Health Impact Assessment
Concepts, theory, techniques, and applications

Edited by

John Kemm
HIA Research Unit, Department of Epidemiology
and Public Health
University of Birmingham, Birmingham

Jayne Parry
HIA Research Unit, Department of Epidemiology
and Public Health
University of Birmingham, Birmingham

Stephen Palmer
Department of Epidemiology and Public Health
University of Wales College of Medicine, Cardiff

D1337897

OXFORD
UNIVERSITY PRESS

OXFORD

UNIVERSITY PRESS

Great Clarendon Street, Oxford OX2 6DP

Oxford University Press is a department of the University of Oxford.
It furthers the University's objective of excellence in research, scholarship,
and education by publishing worldwide in

Oxford New York

Auckland Cape Town Dar es Salaam Hong Kong Karachi
Kuala Lumpur Madrid Melbourne Mexico City Nairobi
New Delhi Shanghai Taipei Toronto

With offices in

Argentina Austria Brazil Chile Czech Republic France Greece
Guatemala Hungary Italy Japan South Korea Poland Portugal
Singapore Switzerland Thailand Turkey Ukraine Vietnam

Published in the United States
by Oxford University Press Inc., New York

© Oxford University Press 2004

The moral rights of the authors have been asserted

Database right Oxford University Press (maker)

First published 2004
Reprinted 2005

British Library Cataloguing in Publication Data

Data available

Library of Congress Cataloging in Publication Data

Data available

ISBN 0 19 852629 6 (Pbk)

10 9 8 7 6 5 4 3 2

Typeset by Newgen Imaging Systems (P) Ltd., Chennai, India
Printed in Great Britain
on acid-free paper by
Biddles Ltd., King's Lynn, Norfolk

Preface

Calls for health impact assessment of policies, projects, and programmes are to be found in documents published by many national governments and by supranational bodies such as the European Union. There is increased awareness that decisions in 'non-health' areas such as the economy, transport, agriculture, manufacture, housing, and law and order have far more effects on the health of the population than decisions in 'health' areas such as health services. It therefore seems reasonable to include consideration of the effects on health in any policy analysis. Few would contest the general desirability of considering health, and a great deal of work, which claims to be health impact assessment, has been done. However there is very little agreement as to precisely what is meant by the term health impact assessment or on the methods, by which it should be done. More recently there have been questions whether health impact assessment had any substance and whether the effort and expense of such assessment added anything to the process of decision making. This book considers what may be expected of health impact assessment and what progress has been made in meeting those expectations.

Health impact assessment is intended to help decision makers in all areas to foresee the health consequences of their decisions, to ensure that these consequences are considered, and to reduce the risk of population health being damaged through some indirect and unintended consequence of a decision. Two main intellectual roots of health impact assessment may be identified: one in environmental impact assessment and one in healthy public policy analysis. These in turn draw on a wide range of disciplines including, economics, ecology, epidemiology, planning, political science, risk assessment, sociology, and toxicology. Each of these contributes to an understanding of the causal chain linking policies, programmes, and projects to the ultimate health outcomes. Knowledge of this chain is the rational base for prediction of consequences in health impact assessment

Theories of health impact assessment have tended to polarize around different views on the nature of evidence. Some groups place heavy emphasis on the use of epidemiology and related sciences, while others place most emphasis on participation and democratic processes. In practice the differences are less marked as all struggle with inadequate data on which to base predictions. Groups also differ on the main purpose of health impact assessment. For some it is a process used to choose between well-defined options while for others it is primarily a process to involve communities in thinking about their own health.

Health impact assessment is a young field of endeavour. There are relatively few examples of completed assessments and even fewer examples, which can be shown to have contributed usefully to decision making. This book does not claim to have a single true theory of health impact assessment or a correct method of doing it. Rather it attempts to describe the work done so far, drawing on examples and thinking from many different disciplines and many parts of the world. It tries to identify the areas of agreement and the questions remaining unanswered. It seeks to map a confused field and signpost possible directions for future progress.

John Kemm
Jayne Parry
Stephen Palmer

Contents

Abbreviations

AHIG	Airport Health Impact Group	EPBC	Environment Protection and Biodiversity Conservation Act
ATSDR	Agency for Toxic Substance and Disease Registry	EPIP	Environment Protection Impact of Proposals Act
BME	Black and Minority Ethnic Communities	EU	European Union
BSE	Bovine spongiform encephalopathy	FAO	Food and Agriculture Organisation
BSL	British sign language	FPTCEOH	Federal, Provincial, Territorial Committee on Environmental and Occupational Health
Btk	Bacillus thuringiensis var. kurstaki		
CAP	Common Agricultural Policy	FPTACPH	Federal, Provincial, and Territorial Advisory Committee on Population Health
CDSR	Cochrane Database of Systematic Reviews		
CHIAT	Community HIA tool	GBD	Global Burden of Disease
CPHM	Consultant in Public Health Medicine	GDP	Gross domestic product
DALY	Disability Adjusted Life Year	GIS	Geographic Information Systems
DHSC	Department of Health and Social Care	GLA	Greater London Authority
DIN	Disease impact number	GP	General Practice
DMBC	Doncaster Metropolitan Borough Council	HA	Health Authority
		HAZs	Health Action Zones
DPH	Director of Public Health	HEBS	Health Education Board for Scotland
EA	Environment Agency		
EC	European Commission	HFA	Health For All
ECHP	European Centre for Health Policy	HIA	Health impact assessment
		HIA-CD	HIA for community development
EHIA	Environmental health impact assessment	HIAG	Health impact assessment guidelines
EIA	Environment impact assessment	HIARU	Health Impact Assessment Research Unit
EIS	Environmental impact statement	HIOs	Health Improvement Officers
EMPCA	Environmental Management and Pollution Control Act	HIS	Health impact screening
		HPP	Healthy public policies
Enhealth	National Environmental Health Council	IAIA	International Association for Impact Assessment
EPA	Environmental Protection Agency	IEH	Institute for Environment and Health

IPC	Integrated Pollution Control	RBI	Reducing Burglary Initiative
IPO	Intersectoral Policy Office	RCTs	Randomized controlled trials
ISD	Information and Statistics Division	RIVM	National Institute of Public Health and Environment
LFG	Landfill gas		
LHC	London Health Commission	SASHU	Small Area Health Statistics Unit
LHCE	Local Healthcare Cooperative	SCIEH	Scottish Centre for Infection and Environmental Health
LHO	London Health Observatory		
LHS	London Health Strategy		
LWMS	Local Welfare Management System	SEA	Strategic Environmental Assessment
MAFF	Ministry of Agriculture, Food and Fisheries	SIA	Social impact assessment
		SNAP	Scottish Needs Assessment Programme
NEHAP	National Environmental Health Action Plans	SRB	Single Regeneration Budget
NEHS	National Environmental Health Strategy	STOP	Society Targeting Overuse of Pesticides
NHMRC	National Health and Medical Research Council	SWSHD	South West Stockholm Health District Authority
NHS	National Health Service	T*f*L	Transport *for* London
NLR	National Aerospace Laboratory	TH	Target Hardening
NOI	Notice of Intention	TRIPS	Trade-related aspects of intellectual property rights
NSPH	Netherlands School of Public Health		
ONS	Office for National Statistics	UBMS	University of Birmingham Medical School
PATH	People Assessing Their Health	UPA	Under Privileged Area
PEEM	Panel of Experts on Environmental Management	UVPG	Umweltverträglichkeitsprüfungsgesetz
PER	Public Environment Report	USEPA	UN Environmental Protection Agency
PHA	Public Health Alliance		
PHIS	Public Health Institute of Scotland	vCJD	Variant Creutzfeld–Jakob Disease
PHP	Public Health Practitioner	VOC	Volatile organic compounds
PIN	Population impact number	WEHAB	Water, energy, health, agriculture, and biodiversity
PRAM	Policy/Risk Assessment Model		
PTSD	Post-Traumatic Stress Disorder	WHO	World Health Organisation
QALY	Quality Adjusted Life Year	WTO	World Trade Organisation
RANCH	Road Traffic and Aircraft Noise Exposure and Children's Cognition and Health		

Contributors

Kate Ardern
South Liverpool PCT
Pavillion 6, The Matchworks
Garston Road
Liverpool L19 2PH

Muna I Abdel Aziz
South Yorkshire Health
Protection Service
5 Old Fulwood Road
Sheffield, S10 3TG

Reiner Banken
Agence d'évaluation des technologies
et des modes d'intervention en santé
(AETMIS)
2021 Avenue Union
Bureau 1040, Montréal
Québec, Canada

Jan JM Barendregt
Department of Public Health
Erasmus MC, University Medical
Center Rotterdam
PO Box 1738, 3000 DR
Rotterdam
the Netherlands

Ruth Barnes
Independent HIA Consultant
Queen's Park, London

Karin Berensson
The Swedish Federation of County
Councils
Stockholm

Martin Birley
International Health Impact
Assessment Consortium
University of Liverpool
156 Meols Parade, Meols
Wirral CH47 6AN

Alan Bond
School of Environmental Sciences
University of East Anglia
Norwich NR4 7TJ

Caron Bowen
188/25 Best Street
Lane Cove
New South Wales 2066
Australia

Ceri Breeze
Public Health Strategy
Division
Welsh Assembly Government
Cathays Park, Cardiff CF10 3NQ

Donald M Campbell
Mid Central Health
PO Box 256
Palmestone North
New Zealand

Carlos Dora
Health impact Assessment
Programme
World Health Organization
Avenue Appia 20
CH-1211 Geneva 27, Switzerland

Margaret Douglas
Department of Health and Health
Policy
Lothian NHS Board
Deaconess House
148 Pleasance, Edinburgh EH8 9RS

Eva Elliott
Welsh HIA Support Unit
School of Social Sciences
Glamorgan Building
Cardiff University
King Edward VII Avenue
Cardiff CF10 3WT

Rainer Fehr
Logd, Landesinstitut fur die
Offentlichen Gesundheit NRW
Westerfeldstr 35/37
33611 Bielefeld, Germany

Ellis AM Franssen
Department of Environment and
Health Research
National Institute of Public Health
and the Environment
PO Box 1, 3720 BA Bilthoven
the Netherlands

Mojca Gabriljecic-Blenkus
Institute of Public Health
Trubarjeva 2, SI-1000 Ljubljana
Republic of Slovenia

Doris E Gillis
St Francis Xavier University
Antigonish, Nova Scotia,
Canada

Alex Hirschfield
Department of Civic Design
University of Liverpool
Liverpool

Danny Houthuijs
Department of Environment and
Health Research
National Institute of Public
Health and the Environment
PO Box 1, 3720 BA Bilthoven
the Netherlands

Bridget Hsu Hage
University of Melbourne
Shepparton, Vic 3632,
Australia

Erica Ison
Institute of Health Sciences
Old Road, Headington
Oxford OX3 7LF

Mike Joffe
Department of Epidemiology and
Public Health
Imperial College
Norfolk Place
London W2 1PJ

John Kemm
HIA Research Unit
Department of Epidemiology and
Public Health
University of Birmingham
Birmingham B15 2TT

Anton E Kunst
Department of Public Health
Erasmus MC, University Medical
Center Rotterdam
PO Box 1738, 3000 DR Rotterdam
the Netherlands

Roy E Kwiatkowski
Chief Office of Environmental
Health Assessment

Health Canada
PL 6604M Room C486
Tupper Building
2720 Riverside Drive
Ottawa, Ontario
Canada

Erik Lebret
Department of Environment and
Health Research
National Institute of Public Health
and the Environment
PO Box 1, 3720 BA Bilthoven
the Netherlands

Juhani Lehto
Department of Social and Health
Policy
Tampere School of Public Health
FIN-33014
University of Tampere Finland

Karen Lock
ECOHOST
London School of Hygiene and
Tropical Medicine
Keppel Street
London WC1E 7HT

Johan P Mackenbach
Department of Public Health
Erasmus MC, University Medical
Center Rotterdam
PO Box 1738, 3000 DR
Rotterdam
the Netherlands

Jonathan Mathers
HIA Research Unit
Department of Epidemiology and
Public Health
University of Birmingham
Birmingham B15 2TT

Ian Matthews
Department of Epidemiology and
Public Health
University of Wales College of
Medicine
Heath Park, Cardiff CF4 4XN

John McCabe
North Sheffield Primary
Care Trust
c/o Welfare House
North Quadrant, Sheffield S5 6NU

Mark McCarthy
Public Health Research Group
Department of Epidemiology and
Public Health
University College London
1-19 Torrington Place
London WC1E 6BT

Sally Macintyre
MRC Social and Public Health
Sciences Unit
University of Glasgow
Glasgow G12 8RZ

Odile Mekel
Logd
Landesinstitut fur die Offentlichen
Gesundheit NRW
Westerfeldstr 35/37
33611 Bielefeld
Germany

Sue Milner
North Liverpool PCT
New Hall Campus
Longmore Lane
Liverpool L10 1LD

Jennifer Mindell
London Health Observatory
Kings Fund

11-13 Cavendish Square
London W1G 0AN

Maurice B Mittelmark
University of Bergen
Christiesgt
13, 5015 Bergen
Norway

Jill Muirie
Department of Health and Health
Policy
Lothian NHS Board
Deaconess House, 148
Pleasance
Edinburgh EH8 9RS

Stephen Palmer
Department of Epidemiology and
Public Health
University of Wales College of
Medicine
Heath Park
Cardiff CF4 4XN

Jayne Parry
HIA Research Unit
Department of Epidemiology and
Public Health
University of Birmingham
Birmingham B15 2TT

Mark Petticrew
MRC Social and Public Health
Sciences Unit
University of Glasgow,
Glasgow G12 8RZ

John Radford
North Sheffield Primary
Care Trust
c/o Welfare House
North Quadrant
Sheffield S5 6NU

Anna Ritsatakis
14 Tsangaris St.
Melissia 15127
Athens, Greece

Ben Rolfe
Welsh HIA Support Unit,
School of Social Sciences
Glamorgan Building
Cardiff University
King Edward VII Avenue
Cardiff CF10 3WT

Ernst W Roscam Abbing
Department of Social
Medicine
University of Nijmegen
PO Box 9102; 6500 HCNijmegen
the Netherlands

Brigit AM Staatsen
Department of Environment and
Health Research
National Institute of Public Health
and the Environment
PO Box 1, 3720 BA Bilthoven
the Netherlands

Hilary Thomson
MRC Social and Public Health
Sciences Unit
University of Glasgow
Glasgow G12 8RZ

Martin Utley
Deputy Director
Clinical Operational
Research Unit
University College London
Gower Street
London
WC1E 6BT

J Lennert Veerman
Department of Public Health
Erasmus MC
University Medical Center
Rotterdam
PO Box 1738, 3000 DR
Rotterdam
the Netherlands

Rudolf Welteke
Logd, Landesinstitut fur die
Offentlichen Gesundheit NRW
Westerfeldstr 35/37
33611 Bielefeld, Germany

Carla MAG van Wiechen
Department of Environment and
Health Research
National Institute of

Public Health and the
Environment
PO Box 1, 3720 BA Bilthoven
the Netherlands

Gareth Williams
Welsh HIA Support Unit
School of Social Sciences
Glamorgan Building
Cardiff University
King Edward VII Avenue
Cardiff CF10 3WT

John Wright
HIA Research Unit
Department of Public Health and
Epidemiology
University of Birmingham
Birmingham B15 2TT

Chapter 1

What is HIA? Introduction and overview

John Kemm and Jayne Parry

Definition of HIA

Health impact assessment (HIA) has been succinctly described 'as any combination of procedures or methods by which a proposed policy or program may be judged as to the effects it may have on the health of a population' [1]. Other definitions (shown in Table 1.1) make clear that HIA is concerned with the health of populations and attempts to predict the future consequences of health decisions that have not yet been implemented.

The purpose of HIA

Health impact assessment is intended to influence decision making so that policies, projects, and programmes in all areas lead to improved population health or at least do not damage population health. There are three ways in which HIA might influence decision making.

◆ By raising awareness among decision makers of the relationship between health and the physical, social, and economic environments, thereby ensuring that they always include a consideration of health consequences in their deliberations.

◆ By helping decision makers identify and assess possible health consequences and optimize overall outcomes of the decision.

◆ By helping those affected by policies to participate in policy formation and contribute to decision making.

The ways in which thinking about the purpose of HIA has developed are discussed further in Chapter 2.

Terminology

The terms retrospective, concurrent, and prospective are used to describe the temporal relation of an HIA to the intervention being assessed. Prospective

Table 1.1 Definitions of HIA

Health impact assessment is 'a combination of procedures, methods and tools by which a policy, a program or project may be judged as to its potential effects on the health of a population and the distribution of effects within the population'.
WHO Gothenburg Consensus Paper [2]
Health impact assessment is 'a methodology which enables the identification, prediction and evaluation of the likely changes in health risk, both positive and negative, (single or collective), of a policy, programme plan or development action on a defined population. These changes may be direct and immediate or indirect and delayed'.
BMA board of science and education [3]
'Health impact assessment is the estimation of the effects of a specified action on the health of a defined population'.
Scott Samuel [4]
'Health impact assessment is a method of evaluating the likely effects of policies, initiatives and activities on health at a population level and helping to develop recommendations to maximize health gain and minimize health risks. It offers a framework within which to consider, and influence the broad determinants of health'.
Scottish Office [5]
Health impact assessment is 'a combination of procedures or methods which enable a judgement to be made on the effect(s)—positive or negative of policies, programmes or other developments on the health of a population or on parts of the population where health are concerned'.
National Assembly for Wales [6].

HIA attempts to predict the consequences of a decision before it has been implemented. Retrospective HIA attempts to identify and describe the consequences of a decision or event that has already occurred. Concurrent HIA attempts to monitor situations and identify and describe the consequences of an intervention as it is implemented.

Studies that have been described as retrospective HIA could equally well be described as outcome evaluation. The distinction that some [7] draw between the two terms is not helpful. Similarly exercises described as concurrent HIA could just as well be described as monitoring or surveillance. In the policy appraisal literature and in European writings on HIA the terms ex ante and post ante are widely used. These terms have the same meaning as prospective and retrospective.

Prediction of the future has to be based on knowledge and experience of the past. Retrospective assessments provide the knowledge and understanding of the relation between interventions and their consequences, which is essential for prospective HIA. Concurrent assessments are chiefly used where consequences are expected but their nature and severity is uncertain. Monitoring these consequences allows early introduction of preventive measures or modification of the intervention.

According to the definitions given at the start of this chapter HIA is concerned with both the prediction of consequences and the subsequent modification of decisions to mitigate harm and to maximize health. If HIA is to achieve this then concurrent and retrospective activities cannot strictly be called HIA for they are undertaken post implementation. Although the terms retrospective and concurrent HIA are well embedded in the literature, in Chapter 36 we argue that it is time to stop using them.

The conceptual roots of HIA

Health impact assessment draws on a wide range of disciplines but the two main conceptual roots lie in

◆ Impact assessment especially environmental impact assessment

◆ Policy appraisal and the promotion of healthy public policy (HPP).

The development of environmental impact assessment was encouraged by increased awareness of environmental damage and is now embedded in a legislative framework in many countries. It draws on cost–benefit analysis, ecology, biological sciences, epidemiology, toxicology, risk assessment, and increasingly sociological disciplines. However at present environmental impact assessments, particularly in the United Kingdom, frequently pay inadequate attention to possible consequences for human health [8]. Many suggest that extension of environmental impact assessment so that health issues were properly covered would be a logical way to develop HIA [9].

Health impact assessment could also be seen as a specialized form of policy appraisal, which seeks to analyze the content of policies and of the policy-making process. Policy appraisal draws on political science, political economy, and social sciences. Until recently however discussion of health consequences was largely limited to policies concerned with the provision of medical facilities. The call for HPP was a response to this restricted view and tried to extend the applications of policy appraisal to health consequences of all policies.

What is health?—Outcomes for HIA

Health impact assessment is concerned with health but the meaning of that term is contested. Frequently the term health is operationalized as the absence of disease and a great deal of 'health' policy is concerned with the provision of health services for those with disease. Sometimes health policy is extended to cover the prevention of disease through measures such as immunization, health education, and even provision of safe food and water. The constitution of WHO [10] however proclaims a very different view of health. 'Health is a state of complete physical, mental and social well-being and not merely the absence of disease or infirmity'. Such a view has been criticized as hopelessly utopian [11] and so boundless as to be meaningless [12]. However it is a better vision against which to compare policy outcomes than one that focuses solely on disease.

In contrast, many HIA practitioners emphasize that they take a much broader health perspective [13] and rightly insist that an analysis limited to impacts on death and frequency of medically defined disease is inadequate. A satisfactory assessment considers the impact on all aspects of physical, mental, and social well-being including objectively assessed states, subjective feelings, sense of well-being, and quality of life considering positive aspects of health as well as negative. Unfortunately epidemiology and related sciences, which could contribute to HIA, are currently limited in their ability to explore outcomes other than death or frequency of objectively assessed disease. This is likely to improve and considerable progress is being made in developing measures of subjective health such as the Nottingham Health Profile [14], the Euroqol [15], and the Sickness Impact Profile with its derivatives such as SF36 [16] and these tools will undoubtedly contribute to the further development of HIA.

The determinants of health

Recent analyses have emphasized the relatively small role played by health services and medical procedures in influencing the health of populations. McKeown [17] in his influential book *The role of medicine* pointed out how the dramatic decline in deaths from infectious disease that took place in the latter half of the nineteenth and first half of the twentieth century was associated more with increasing prosperity, better nutrition, better housing, and improved sanitation than it was with any development in medicine. The Black Report [18] and 20 years later the Acheson enquiry [19] into health inequalities in the United Kingdom emphasized the powerful influence that conditions of living exercise over health. The overarching importance of general socio-economic, cultural and environmental conditions, housing and working conditions, and social and community influences on health has been emphasized [20,21] as has the effect of income distribution [22]. The terms atomistic fallacy [23] and individualistic fallacy [24] have been coined to describe the futility of attempting to understand the health of individuals in isolation without considering the communal context within which they exist. A hierarchical framework that recognizes individual, communal, and macro-political level factors is needed to understand the relationship between an individual's health and their environment. The development of multi-level modelling techniques to explore these relationships offers an additional tool to the HIA armoury.

The realization that virtually every area of human activity influenced health leads to the conclusion that most public or political decisions have the potential to impact health for good and ill. Major improvements in population health are more likely to be achieved through interventions in economic, industrial, housing, transport, agriculture, education, law and order, and other 'non-health' areas than in the policy areas with which ministries of health are usually concerned. Milio [25] argued that public policy should be used and assessed for the way it affected health and the World Health Organisation (WHO) includes 'healthy public policy' as one of the key health promotion actions in the Ottawa Charter [26]. HIA provides a framework within which the healthiness or otherwise of public policy can be assessed [27].

Issues requiring coordinated policy responses

Governments are concerned with the overall well-being of their populations but responsibility for each issue is segmented between the many different ministries or branches of government. While this may make for efficient government it raises problems when the actions of one branch have consequences for the concerns of another. There are many issues, of which health is only one, that are affected by the decisions of numerous different branches of government and require a coordinated policy. Environmental sustainability, inequalities within society, social inclusion, well-being of families, law and order, and fiscal balance are other examples of issues that cut across the concerns of several different ministries. Given the requirement to consider all these issues policy makers may be concerned that excessive concentration on one issue such as health could detract attention from other equally important issues.

Environmental sustainability is a prerequisite for long-lasting health and there is a need to find forms of economic development that meet 'the needs of the present without compromising the ability of future generations to meet their own needs' [28]. There is concern that depletion of non-renewable resources, loss of biodiversity, overloading of pollution sinks, and growth of human population could lead to a situation in which the capacity of the biosphere to support human life and health was exceeded [29]. A programme of action commonly referred to as Agenda 21 [30,31] was produced at an international conference of heads of government in Rio de Janeiro. National and local governments frequently treat this as a 'cross-cutting' issue and have set up procedures to ensure that decisions in all areas are congruent with this overall goal. Health Impact Assessment procedures could develop along similar lines.

Health inequalities are another cross-cutting issue for many governments. The fact that health inequalities exist and are getting wider in many countries has been noted [18,19]. Their existence is recognized as inequitable and their reduction has been made another overarching goal for government. It has been suggested that all decisions should be assessed for their impact on health inequalities [32] but since HIA has to be concerned with the distribution of health impacts across populations it should provide the information needed to address inequality issues [33].

Positivistic and relativistic bases for predictions in HIA

Health impact assessment involves making predictions about the future consequences of decisions. Predictions based on natural science disciplines such as epidemiology and toxicology assess for each hazard the number of people exposed to particular levels and the likelihood of an individual experiencing harm when exposed to these levels (dose–response curve). In theory this approach could be extended to all determinants of health and all outcomes but the current level of knowledge means it is usually limited to physico-chemical hazards and the frequency of death and disease outcomes. Sometimes this approach may be built into complex models, which take into account changing levels of exposure and delays between exposure and outcome.

All these approaches share a world view that the links between an intervention and its consequences are describable by scientific laws. The validity of these laws is tested by following principles described by Popper in which extensive attempts are made to falsify through experiment hypotheses derived from the laws. The more that hypotheses derived from it resist falsification the more is a law considered a sound basis for prediction. This is part of a wider positivist philosophy, which asserts that there is an objective reality of which law like generalities (scientific laws) are a part, which can be imperfectly known through observation. Logical positivism is an extreme variant of positivism, which holds not only that there are objective realities but also that there are no realities that could not be known by observation.

Biomedical scientists adopting positivist approaches to HIA and other problems sometimes claim that it is a value free approach, which provides no basis for preferring one value over another. While measuring the concentration of a particular substance or studying its physiological effect might be value free, questions such as which physiological reactions to study, how to interpret the data, and what weight to give to different types of evidence can never be value free.

While immensely powerful this positivist approach is limited in application. The links between health and its determinants are frequently characterized by complexity. Slight uncertainty in initial conditions is quickly and enormously magnified and the system becomes unpredictable because the initial conditions can never be specified with sufficient precision [34]. Furthermore it does not provide any basis for handling so called 'irrational fears'. Health impact assessment needs to consider all these things.

A totally different approach is commonly used in the humanities and social sciences. This emphasizes that interpretation of events depends on their context and on the person perceiving them. It pays great attention to the understanding of underlying meanings. The basis for prediction is deemed to be held by the people, who would be affected by the policy, and understanding of the situation comes from these people and from unravelling the meanings that events hold for them. By emphasizing the difficulty of knowing underlying reality and the differences in perception it provides a framework in which to discuss values, anxieties, and 'irrational' fears. This approach is informed by a worldview which holds that much truth is socially constructed, and that positivist talk of objective reality is misleading since all perception is influenced by the perceiver and all knowledge is subjective. A few take the extreme position of relativism holding that all perceptions of the truth are equally valid and that there are no grounds for preferring one version over another but this extreme offers no basis for prediction.

Neither positivistic nor relativistic perspectives alone provide a satisfactory basis for HIA. Prediction of the effects such as those of chemical pollutants on mortality or physical disease is probably best understood from a positivistic perspective. On the other hand prediction of outcomes such as anxiety, amenity, social inclusion, and quality of life may well be more amenable to relativistic approaches.

Values of HIA

The Gothenberg Consensus Paper [2] described the values of HIA as being democracy, equity, sustainable development, and ethical use of evidence. Two of these equity and sustainable development seem to be criteria by which impacts should be judged. Apart from guiding how impacts should be investigated (looking at distribution of impacts as well as aggregate impact for equity, looking at long- and very long-term as well as short-term impacts for sustainability) they say little about how assessors should behave. The other two values, democracy and ethical use of evidence, have clear implications for the conduct of assessors.

Democracy 'emphasizes the right of people to participate in the formulation, implementation and evaluation of policies that affect their life'. It underpins the importance of participation and openness in HIA. It implies that people should have the opportunity to know and influence the questions being asked and the issues being investigated in an HIA. It implies that they should be able to see all the evidence collected and place different interpretations on it if they so wish. The commitment to openness may sit uncomfortably with the desire to influence decision makers. The value of democracy raises difficult questions as to how disagreements between 'technical experts' and citizens should be resolved.

Ethical use of evidence is clearly desirable but begs the question what is ethical? It certainly involves adopting the highest scientific standards of rigour, using clear criteria to select and judge evidence, and not neglecting evidence because it does not fit the argument being made. It implies diligence in searching for evidence though there virtually always has to be a compromise between the ideal and what can be done with the time and resources available. Ethical use of evidence requires impartiality so that the decision maker and others can trust the assessor to present an honest assessment of all the evidence available to them on the options under consideration. Public health workers may be more accustomed to advocacy, which involves putting the case for the option they consider best for health, playing down any weaknesses in their own argument and emphasizing any weaknesses in the oppositions' argument. The dilemma is worsened because the opposition is frequently not impartial. None the less HIA is different from advocacy and the assessor has a duty to be impartial. An individual can be both advocate and impartial assessor, but they cannot be both at the same time.

Levels at which HIA takes place

Decisions are made at many levels from supranational to local. International bodies such as the WHO, the World Bank, and the European Union (EU), national governments, regional governments, local authorities, health authorities, transport authorities, non-governmental organizations, and many other bodies make decisions at their own levels. The ways in which decisions are taken and the capacity to make impact assessments may vary between levels and organizations so that different approaches to HIA may be appropriate for each context.

Similarly decisions vary in their scope. Policy decisions produce overall frameworks setting the goals for a particular area, laying out general directions, and guiding how issues within that area should be determined. Decisions on programmes set in motion, linked activities contributing to a particular goal. Project decisions cover a limited action such as construction of a building or piece of infrastructure, provision of a new service, changing management structures, or mounting a communication campaign. All of these may have impacts on health and could benefit from HIA though different applications may well need different assessment processes.

Linking HIA and decision making

Health impact assessment like policy evaluation has to take place at several levels. Fischer [35] has classified these levels as programme verification, situational validation, societal validation, and social choice. Programme verification considers whether the intervention will achieve its stated objectives and whether it uses resources efficiently but an impact assessment has to ask much more. Situational validation considers whether the stated objectives will be relevant to the problem. Societal vindication asks whether the intervention has instrumental value for the health of society as a whole while social choice asks whether the fundamental ideology of the policy will be compatible with health. Epidemiology is a very powerful tool for programme verification but any analysis restricted to this level is incomplete. Other disciplines such as sociology and philosophy have more to contribute to HIA at the higher levels of discourse.

This book repeatedly emphasizes that HIA is intended to be useful and to assist decision makers. It follows that the HIA has to be closely linked to the decision-making process and structured so that it is possible for the assessment to influence the decision. Many descriptions of the HIA procedures intermingle assessment and decision-making steps [13]. Descriptions of HIA often seem to assume a linear (rational deductive) mode of policy making in which there is a logical point to start the assessment and another point at which to present and consider it. In practice most policy making is incremental consisting of no more than marginal adjustments to existing policies and structures limited to what is deemed possible on the basis of value judgements and careful negotiations with interested parties [36]. In this model of policy making there is no obvious point at which HIA should be commissioned or presented. None the less if it is to be useful HIA must fit into this untidy process. One characteristic of incremental policy-making is that decisions are often made very rapidly as windows of opportunity arise and consequently HIA may also have to be made with corresponding haste. Assessments that do not conform to decision-making timetables will not influence that decision. Furthermore the findings of HIA have to be presented in a form that is policy relevant and addresses the concerns of the policy makers [37]. Decision makers will need to know not only the predicted consequences of different options but also the degree of certainty that attaches to those predictions.

Another requirement of HIA is that it should be proportionate to the decision it is intended to influence. HIAs have a cost in time and possibly other resources. It is clearly unreasonable to use £10,000 worth of resources to assess a decision about the use of £20,000 worth whereas a much greater resource expenditure would be justified for an assessment linked to a decision involving many million pounds worth of resource.

In many ways the ideal solution to the problem of HIA being external to the decision-making process would be for the decision makers to own and control the HIA process. This would then ensure that it addressed issues relevant to them and occurred at relevant times. The argument against this arrangement is that it would be too easy for HIA to become tokenistic and avoid health issues where these were inconvenient for other policy goals.

Quality criteria for HIA

Those practising or commissioning HIA have to be concerned with its quality. Many descriptions of HIA include an audit step that involves reflection on the process. One set of quality criteria relates to the utility of the process and whether it can be said to have influenced or assisted the decision-making process in any way. For example, a useful HIA might have made the decision maker aware of health impacts of which they were previously unaware, or it might have allowed them to assess the size of an impact or given information relevant to trade offs in the decision, or it might have allowed stakeholders to participate more meaningfully in formation of the policy. An HIA would also be useful if it resulted in better mitigation of harmful impacts or enhancement of beneficial ones or it might be useful by making the decision-making process more transparent. It would be difficult to say that an HIA that had done none of these things was of acceptable quality [38].

A second criterion of quality is predictive accuracy. Health impact assessment predicts a set of consequences if a particular decision is implemented. It is reasonable to ask whether consequences predicted for the short term subsequently occurred. This criterion has limited applicability since it is not possible to follow up on very long-term consequences. Nor is it possible to verify the counterfactual and check predictions made for options that were not chosen would have been correct if that option had been implemented.

A third set of criteria relates to the process of the HIA [12]. Has the assessment been given a brief and addressed issues relevant to the decision in question. Were those who should have been consulted about the decision (stakeholders) involved in an appropriate and timely way? Was the evidence from the literature on consequences of similar decisions searched in a systematic manner [39]? Has evidence been gathered from appropriate key informants? Have alternate options been adequately explored and have efforts to mitigate negative effects concentrated on the largest impacts.

Evaluation of HIA may be said to be the least satisfactory aspect of this subject and it will be discussed further in Chapter 36.

Organization of this book

Health impact assessment is an evolving discipline and there is as yet little shared understanding of what it is or how it should be done. This book does not attempt to define boundaries or prescribe methods. It does seek to map the progress that has been made so far, outline the different positions, and identify the issues on which further work is required. Chapters 1–3, the first section, review the history of HIA, its development from environmental impact assessment and policy appraisal, and how it is now being applied to the problem of reducing health inequalities. Chapters 4–6, the second section, look at how epidemiology, social sciences, and modelling contribute to HIA. Chapters 7 and 8, also in the second section, look at the sources of evidence and the role of participation in HIA. Chapters 9–13, the third section, look at various practical aspects of undertaking an HIA.

Chapters 14–23, the fourth section, describe how HIA has been used in various international, national, and local government levels. Chapters 24–35, the fifth section, are case studies describing the application of HIA to policies and projects at local and national levels. This section covers subjects as diverse as a moth eradication programme, decisions on landfill sites, city transport strategies, construction of new airports, estate refurbishment projects to national policies on agriculture, alcohol, or smoking. Chapter 36 draws together this diverse experience, summarizing the points of agreement, the undecided questions, and possible directions for future work.

References

1 Ratner PA, Green LW, Frankish CJ, Chomik T, and Larsen C. Setting the stage for health impact assessment. *Journal of Public Health Policy* 1997; **18**:67–79.

2 WHO European Centre for Health Policy. Health Impact Assessment: main concepts and suggested approach Gothenburg Consensus paper. 1999.

3 BMA Board of Science and education. *Health and Environmental Impact Assessment.* London: Earthscan, 1998.

4 Scott-Samuel A. Health impact assessment—theory into practice. *Journal of Epidemiology and Community Health* 1998; **52**:704–705.

5 Scottish office. *Towards a Healthier Scotland (A white paper on health).* Edinburgh: The Stationery Office, 1999.

6 National Assembly for Wales. Developing health impact assessment in Wales. Cardiff: National Assembly for Wales, 1999.

7 Milner S and Marples G. *Policy Appraisal and Health Project: Phase 1 a Literature Review.* Newcastle: University of Northumbria, 1997.

8 Arquiaga MC, Canter LW, and Nelson DI. Integration of health impact considerations in environmental impact studies. *Impact Assessment* 1994; **12**:175–197.

9 Joffe M and Sutcliffe J. Developing policies for a healthy environment. *Health Promotion International* 1997; **12**:169–173.

10 World Health Organisation Constitution WHO Geneva 1946.

11 Siracci R. The World Health Organisation needs to reconsider its definition of health. *British Medical Journal* 1997; **314**:1409–1410.

12 Seedhouse D. *Health Promotion. Philosophy, Prejudice and Practice.* Chichester: John Wiley, 1997.

13 Scott-Samuel A, Birley M, and Ardern K. *The Merseyside Guidelines for Health Impact Assessment.* Liverpool: Merseyside Health Impact Assessment Steering Group, 1998.

14 Hunt SM, McEwan J, and McKenna SP. *Measuring Health.* London: Croom Helm, 1986.

15 EuroQol Group. EuroQol: a new facility for the measurement of health related quality of life. *Health Policy* 1990; **16**:199–208.

16 Ware JE and Sherbourne CD. The MOS 36 items short form health survey (SF36) Conceptual framework and item selection. *Medical Care* 1992; **30**:473–480.

17 McKeown T. *The Role of Medicine* Basil. Oxford: Blackwell, 1979.

18 Black D, Morris JN, Smith C, and Townsend P. Inequalities of health: The Black Report 1980—(Reprinted London, Penguin).

19 Acheson D. *Independent Inquiry into Inequalities in Health: Report.* London: The Stationery Office, 1998.

20 Whitehead M. Tackling inequalities a review of policy initiatives. In Benzeval M, Judge K, and Whitehead M (eds.), *Tackling Inequalities in health.* London: Kings Fund, 1995.

21 Hertzman C, Frank J, and Evans RG. Heterogeneities in health status and the determinants of population health. In Evans RG, Barer ML, and Marmor TR. (eds.) Why are some people healthy and others not? New York: de Gruyter, pp 67–92, 1994.

22 Wilkinson R. *Unhealthy Societies: the Afflictions of Inequality.* London: Routledge, 1996.

23 Marmot MG. Improvement of social environment to improve health. *Lancet* 1998; **351**:57–60.

24 Krieger N. Epidemiology and the web of causation: has anyone seen the spider? *Social Sciences and Medicine* 1994; **39**:887–903.

25 Milio N. *Promoting Health through Public Policy.* Ottawa Canadian Public Health Association, 1986.

26 World Health Organisation. The Ottawa charter of health promotion. *Health Promotion* 1986; **1**:i–v.

27 Kemm J. Health impact assessment: a tool for healthy public policy. *Health Promotion International* 2001; **16**:79–85.

28 World Commission on Environment and Development. Our common future. Oxford: Oxford University Press, 1988.

29 McMichael AJ. *Planetary Overload. Global Environmental Change and the Health of the Human Species.* Cambridge: Cambridge University Press, 1993.

30 Keating M. The earth summit agenda for change: A plain language version of Agenda 21 and other Rio agreements. Geneva: Centre for our common future, 1993.

31 Dodds F. *The Way Forward Beyond Agenda 21*. London: Earthscan publications, 1997.

32 Lester C, Hayes S, Griffiths S, Lowe G, and Hopkins S. Implementing a strategy to address health inequalities: a Health Authority approach. *Public Health* 1999; 1:90–93.

33 Douglas M and Scott-Samuel A. Addressing health inequalities in health impact assessment. *Journal of Epidemiology and Community health* 2001; 55:450–451.

34 Kolata G. Theory of chaos science. 1986; 231:1068–1070.

35 Fischer F. *Evaluating Public Policy*. Chicago, IL: Nelson-Hall, 1995.

36 Ham C. *Health Policy in Britain* (3rd edn). Basingstoke; Macmillan, 1992.

37 O'Neill M and Pederson AP. Building a methods bridge between policy analysis and healthy public policy. *Canadian Journal of Public Health* 1992; 83(Suppl 1):S25–S30.

38 Kemm JR. Can health impact assessment fulfill the expectations it raises? *Public Health* 2000; 114:431–433.

39 McIntyre S and Pettogrew M. Good intention and received wisdom are not enough. *Journal of Epidemiology and Community Health* 2000; 54:802–803.

Chapter 2

The development of HIA

John Kemm and Jayne Parry

Early origins

In 1996 a paper appeared with the title 'Health impact assessment—an idea whose time has come' [1] and this is a convenient point to start discussion of the recent development of health impact assessment (HIA). However, the ideas that have been incorporated in HIA start much earlier. In the sixteenth century Machiavelli was practising sophisticated policy analysis though health was not one of his concerns. Health and welfare of populations was an important consideration for policy makers by the nineteenth century and political science was well developed by the middle of the twentieth century.

Cost–benefit analysis of policies with the notion that the consequences of policies and project options could be compared using a common metric was being applied in the 1970s. Considerable progress was made in estimating the benefits to be expected with different options. The consequences of catastrophic events in large and complex structures and the increasing number of such structures (e.g., nuclear industry, petroleum industry, air transport) lead to the development of very sophisticated methods of modelling safety critical systems in order to reduce risk to very low levels.

Concern over unforeseen environmental consequences of large engineering projects especially in the Third World became widespread. This led project funders, notably the World Bank, to require thorough examination of how proposals would affect flora, fauna, and the physical environment. Projects were often associated with massive disruption of human communities, which would be displaced and have their culture and lifestyle radically changed. It became clear that anthropologists and sociologists had much to offer environmental impact assessment teams [2] and social impact assessment [3] developed. Social impact assessment and HIA have much in common. It has been suggested, 'it is difficult to envisage what a social impact would be that did not have an impact on health or what territory could be covered by a social impact assessment that was not enveloped within a health impact assessment' [4].

Development of methods

Early descriptions of HIA spoke of 'a methodology' [5] but numerous different methods have been described. Maturing of the field has brought recognition that HIA is 'a combination of procedures methods and tools' [6]. Health impact assessments primarily focussed on the needs of decision makers tend to use methods different from those that are primarily intended to make decisions more participatory. Health impact assessments applied to projects has tended to use methods different from HIA applied to policy. Health impact assessment in the context of an environmental impact assessment or a strategic environmental assessment tends to be done rather differently from HIA separate from other impact assessments. In the United Kingdom, the different settings of government departments and health authority (Primary Care Trust) also require different methods and HIA as part of a planning enquiry or Integrated Pollution Prevention and Control (IPPC) determination needs methods different from those done without a legal framework. In other countries there will be similar differences in the methods needed for HIA in different contexts.

The methods used also tend to reflect the disciplines of those executing the HIA. Epidemiologists tend to use 'epidemiological' methods, social scientists to use 'sociological' methods, community developers to use 'community development' methods, and so on. The response to this diversity should not be to engage in futile argument as to which is best, but to identify the strengths and weaknesses of each and find which methods are appropriate for which purpose. Amid all the different accounts, the basic steps identified in the earliest descriptions of HIA have remained unchanged (Fig. 2.1). Screening, scoping, assessment, communication to decision makers, implementation, and monitoring are consistent features in all descriptions.

It has also become clear that the intensity of effort that is possible or reasonable in different settings varies. A small voluntary sector organization does not have

Fig. 2.1 Stages of HIA.

the same resources for HIA as a government department. A decision as to whether to hold a farmers' market does not justify the same level of assessment as a multi-million-pound project to build a motorway. Many HIAs involved only a few people spending a few hours without any recourse to literature searches or professional epidemiologists or sociologists. These small-scale efforts were recognized as having their place. It has been suggested that HIA be categorized into three levels of intensity: mini, standard, and maxi HIA ranging from an exercise taking two or three people 2 or 3 hours to an extensive investigation involving many people and requiring many months [7]. It should be noted that the terminology used to describe small or quick HIAs is extremely varied, and confusing. They have been variously termed 'rapid', 'mini', and 'desk-top' assessments. These terms are not used consistently by different chapter contributors but each has defined how the terms are used in their chapters. However, the need for a more consistent terminology is clear and this is discussed further in Chapter 36.

Development of HIA of projects

In the context of major projects in developing countries the main health impacts resulted from water borne disease, vector borne disease, displacement of populations, and disturbed culture and lifestyle [8] (see Chapter 32). Interest in health and environmental impact assessment was also growing in developed countries. In New Zealand the Resource Management Act 1991 required authorities to make an 'assessment of any actual or potential effects on the environment', which includes 'any effect on those in the neighbourhood or wider community including socio-economic and cultural effects'. A guide to health impact assessment [9] was published in 1995 to assist authorities with this task. In Australia development planning and resource issues are regulated by the individual states but the national government produced a report [10] to guide states on how they could involve impact assessment in their planning and development (see Chapter 20). In Germany health aspects were considered in the context of environmental impact assessment [11]. In Holland various large development projects, such as the expansion of Schiphol airport, were the subject of assessments that covered both health and environmental impacts (see Chapter 24). In Ottawa, Canada, projects requiring environmental impacts were scrutinized for possible health impacts and those with greatest potential subjected to fuller assessment (see Chapter 27). In the United Kingdom an HIA was submitted as evidence to the planning enquiry on a third runway for Manchester airport [12] and the British Medical Association published a guide to linked health and environmental impact assessment [5].

The development of environmental impact assessment and of HIA within it is discussed in Chapter 12.

Development of HIA of policy

In the 1980s within health promotion circles there was considerable interest in 'healthy public policy' [13,14]. In the 1990s the UK government acknowledged the existence of 'Inequalities in Health' [15] and adopted their reduction as an overarching policy goal. This interest required an ability to predict the health consequences of policy. In British Columbia, Canada, HIA was a requisite in the preparation of policy and guidance [16] was published on how to do this (see Chapter 15). In Holland the government gave considerable thought as to how health consequences of policies could be assessed and developed procedures to undertake this [17] (see Chapter 16). In Sweden the focus fell on local rather than national authority policy and guidance was published as to how they could assess the health impacts of their policies [18] (see Chapter 19). In England the Department of Health published a discussion of how health aspects of policy could be appraised though the discussion was theoretical and heavily biased toward economic outcomes [19]. The Greater London Assembly has developed a system for assessing the health impacts of all its strategies (Chapter 21).

Meanwhile in Europe and particularly in the United Kingdom development of HIA was encouraged by a series of supra-national and national government statements. In England the green paper 'Our healthier nation: A contract for the nation' [20] stated 'the Government will apply health impact assessments to its relevant key policies, so that when they are being developed and implemented, the consequences of those policies for our health is considered'. This commitment was renewed in the subsequent white paper 'Saving lives: our healthier nation' [21]. Governments in Scotland [22], Wales [23], and Northern Ireland [24] made similar commitments. WHO Europe [25] included 'Member States should have established mechanisms for health impact assessment and ensured that all sectors become accountable for the effects of their policies and actions on health' as one of its Health 21 targets. The High Level Committee on Health of the European Union recommended the development of an easy-to-use checklist of steps in policy appraisal for health impact to be used for policy development [26]. Supported by this wealth of commitment one might have expected the rapid introduction of HIA into government policy-making, but this has not been the case. It is also interesting to speculate how and why HIA achieved such high-level acceptance at a time when it could be argued that no one was clear what it was, let alone how to do it.

Why do HIA

The first HIAs were undertaken by enthusiasts apparently to be discussed and admired by other HIA enthusiasts—including the authors of this chapter. It was not clear who else was supposed to read the reports or how they were to contribute to public health in any specific way. There was an implicit, albeit erroneous, assumption that the assessment and decision-making steps would be directly and automatically linked. More recently, HIA has become more purposive and two streams of thought as to its utility have emerged.

One stream encouraged by the statements of governments and authorities that they wished to include HIA in their decision-making processes set out to shape HIA to engage decision makers. A new emphasis on the relation between health impact assessors and decision makers appeared in guides to HIA [7,27]. Health impact assessment paid more attention to the real world of decision making and realized that it had much to learn from political science. Some saw HIA as a tool for meeting the call of the Ottawa charter to 'put health on the agenda of policy makers in all sectors and all levels, directing them to be aware of the health consequences of their decisions and to accept their responsibilities for health' [14]. While this worked well with decision makers already committed to health, those less committed did not welcome an approach that they interpreted as an attempt to force them to adopt the priorities and agenda of another interest group. The contrast between EIA with its clear and legislatively defined place in decision making and HIA, which had no legal status in addition to frustration at the reluctance of many decision makers to adopt HIA, led some to urge that HIA should become obligatory.

An alternative approach seeks a less directive style to engage decision makers with the health agenda. HIA is presented as a tool for use by decision makers to help them gain better insight of outcomes, balance health against other policy considerations, appraise options, and improve the trade offs. HIA is presented as a means of dealing with many of the inherent but more challenging aspects of decision making. This approach was often more acceptable to decision makers. Realization that HIA had to find a place beside numerous other impact assessments on the policy maker's desk prompted some to argue for HIA to evolve from a stand alone assessment to become one component of a multi-purpose impact assessment. Supporters of integration hope that by making the process less burdensome it will result in an increased use of HIA-like approaches. Others fear that the focus on health will be unacceptably diluted.

Some bold enthusiasts, such as those who advocate comparative risk assessment [28], have even suggested that HIA could reduce different outcomes to a common metric. It would then be possible to make the trade offs by simply summing the different outcomes for each option and then directly comparing the options. One possibility might be to assign a monetary value to all utilities [19]. Others have tried to use measures such as QALYs (Quality Adjusted Life Year) [29] or DALYs (Disability Adjusted Life Year) [30]. However the assumptions underlying these measures are contentious and they are probably not capable of satisfactorily combining outcomes as disparate as those with which HIA is concerned. The notion that some elaborate calculation could remove the need for judgement in policy making is fanciful. HIA can only be an aid to and not a substitute for judgement in decision making.

The other stream driven by notions of participative democracy emphasizes the capacity of HIA to bring the different stakeholders into the process and make decision making more transparent and inclusive. How decision makers view this approach will depend on their willingness to share power with those affected by policy decisions. This type of HIA found much in common with community development and social education [31]. In some cases the focus shifted so far towards community participation that decision making was forgotten. (See, e.g., Chapter 13.) An emphasis on participation focuses attention on the recommendation step of HIA seeing itself as a way of giving a voice to the voiceless. But claims of HIA to be a means of advocacy and a channel for participation raise uncomfortable questions about legitimacy and control (how is it decided who participates? can local and health authority staff be the voice of their community? who sets the agenda?). These questions have not yet been properly addressed in HIA.

Addressing the needs of decision makers and participation are both important in HIA and this book contains examples of both streams. The different streams of HIA are not incompatible and different combinations are likely to be appropriate for different situations.

Development of the evidence base of HIA

As HIA came to be noticed its evidence base attracted criticism. With many HIAs the justification for the predictions and recommendations was far from clear. Parry and Stevens [32] wrote 'limitations compromise predictions to such an extent that the standard model of health impact assessment should be abandoned'. Others pointed out the dangers of basing recommendations on incomplete understanding [33] and argued that HIA should be built on the same sort of evidence as was being urged for clinical medicine. Proponents of participatory HIA objected that epidemiological and toxicological evidence forced HIA into a very narrow view of health, excluded most stakeholders from the discussion, and worst of all failed to recognize the richness and complexity of human societies.

The basis for prediction is understanding of causal connections built up from past experience (evaluations of previous interventions) using epidemiological and sociological methods. Here much progress has been made in building up databases of systematic reviews and there is scope for a great deal more (see Chapter 7). The evidence debate will not be advanced by parroting either the strengths or the shortcomings of randomized controlled trials and other epidemiological methods. Rather there is a need to recognize on the one hand that many interventions are necessarily one offs in unique situations for which no true control is possible and on the other that in such situations we need rigorous methods of causal attribution. Critiques of HIA need to accept the reality that certainty is unattainable and HIA cannot replace judgement.

Development of the HIA community

The number of people interested in HIA has grown over the years. The first UK Health Impact Assessment Conference was held in Liverpool, England, in 1998 and has now been held five times with the fifth and latest UK and Ireland Conference in Birmingham, England, in 2003. The growth of European HIA networks is described in Chapter 14 and the growing interest of the IAIA (International Association for Impact Assessment) is described in Chapter 32. HIA reports do not adapt easily to produce the very concise papers sought by scientific journals but the number of HIAs published in the literature is growing. More importantly a number of easily accessible depositories for HIA reports have been developed such as the HDA HIA website [34] and more recently the WHO HIA website [35]. Unlike environmental impact assessment, HIA remains the preserve of non-specialists and has not developed into an industry with paid consultants.

Interest in HIA is apparent in many parts of the world. Experience with HIA has been gathered in the United Kingdom, Scandinavia, Germany, Netherlands, Australia, New Zealand, and Canada. In the accession countries (countries about to join the EU) and in South Eastern Asia interest has been stimulated by WHO.

This chapter has attempted to trace how HIA has developed over the past 10 years (up to 2003). During that time interest in HIA has risen exponentially and there is ample scope for further development over the next 10 years.

References

1 Scott-Samuel A. Health impact assessment: an idea whose time has come. *British Medical Journal* 1996; **313**:183–184.

2 Goodland R and Mercier JR. The Evolution of Environmental Assessment in the World Bank: from 'Approval' to Results. Environment Department Papers No 67. New York: World Bank, 1999.

3 **Burdge RJ and Vanclay F.** Social impact assessment. In Vanclay F and Brinstein DA (eds.), *Environmental and Social Impact Assessment.* Chichester: John Wiley, 1995.

4 **Hendley J, Barnes R, Hirschfield A, and Scott-Samuel A.** What is HIA and How can it be Applied to Regeneration Programmes? Working paper series No 1. Health Impact Assessment Research and Development Project. Liverpool: University of Liverpool, 1999.

5 BMA Board of Science and Education. *Health and Environmental Impact Assessment.* London: Earthscan, 1998.

6 WHO European Centre for Health Policy. Health Impact Assessment: Main Concepts and Suggested Approach. Gothenburg Consensus paper, 1999.

7 West Midlands Directors of Public Health Group. Using Health Impact Assessment to make better decisions: A simple guide. Birmingham: NHS Executive West Midlands, 2001.

8 **Birley MH.** *The Health Impact Assessment of Development Projects.* London: HMSO, 1995.

9 Public Health Commission. *A Guide to Health Impact Assessment.* Wellington, New Zealand: Public Health Commission, 1995.

10 **Ewan C, Young A, Bryant E, and Calvert D.** *Framework for Environmental and Health Impact Assessment.* Canberra: National Health and Medical Research Council, Australian Government Publishing Service, 1994.

11 **Fehr R.** Environmental health impact assessment—evaluation of a 10 step model. *Epidemiology* 1999; **10**:618–625.

12 **Will S, Ardern K, Spencely M, and Watkins S.** Proof of evidence of Stockport Health Commission to Town and Country Planning Act Enquiry on Application by Manchester Airport PLC for the development of a second main runway (part) and associated facilities. 1994.

13 **Milio N.** *Promoting Health through Public Policy.* Ottawa: Canadian Public Health Association, 1986.

14 World Health Organisation. The Ottawa Charter for health promotion. Health Promotion 1, iii–v.

15 **Acheson D.** *Independent Inquiry into Inequalities in Health.* London: The Stationery Office, 1998.

16 Population Health Resource Branch. *Health Impact Assessment Toolkit.* Vancouver, British Columbia: Ministry of Health, 1994.

17 **Putters K.** *Health Impact Screening.* Rijswijk: Ministry of Health, Welfare and Sport, 1994.

18 Landstings Forbundet and Svenska Kommunforbundet. *Focussing on Health.* Stockholm: Landstings Forbundet. www.lf.se/hkb

19 Department of Health. *The Health of the Nation; Policy Appraisal and Health; A Guide from the Department of Health.* London: Department of Health, 1995.

20 Department of Health. *Our Healthier Nation: A Contract for the Nation.* London: The Stationery Office, 1998.

21 Department of Health. *Saving Lives: Our Healthier Nation.* London: The Stationery Office, 1999.

22 Scottish Office. *Working together for a Healthier Scotland.* Edinburgh: The Stationery Office, 1998.

23 Welsh Office. *Better Health; Better Wales.* Cardiff: The Stationery Office, 1998.

24 Secretary of State for Northern Ireland. *Well into 2000*. Belfast: Department of Health and Social Services, 1997.

25 WHO. *Health 21: An introduction to the Health for All Policy Framework for the WHO European Region*. Copenhagen: WHO, 1999.

26 Hubel M. Evaluating the health impact of policies: a challenge. *Eurohealth* 1998; **4**:27–29.

27 Taylor L and Blair-Stevens C (eds.). Introducing health impact assessment (HIA): Informing the decision making process. London: health development Agency 2002 http://www.phel.gov.uk/hiadocs/Full_copy_of_HDA_short_guide.pdf

28 Ezzati M. Comparative Risk Assessment in the Global burden of disease study and the environmental health risks. In Kay D, Pruss A and Corvalan C (eds.), *Methodology for assessment of environmental burden of disease*. WHO/SDE/WSH/)).7. Geneva: World Health Organisation, 2000.

29 Williams A and Kind P. The present state of play about QALYs. In Hopkins A. (ed.), *Measures of Quality of Life and the Uses to which such Measures may be put*. London: Royal College of Physicians, 1992.

30 Murray CJL. Quantifying the burden of disease: the technical basis for disability adjusted life years. *Bulletin of World Health Organisation* 1994; **72**:429–445.

31 Mittelmark M. Promoting social responsibility for health: health impact assessment and public policy at the community level. *Health promotion International* 2001; **16**:269–274.

32 Parry J and Stevens A. Prospective health impact assessment: pitfalls, problems and possible ways forward. *British Medical Journal* 2001; **323**:1177–1182.

33 MacIntyre S and Petticrew M. Good intentions and received wisdom are not enough. *Journal of Epidemiology Community Health* 2000; **54**:802–803.

34 HDA health impact assessment gateway http://www.hiagateway.org.uk/resources/completed_hia_database/completedhialist.asp

35 WHO website—Health Impact Assessment http://www.who.int/hia/examples/en/

Chapter 3

Health inequalities and HIA

Johan P Mackenbach, J Lennert Veerman,
Jan JM Barendregt, and Anton E Kunst

Introduction

At the start of the twenty-first century all European countries are faced with substantial socio-economic inequalities in health with systematically higher morbidity and mortality rates in people with lower as compared to those with higher socio-economic status. There is no sign that these inequalities will spontaneously disappear, on the contrary all the evidence suggests that inequalities have increased over the last decades of the twentieth century [1–3].

The emphasis of research in the area of socio-economic inequalities in health has gradually shifted from description to explanation, and a general understanding of the factors involved has emerged. Childhood circumstances, material factors, health-related behaviours, and psychosocial factors have all been shown to contribute significantly to the explanation of socio-economic inequalities in health [4–9]. A better understanding of the explanation has laid the foundation for the systematic development of strategies to reduce such inequalities. During the 1990s important progress has been made in this field, and around the year 2000 national advisory committees in several European countries (Britain, Sweden, Netherlands) have laid out complete 'blueprints' for strategies to tackle socio-economic inequalities in health [10–12].

All these strategies acknowledge the fact that interventions and policies that can be expected to reduce health inequalities will mainly have to be found outside the health care system in a narrow sense of the word. Most of the factors involved in the explanation of health inequalities can only be tackled by 'intersectoral' policies such as income and housing policy, taxation of products like tobacco and alcohol, or regulations of working conditions. It is within this context that a plea has been made for the application of health impact assessment (HIA). Indeed, all the national reports mentioned here recommend HIA as a method to steer policies in these other areas towards contributing to a reduction in health inequalities [10–12].

Health impact assessment has been defined as 'a combination of procedures, methods and tools by which a policy, programme or project may be judged as to its potential effects on the health of a population, and the distribution of those effects within the population' [13]. This definition testifies to the awareness among those involved in developing HIA that it is important to assess the impact of policies not only on the health of the 'average' citizen, but to also look carefully at the health impact in vulnerable groups, such as those with low socio-economic status. Indeed, there is a general consensus among HIA experts that all HIA methods and procedures should incorporate a focus on health inequalities [14].

In this chapter the current state of affairs on this intersection between the older health inequalities tradition and the evolving HIA field is reviewed. Then the available evidence on health inequalities is summarized. Next possible approaches for assessing the impact of policies on health inequalities are outlined, and a few HIA examples are discussed in which health inequalities were explicitly considered. Finally, a research agenda is laid out to develop the tools that will in the future help us to reliably and validly assess the impact of policies on health inequalities.

Health inequalities: what do we know

Many countries have nationally representative surveys with questions on both socio-economic status (education, occupation, income) and self-reported morbidity (e.g., self-assessed health, chronic conditions, disability). These data show that inequalities in self-reported morbidity are substantial everywhere, and nearly always in the same direction: persons with a lower socio-economic status have higher morbidity rates [15,16]. For mortality too, a harder but rarer outcome measure, socio-economic inequalities of considerable magnitude have been found in all countries with available data. For example, the excess risk of premature mortality among middle-aged men in manual occupations compared to those in non-manual occupations ranged between 33 and 71 per cent in a comparative study of the situation in the 1980s in Western Europe [17].

Early debates about the explanation of socio-economic inequalities in health focused on the question whether 'causation' or 'selection' was the more important mechanism. Social selection explanations imply that health determines socio-economic position, instead of socio-economic position determining health. There is indeed some evidence that during social mobility selection on (ill)-health may occur, with people who are in poor health being less likely to move upward and/or more likely to move downward [18,19]. Although the occurrence of health-related selection as such is undisputed, its contribution to the explanation of socio-economic inequalities in health is less clear. Only a few studies have investigated this directly, and these have shown that the contribution to inequalities in health by occupational class is small [18,19].

Longitudinal studies, in which socio-economic status has been measured before health problems are present and in which the incidence of health problems has been measured during follow-up, show higher risks of developing health problems in the lower socio-economic groups. This clearly shows that 'causation' instead of 'selection' is the main explanation for socio-economic inequalities in health [20,21]. This 'causal' effect of socio-economic status on health is likely to act through a number of more specific health determinants, which are differentially distributed across socio-economic groups. There is no doubt that 'material' factors, that is, exposure to low income and to health risks in the physical environment, are part of the explanation [5]. How low income affects health, and the relative importance of different pathways related to low income is, however, far from clear. Psychosocial factors also play an important role. Being in a low socio-economic position may be a psychosocial stressor, which, through biological or behavioural pathways, could lead to ill-health. Psychosocial factors related to work organization, such as job strain, have been shown to play an important role in the explanation of socio-economic inequalities in cardiovascular health [8,9]. Health-related behaviours, such as smoking, diet, alcohol consumption, and physical exercise, are also important 'proximal' determinants of socio-economic inequalities in health. In most countries smoking is more prevalent in the lower socio-economic groups [7]. The contribution of diet to inequalities in health is less clear. In many countries people in lower socio-economic groups consume fewer fresh vegetables and fruits [6], but data on fat consumption do not suggest consistent differences between socio-economic groups [22]. On the other hand, obesity is very strongly associated with socio-economic status, with higher prevalence rates of obesity in the lower socio-economic groups, particularly in richer countries [6]. Data on socio-economic differences in alcohol consumption are also not always consistent, but frequently lower socio-economic groups have higher rates of both abstinence and excessive alcohol consumption [6].

It was once thought that inequalities in health would disappear with greater economic prosperity and the development of the welfare state. There is some evidence that inequalities in health are indeed responsive to this. In Britain income inequalities were deliberately reduced during the Second World War to 'gain the cooperation of the masses' in the war effort [23]. Perhaps as a result inequalities in mortality reached their narrowest in England and Wales around 1950 [24]. In the United States, the Second World War was also one of only two periods in the last century showing an abrupt decline in the racial gap in infant mortality, coinciding with the only significant decline in the racial wage gap for men [25]. These examples, and the evidence on between-country variations in the size of health inequalities, suggest that socio-economic inequalities in health are not fixed but can be reduced by deliberate action.

In recent years, researchers in Europe have developed promising conceptual frameworks to support the development of policies and interventions to reduce inequalities in health. Margaret Whitehead's paper 'The concepts and principles of equality in health' has laid the normative foundations for action in this area [26]. Margaret Whitehead's and Göran Dahlgren's paper 'Policies and strategies to promote inequalities in health' developed a systematic approach to identifying entry-points for policies and interventions to reduce inequalities in health [27]. The British King's Fund report on 'Tackling inequalities in health', of which Margaret Whitehead was a co-author, filled in this approach by listing sets of possible actions in four different policy areas: housing, social security, smoking, and health care [28].

While slightly different models of entry-points for policies to reduce inequalities in health have been developed, there is general agreement that at least the following entry-points should be considered [29]:

- reducing inequalities in power, prestige, income, and wealth linked to different socio-economic positions, for example, by reducing income inequalities and promoting educational achievement for children from families with lower socio-economic status;

- reducing the effect of health on socio-economic position, particularly by reducing the economic consequences of ill-health, for example, by providing decent illness benefits and supporting the school career of children with chronic diseases;

- reducing the effect of socio-economic position on the risk of being exposed to specific health dangers, such as unfavourable material, psychosocial and behavioural factors, for example, by providing good quality housing regardless of income, by improving working conditions, and by developing health promotion programmes that are sensitive to the perceptions and needs of people in lower socio-economic groups;

- reducing the health effects of being in a low socio-economic position by providing health care, for example, by strengthening primary care in deprived neighbourhoods and by securing accessibility to good quality health care regardless of income.

This conceptual development recently culminated in official government reports with specific policy proposals in three European countries: Britain, the Netherlands, and Sweden. In Britain, an 'Independent inquiry into inequalities in health' was held by the Acheson committee. This reviewed all the evidence, with the help of a large number of experts, and came up with 123 recommendations in 11 areas for future policy development [10]. In the Netherlands consecutive governments during the 1990s have pursued a systematic research-based strategy to cope with the issue of socio-economic inequalities in health [30]. In a final research programme a number of interventions and policies was evaluated as to their (potential) effectiveness, mostly in quasi-experimental designs. On the basis of the results of these studies, a government committee developed a set of 26 recommendations which cover all the entry-points mentioned in the previous paragraph [12]. In Sweden a parliamentary committee recently suggested targets and strategies for a new policy to reduce inequalities in health, 'Health on equal terms' [11]. The committee gave priority to tackling more than a dozen determinants, ranging from housing segregation and work organization to alcohol and smoking. All three policy reports also made strong pleas for the application of HIA to involve other policy areas in strategies to reduce health inequalities.

Steps in, and requirements for, 'Health Inequalities Impact Assessment'

According to the Gothenburg Consensus Paper, the first stage of any HIA is to select which policies or programmes could have an impact on health and what kind of impact ('screening' process), and to determine what further work should be carried out, by whom and how ('scoping' process). Screening and scoping are usually done on the basis of careful study of available documents and consultation of experts. If screening and scoping suggest that further work should be done, then the consensus paper suggests that one of three broad categories of action could be taken; rapid health impact appraisal based on existing knowledge; a more in-depth health impact analysis requiring the compilation and analysis of new information; and in the case of broad policies that make an in-depth analysis infeasible, a broad-brush health impact review [31].

It the first stage of screening and scoping it should also be determined whether any further work on health inequalities would need to be carried out. Is it possible that the policy will have an effect on health inequalities? If the answer to this question is affirmative, then the next stage, be it in the form of rapid health impact appraisal, health impact analysis, or health impact review, will need to have an explicit focus on health inequalities.

Policies in areas like income, education, housing, transport, or working conditions will impact on health through specific determinants such as purchasing power, health-related knowledge, house dampness, accident risks, and job control. Likewise, these policies will impact on health inequalities through their effect on the distribution of specific determinants. Only if their effect on these determinants is limited to, or stronger in, certain socio-economic groups will they have an effect on the size and pattern of socio-economic inequalities in health.

This suggests that, in order to determine whether the proposed policy could possibly affect health inequalities, the first stage of screening and scoping should include the following steps:

1. identify the health determinants on which the policy may impact;
2. check the current socio-economic distribution of these determinants, and their contribution to current socio-economic inequalities in health; and
3. judge whether the policy could differentially affect the prevalence of these determinants in different socio-economic groups, and thereby contribute to widening or narrowing of socio-economic inequalities in health.

The first step is common to all HIAs, and is part of all screening and scoping exercises. Whether or not the policy can reasonably be expected to have an effect on important determinants of health, will of course largely determine whether a 'full' HIA has to be carried out or not. This will usually be based on a careful analysis of the content (intended effects, target groups, proposed implementation strategy, accompanying measures, time-schedule, etc.) of the policy, as well as a brief review of previous experiences with similar policies (implementation problems, side-effects, etc.).

The second and third steps are specific for the health inequalities focus, but should be incorporated in general procedures for screening and scoping. The possible effects of policies on health determinants can only be assessed against the background of their current prevalence and contribution to the occurrence of ill-health. While such background information needs to be available any-how, a differentiation by socio-economic group is essential for step 2. Similarly, the usual assessment of whether the proposed policy could affect the preva-lence of health determinants should also be differentiated by socio-economic group. Existing instruments for screening and scoping, such as checklists, could easily be extended with a few questions related to such a health inequalities focus, but application is far from trivial and will require availability of specific data and availability or consultation of specific expertise.

If the conclusion of the first stage is that a full HIA is necessary, and that this needs to have a health inequalities focus (in addition, mostly, to other foci), then the steps shown in Table 3.1 could provide useful guidance.

Table 3.1 Steps needed to focus on health inequalities in an HIA

A. Estimate the prevalence of health determinants on which the policy may impact

B. Check the current socio-economic distribution of these determinants, and their contribution to current socio-economic inequalities in health

C. Estimate the effect of the policy on the prevalence of these determinants in different socio-economic groups

D. Estimate the effect of changes in the prevalence of these determinants in different socio-economic groups on inequalities in health.

Use of the term 'estimate' in Table 3.1 is intended to show that while the generation of quantitative information should be aimed for, this will not always be feasible under current circumstances. Even if the impact assessment has to be based largely on qualitative methods, however, keeping in mind a quantitative conceptual framework may help to structure the exercise and to identify the main gaps in reasoning.

Step A, again, is common to all full HIAs, but steps B, C, and D add an extra dimension.

Step B requires access to information and expertise in the field of socio-economic inequalities in health. In many countries, the recent surge in the interest in socio-economic inequalities in health has greatly increased the amount of available information, but this may not be readily accessible by those performing HIA. It is therefore advisable to include overviews of socio-economic distributions of health determinants, and of their contribution to socio-economic inequalities in health, to be included in the (Internet) 'libraries' that are being developed to support HIA in various countries.

Step C usually requires a variety of approaches: detailed analysis of the 'content' of the proposed policy; literature search to identify the effect of previous, related policies; systematic consultation of a wide range of experts. While the general approach is common to all HIAs, estimation of the policy's effect not only on the 'average' prevalence of health determinants, but also on the socio-economic distribution of these determinants, will usually complicate this exercise substantially. It requires insights into the sections of the population that will be affected by the policy and into their behavioural responses to the policy, and such insights may not be readily available either from the literature or from experts.

Step D translates changes in the distribution of determinants into changes in the distribution of health, and requires insights into the link between the two. Whereas step B depends on the availability of knowledge on the current contribution of determinants to the explanation of socio-economic inequalities in health, step D requires an understanding of the effect of manipulation of these determinants. The easiest would be to simply assume that the strength of current associations predicts the size of manipulation effects, and perhaps that is all one can do in the absence of better information. Even then, this may require a quantitative model linking determinants to health outcomes. Such models have been constructed (e.g., Prevent) but do not yet include a differentiation by socio-economic status.

Examples

This section briefly describes three examples of HIA studies that include explicit references to policy impacts on health inequalities in their conclusions, based on a structured approach to identify those impacts.

The Edinburgh transport strategy (UK) [32]

Description of the HIA exercise

The study assesses three possible scenarios for an urban transport strategy: (1) emphasis on private motorized transport, (2) accompanying measures like traffic calming, improvement of public transport, more possibilities for cyclists and pedestrians, and (3) an intermediate scenario. The HIA consisted of screening followed by 'rapid health impact appraisal'. The screening involved selection of partners and informants, a literature review, community profiling, and policy analysis. It resulted in the identification of key impacts, and identified the following determinants that could be affected by the policy scenarios: accidents, pollution, physical activity, access to goods and services, and community network. It also identified health inequalities as a possible focus for further HIA.

The rapid health impact appraisal consisted of two half-day discussion sessions within the project group to determine the effect of policy scenarios on determinants/health in two population groups (disadvantaged, predominantly non-car-owning; middle class, affluent, predominantly car-owning). The conclusion was that 'Transport policy could have significant impacts on the health of the people in Edinburgh. Many potential impacts are borne by the groups of people who already have the poorest health. Thus transport has important impacts on inequalities in health. There is clearly a need to give special consideration to the most disadvantaged groups of people when designing transport policy. The scenario with investment in public transport, cycle ways and pedestrian zones predicted the best health impacts for all groups'.

Comments

From the perspective of the framework discussed in this chapter, we can describe the study as follows. The assessment identifies the main determinants influenced by the policy (step A), and for two out of five (road traffic accidents and access to good and services) it gives information on the distribution over different socio-economic groups (step B). The effects of policy are estimated in terms of five ordinal categories of impact size, for a simplified scheme of higher and lower socio-economic groups (step C). The overall effects on the health of the different groups and on health inequalities are stated in the conclusion in general terms (step D). The approach could serve as a basis for a more in-depth health impact analysis, which would need to further specify the distribution of determinants and link this to the available epidemiological evidence.

EU's Common Agricultural Policy on fruits and vegetables (Sweden) [33]

Description of the HIA exercise

The HIA assesses the health effects of the EU's Common Agricultural Policy on fruits and vegetables. It uses available information, and thus falls in the category of 'rapid health impact appraisal'. The appraisal was performed on the basis of policy analysis and a limited literature review. The authors concluded that 'by artificially keeping prices of fruits and vegetables up, the EU Common Agricultural Policy contributes to socio-economic inequalities in health'.

Comments

This HIA is quite specific in its approach. It identifies the health determinant affected by the policy (fruit and vegetable consumption), discusses the socio-economic distribution of this determinant and its contribution to current socio-economic inequalities in health, and pinpoints the mechanism through which the policy may affect the socio-economic distribution of the determinant (i.e., prices). It does remain predominantly qualitative: there is no quantitative framework and no attempt at quantifying the effects.

Tobacco discouragement policy (Netherlands) [34]

Description of the HIA exercise

The study assesses a number of different policy options for the discouragement of tobacco consumption in the Netherlands in 1998. The HIA can be characterized as a 'health impact analysis'. It involved literature study to estimate the effects of various policies on smoking behaviour, epidemiological modelling to link changes in smoking behaviour to health outcomes, and an analysis of the economic effects of the policy measures. One of the conclusions is that 'some policies may affect the socio-economic distribution of smoking, and thereby socio-economic inequalities in health'.

Comments

This HIA follows the 'rational' model outlined in the third section of this chapter, and even includes a quantitative estimate of policy effects on future population health. However, although socio-economic inequalities in smoking and in health are discussed explicitly, the modelling effort did not include a differentiation by socio-economic status, and conclusions about policy impacts on health inequalities remained qualitative.

Discussion

Although health inequalities are mentioned in a relatively large number of HIA reports, relatively few use a structured approach to estimate policy impacts on health inequalities. Three examples that did follow a structured approach were described, and each of these includes elements of the stepwise approach outlined in the third section of this chapter. While none of them attempts to quantify policy impacts on health inequalities, they do suggest that with investments in further methods development and data collection such quantification may come within reach.

In our search through the official and unofficial literature we have seen many other examples of HIA that suggest policy impacts on health inequalities, but use a less structured approach. Many of these were limited to a broad-brush and qualitative assessment on the basis of some literature and expert consultation. While the results of these assessments may have played a useful role in the policy process, for example, by alerting policy makers to the existence of vulnerable groups that might be affected negatively (or positively) by their policies, they are unlikely to fulfil basic standards of reproducibility and validity. This situation is symptomatic of a field in which there is no consensus about methods, and in which methods development may simply not yet have reached a stage in which consensus would be useful. While this may also apply to the HIA field in general, it is certainly true for the even more complex subarea of health inequalities impact assessment.

Our rational approach, with a step-wise and systematic assessment of the impact of policies on determinants, resembles that proposed by Joffe and Mindell [35] as well as others [36,37]. Such approaches may appeal to rational minded researchers and policy makers, but it is unclear whether they can as yet be used in practice. In fact, the three examples described here suggest that full application of such a systematic, partly quantitative approach is currently not feasible. This is certainly true for 'rapid health impact appraisal', and may also currently be true for 'health impact analysis', because availability of more resources in the form of time and money will not solve the more fundamental problems of lack of relevant data and understanding of mechanisms.

We would therefore like to end this chapter with a research agenda. Letters B, C, and D refer to the steps outlined for a full HIA in Table 3.1.

Step B—assess current distribution of determinants and its effect on health inequalities. Increase routine availability of data on the socio-economic distribution of health determinants, not only for national but also for regional and local populations, and include these in databases and libraries accessible to those performing HIA.

Make overviews of the contribution of health determinants to socio-economic inequalities in health, and include these in databases and libraries accessible to those performing HIA.

Step C—from policy to determinants. Make overviews of documented effects on inequalities in health determinants (and, if possible, health) of policy in non-health areas.

Adapt currently available methods, such as policy analysis and Delphi approaches, for prospective estimation of policy impacts on inequalities in health determinants.

Step D—from determinants to health. Make overviews of documented health effects of specific determinants likely to play a role in many HIAs (e.g., food consumption patterns, physical exercise, perceived control, transport, welfare benefits). Ideally, a list of (relative) risk estimates for a wide range of determinants is at the disposal of anyone wishing to perform an HIA.

Development of methods for the quantification of health effects of changes in the distribution of determinants. This is likely to involve epidemiological modelling, but current models (e.g., Prevent) do not include a socio-economic differentiation of the population. Further development of these models is necessary.

References

1 Mackenbach JP, Bos V, Andersen O, Cardano M, Costa G, Harding S, Reid A, Hemstrom O, Valkonen T, Kunst AE. Widening socioeconomic inequalities in mortality in six western European countries. *International Journal of Epidemiology* 2003; **32**:830–837.

2 Harding S. Social class differences in mortality of men: recent evidence from the OPCS longitudinal study. *Population Trends* 1995; **80**:31–37.

3 **Borrell C, Plasencia A, Pasarin I, *et al*.** Widening social inequalities in mortality: the case of Barcelona, a southern European city. *Journal of Epidemiology Community Health* 1997; **51**:659–667.

4 **Barker.** *Mothers, Babies and Health in Later Life.* 2nd edn. Edinburgh: Churchill Livingstone, 1994.

5 **Davey Smith G, Blane D, and Bartley M.** Explanations for socioeconomic differentials in mortality. Evidence from Britain and elsewhere. *European Journal of Public Health* 1994; **4**:131–144.

6 **Cavelaars AE, Kunst AE, and Mackenbach JP.** Socio-economic differences in risk factors for morbidity and mortality in the European Community: an international comparison. *Journal of Health Psychology* 1997; **2**:353–372.

7 **Cavelaars AE, Kunst AE, Geurts JJ, Crialesi R, Grotvedt L, Helmert U, *et al*.** Educational differences in smoking: international comparison. *British Medical Journal* 2000; **320**(7242):1102–1107.

8 **Marmot MG, Bosma H, Hemingway H, Brunner E, and Stansfeld S.** Contribution of job control and other risk factors to social variations in coronary heart disease incidence. *Lancet* 1997; **350**(9073):235–239.

9 **Schrijvers CT, van de Mheen HD, Stronks K, and Mackenbach JP.** Socioeconomic inequalities in health in the working population: the contribution of working conditions. *International Journal of Epidemiology* 1998; **27**(6):1011–1018.

10 **Acheson DC.** *Independent Inquiry into Inequalities in Health.* London: The Stationary Office, 1998.

11 **Ostlin P and Diderichsen F.** *Equality-Oriented National Strategy for Public Health in Sweden.* Brussels: European Centre for Health Policy, 2000.

12 **Mackenbach JP and Stronks K.** A strategy for tackling health inequalities in the Netherlands. *British Medical Journal* 2002; **325**:1029–1032.

13 WHO European Centre for Health Policy. Health Impact Assessment. Main concepts and suggested approach. Gothenburg Consensus Paper, December 1999. Copenhagen: WHO Regional Office for Europe, 1999.

14 **Douglas M and Scott-Samuel A.** Addressing health inequalities in health impact assessment. *Journal of Epidemiology Community Health* 2001; **55**(7):450–451.

15 **Cavelaars AEJM, Kunst AE, Geurts JJM, *et al*.** Morbidity differences by occupational class among men in seven European countries: an application of the Erikson-Goldthorpe social class scheme. *International Journal of Epidemiology* 1998; **27**:222–230.

16 **Lahelma E and Arber S.** Health inequalities among men and women in contrasting welfare states. Britain and three Nordic countries compared. *European Journal of Public Health* 1994; **4**:213–226.

17 **Mackenbach JP, Kunst AE, Cavelaars AE, Groenhof F, and Geurts JJ.** Socioeconomic inequalities in morbidity and mortality in western Europe. The EU Working Group on Socioeconomic Inequalities in Health. *Lancet* 1997; **349**(9066):1655–1659.

18 **Wadsworth MEJ.** Serious illness in childhood and its association with later-life achievement. In Wilkinson RG (ed.), *Class and Health, Research and Longitudinal Data.* London: Tavistock, 1986.

19 **van de Mheen H, Stronks K, Schrijvers CT, and Mackenbach JP.** The influence of adult ill health on occupational class mobility and mobility out of and into employment in the The Netherlands. *Social Science and Medicine* 1999; **49**(4):509–518.

20 Marmot MG, Smith GD, Stansfeld S, Patel C, North F, Head J, *et al.* Health inequalities among British civil servants: the Whitehall II study. *Lancet* 1991; **337**(8754):1387–1393.

21 Schrijvers CT, Stronks K, van de Mheen HD, and Mackenbach JP. Explaining educational differences in mortality: the role of behavioral and material factors. *American Journal of Public Health* 1999; **89**(4):535–540.

22 Davey Smith G and Brunner E. Socio-economic differentials in health: the role of nutrition. *Proceedings of the Nutrition Society* 1997; **56**:75–90.

23 Titmuss RM. War and social policy. In Titmuss RM (ed.), *Essays on the Welfare State.* London: Unwin, 1958.

24 Pamuk ER. Social class inequality in mortality from 1921 to 1972 in England and Wales. *Population Studies (Camb)* 1985; **39**(1):17–31.

25 Collins WJ and Thomasson MA. Exploring the racial gap in infant mortality rates, 1920–1970. Cambridge, Massachusetts: National Bureau of Economic Research, 2002.

26 Whitehead M. *The Concepts and Principles of Equity and Health.* Copenhagen: World Health Organization, 1990.

27 Whitehead M and Dahlgren G. What can be done about inequalities in health? *Lancet* 1991; **338**(8774):1059–1063.

28 Benzeval M, Judge K, and Whitehead M. Tackling inequalities in health; an agenda for action. London: King's Fund, 1995.

29 Mackenbach JP, Bakker MJ, Sihto M, Diderichsen F. Strategies to reduce socioeconomic inequalities in health. In Mackenbach JP, Bakker MJ (eds.), *Reducing Inequalities in Health: a European Perspective.* London: Routledge, 2002.

30 Mackenbach JP. Socioeconomic inequalities in health in The Netherlands: impact of a five year research programme. *British Medical Journal* 1994; **309**(6967):1487–1491.

31 Douglas M (ed.). Health Impact Assessment: from theory to practice. Leo Kaprio Workshop; 2000 September; Gothenburg. Nordic School of Public Health.

32 SNAP. Health Impact Assessment of the City of Edinburgh Council's Urban Transport Strategy. Edinburgh: SNAP, p. 60, 2000.

33 Leather S. The CAP regime for fruit and vegetables. In Dahlgren G, Nordgren P, Whitehead M (eds.), *Health Impact Assessment of the EU Common Agricultural Policy.* Stockholm: Swedish Institute for Public Health, pp. 17–22, 1996.

34 Willemsen MC, Zwart WMd, Mooy JM, Gunning-Schepers LJ, Leeuwen MJv, and Sleur DG. GES Tabaksontmoedigingsbeleid. Utrecht: NSPH, p. 145, 1998.

35 Joffe M and Mindell J. A framework for the evidence base to support Health Impact Assessment. *Journal of Epidemiology Community Health* 2002; **56**(2):132–138.

36 McCarthy M, Biddulph JP, Utley M, Ferguson J, and Gallivan S. A health impact assessment model for environmental changes attributable to development projects. *Journal of Epidemiology Community Health* 2002; **56**(8):611–616.

37 Mindell J, Hansell A, Morrison D, Douglas M, and Joffe M. What do we need for robust, quantitative health impact assessment? *Journal of Public Health Medicine* 2001; **23**(3):173–178.

Chapter 4

Causal mechanisms for HIA: learning from epidemiology

Stephen Palmer

Health impact assessors who wish to advise policy makers and base their health impact assessment (HIA) on the fullest scientific evidence need to appreciate the potential and pit falls of epidemiology. This chapter will explore how epidemiological reasoning can give insight into the causal mechanisms by which decisions impact on health, go some way towards estimating the size of those impacts, and thereby provide a basis for prediction.

Epidemiology is the study of the occurrence of disease in populations and their causes. Whereas in clinical situations the focus of concern is the individual patient, in epidemiology the focus is the relation of disease in the individual to the occurrence or risk of disease in the relevant population. Epidemiological studies produce statistical associations between features of people or their environment, and health or disease states. These associations are interpreted within causal frameworks that presume that these relationships will apply to other populations in the future. This presumption is critical to the use of epidemiology as a basis for prediction in HIA.

The steps in the epidemiological approach are:

- demonstration of an association,
- assessment of its validity,
- interpretation of the association within a causal framework, and
- prediction to future populations.

Validity

Associations may be spurious, resulting from misinformation, from misclassification of people and/or disease, or from the operation of chance. Good design and conduct of epidemiological studies and careful analyses and interpretation of results reduces the risk of finding spurious associations. There is an alarmingly long and growing list of possible biases in epidemiological studies that have to be avoided. Critical appraisal of epidemiological studies requires formal assessment of the possible biases. The main groups of biases in addition to chance are selection biases, measurement biases and other confounding factors.

Even when associations are considered to be true, the most that can be said statistically is that they are unlikely to be due to chance (the likelihood is estimated by the confidence intervals and p values), and that there was not evidence of selection or measurement bias. Specific issues to do with validity in epidemiological studies are discussed under each method later in this chapter.

Causal inference

Rothman and Greenland [1] define a cause of a disease as 'an antecedent event, condition, or characteristic that was necessary for the occurrence of the disease at the moment it occurred, given that other conditions are fixed'. Factors that precede a disease may be insufficient in themselves to cause that disease, but may be necessary parts of a set of components that will be sufficient to cause the disease; or they may be contributing components but only under particular sets of conditions. Thus smoking is a cause of lung cancer, but not all people who smoke get lung cancer, and occasionally non-smokers get lung cancer. Such exceptions create doubts in the public mind about belief in a causal link between smoking and lung cancer but are not at all fatal to the argument. Smoking is still a very strong causal factor.

Bradford Hill published a set of criteria that has been widely used in epidemiology to test whether an association was likely to be causal [2]. It is important to lay out the relative merits of each criterion.

Strength

This refers to the size of the measure of the association (the relative risk or odds ratios) estimated in analytic studies. In infectious disease outbreaks odds ratios greater than 10 are common; that is, people exposed to, say contaminated food, are more than ten times as likely to be cases. The same size of effect is true for smoking and lung cancer. Very often in chronic disease epidemiology, however, the relative risk is less than 2. To many this reduces the power of causal inference because undetected biases could produce these results; but small effects may nonetheless have causal explanations. In multifactor diseases, especially those with long latency, large effects from single factors would not be expected.

Consistency

If many studies in different population groups at different times by different researchers produce consistent results it is unlikely that local biases are the cause. However, a discrepant finding may indeed indicate a specific causal factor that is present only some of the time or in only some populations.

Specificity

This refers to a cause producing a single disease effect, for example, smoking and squamous cell carcinoma of the bronchus. The specificity can lead to clues for further studies of biological mechanisms. However, there is no logical reason why one cause could not have many effects, as in the case of smoking.

Temporality

Rothman points out that this is the only criterion in the list that has to be satisfied for causal inference; the cause must precede the effect. In cross-sectional studies this relationship is not usually possible to prove.

Biological gradient

This refers to a dose-effect; the more a person smokes the higher the risk. Occasionally bias can replicate this, but in most cases this is a strong factor to consider.

Plausibility

This applies to the biological plausibility of the hypothesis or putative mechanisms to explain findings. Very often, however, no mechanism can be suggested and only after prolonged investigation are the unlikely associations of events causally linked (e.g., proximity to a fruit stand and Legionnaires disease—transmitted in water aerosols used to 'freshen' fruit).

Coherence

If the epidemiological evidence conflicts with the 'known' biology of the disease, which evidence should be accepted? Many experienced field epidemiologists would argue that we should believe the epidemiology, and challenge the 'science'. Often the science has had to catch up with epidemiological findings.

Experimental Evidence

Neither plausibility nor coherence can really be considered as criteria of causal inference in epidemiological studies. Experimental evidence is of added value and can test hypotheses raised from epidemiological studies. Of course this is highly desirable.

Types of study

Cross-sectional studies

Cross-sectional studies record the occurrence in a population of possible causal factors and of disease states at a given point of time. These are the weakest of the epidemiological studies that may be used to understand causation.

Because they consider a population at a given point in time, the time order of events, that is, what is cause and what is effect, is often impossible to untangle. For example, a survey of mental health, housing, employment, and income levels may show a strong association between poor mental health, bad housing, unemployment, and low income. But poor mental health may lead to loss of work, lower income, and therefore to residence in poorer housing. Or, poor housing, income, and loss of employment may lead to poor mental health. Or, as is likely there may be a mixture of the two. Cohort studies (see later) will be needed to clarify these relationships.

Correlation studies (small areas)

Correlation studies (sometimes called ecological studies) compare variables summarized over groups of people. In small-area studies, exposure to possible causal factors is related to the frequency of disease in small areas. For example, people living in areas close to a point source of pollution may appear to have an increased rate of congenital malformations [3]. The public, especially those suffering from the nuisance of smells and sight from the source, may understandably make a causal connection and believe that toxins from the source are causing the ill-health. But the areas may differ in factors other than the source of the pollution. The population living nearer the site may be of higher or lower socio-economic level, of different age structure, or have different lifestyles. Each of these may be associated with other causes of congenital malformations such as smoking, poor diet, or use of recreational drugs. These differences, beside the one of interest, are known as confounding factors.

The possibility of confounding factors is a common problem of epidemiology [4]. In small-area epidemiology it is compounded by the fact that individual-level data are seldom available for income, lifestyle, or other relevant factors. Often one has to rely on area-based measures of socio-economic deprivation (i.e., assume all residents in the area have the average area characteristic). This lays one open to the ecological fallacy—assuming that the association found at the population level pertains to individuals also. The possibility cannot be excluded that within the poorer areas it is the better off individuals who have the worse reproductive outcome for other reasons. Nevertheless ecological studies can give important clues to aetiology. For example, the correlation of average salt-intake level in a population and average values in blood pressure [5] was an important indicator that salt intake influences blood pressure. While population-level studies are useful in generating hypotheses they cannot provide quantitative evidence to predict the size of the effect at the individual level (i.e., how much does blood pressure rise for a given rise in salt intake).

Correlational studies (time series)

Time-series studies look at changes in the frequency of disease and exposure to possible causal factors in a population over a period of time. Introduction of a time element can strengthen causal inference in correlation studies. For example, in infectious disease epidemiology such approaches are used to detect outbreaks, to analyze seasonality, and to evaluate the effectiveness of interventions such as vaccine programmes. Little in the way of statistical analysis is needed because the causal pathways for microbiological disease are usually well understood. There is a sound biological basis for believing, for example, that HIB vaccine will reduce population incidence of haemophilus influenza and if incidence rises after starting a vaccine programme one may conclude that there is a problem with the programme. When this actually happened recently it was shown that the rise in incidence was due to lower vaccine efficacy following the introduction of a new combined vaccine [6].

Consider the public health challenge of advising on point sources of pollution such as a landfill site [7]. Comparison of the rate of disease before and after opening of a site, taking into account the latent period of the disease, would test the hypothesis that the source increases disease. However, the approach cannot exclude that any changing rate of ill-health might be due to other secular trends such as the opening of other sources of pollution, or that as a result of opening the site healthier (and wealthier) people have migrated out of the area.

Causal inference will be strengthened if it is possible to demonstrate a dose–response between exposure and ill-health (i.e., risk of disease increase as level of exposure increases). Too often accurate individual-level or even area-level measures of exposure to specific pollutants are not available. Proxy measures are therefore taken, which may be as crude as using geographical distance from the site as a measure of exposure [7]. If the study is carried out at the area level no allowance can be made for the variability of duration of residence, time spent out of doors, place of employment, and other such factors. However, these factors are likely to work against showing an association if the source is causing ill-health. Studies that combine measures of exposure (or proxies for exposure) and compare before and after, are relatively strong methodologically. Importantly, however, the statistical power of such studies will usually be low because the number of people residing around the sites is small. Combining data from many sites will increase the power, but this will also suffer from a dilution bias if there is heterogeneity between populations and sites.

One further caution needs to be added. Such studies are often done because of public concern that rates of ill-health are high in an area. Variations in rates will occur by chance, and demonstrating by statistical comparison with other areas that the rate is high, does not prove causality. Such a study may generate hypotheses that can then be independently tested in other sites and other populations. For example, when an association between incidence of leukaemia and residence near one television transmitter was found at one site, this finding could not be replicated at other sites across the country [8].

How might these findings be used in HIA? How can they inform an HIA of a proposal to extend or change the use of a landfill site? The most that can be said from the evidence to date is that the findings are compatible with the belief that, in the past, as yet unspecified toxic substances given off or leaked by landfills may have exposed pregnant women to a slightly increased risk of having a congenital malformation. The relative risk estimates with confidence intervals could be used to calculate the number of excess cases that would occur if that risk pertained to the new proposed site, all other things being equal. There would however be no certainty that the landfill in question was of a similar risk, or that improvements in technology may not significantly reduce levels of pollution. There would be more confidence in the prediction if a particular chemical was associated with a specific congenital malformation and that specific chemical was predicted to be present in specific amounts over specific distances from the site. It is highly unlikely that such data will be available in the foreseeable future. At the very least, the qualitative inference is that pollution from landfills of neighbouring areas could increase congenital malformations and therefore prevention through good engineering practice and close environmental monitoring would be a high priority. Waste disposal and HIA is discussed further in Chapter 28.

Perhaps the best example of using correlational studies to infer causation and then to estimate predicted health effects, is provided by studies of fine air particles and cardio-respiratory disease [9]. Many time-series studies have shown strong relationships over short-time periods between levels of air pollution of particles and the rate of cardio-respiratory disease mortality, which remain after taking account of confounding and other possible biases. The link is biologically plausible since small particles enter the lungs and have toxic effects on the airways, although specific mechanisms to cause mortality are not yet agreed on. However, the size of the effect is low, a 1 per cent increase in mortality for each 10 μg/m^3 increase in PM$_{10}$. There is a dose-response effect, and, mortality increases follow increases is air pollution and not vice versa. When the estimates of risk across different studies vary (as they do) one has to decide which estimate should be used in prediction? The usual approach to this problem is to combine results in a meta analysis in order to develop a 'consensus' risk estimate.

Causal inference from time-series studies is strengthened by results from cohort studies. In several cities the long-term average concentrations of particles were consistently predictive of relative risk of death, with a linear dose-response effect though the risk estimates were larger than in the time-series studies [10]. Work done by the UK Department of Health has used the risk estimates to assess the impact of reducing air pollution, expressing the findings in quality-adjusted life years [11]. However caution is needed in using evidence from one context to predict impacts in another. The Department of Health concludes 'There is evidence from the United States for a chronic effect of particles on mortality, although it is uncertain whether this evidence would apply in the UK, and, if so, what the size of the impact would be'.

Case–control studies

In a case–control study, historical data, usually recalled by the patient, in cases with disease are compared with data from well people. The definition of cases, accuracy of data recall, and appropriateness of controls are critical to validity of the results. Case–control studies are major tools in detecting the causes of outbreaks of infectious diseases and have also been used extensively in environmental epidemiology and cancer epidemiology. In more recent years they have become the preferred method of genetic epidemiology [12]. Case–control studies are particularly suited to identify exposures that may be causes of rare diseases. They can examine multiple exposures of interest in one study, and are often said to be relatively quick to undertake.

When interpreting the results of case–control studies the following points have to be considered. Since they rely on recall or records for information on past exposures they may be subject to significant measurement bias. Validation of information is often difficult or impossible especially for diseases with long incubation periods. Control of extraneous variables may be incomplete and selection of an appropriate comparison or control group may be difficult.

Case–control studies allow the calculation of odds ratios (and confidence intervals)—the ratio of the odds of the exposure in the case group compared to that in the control group. Statistically, this turns out to be the same as the odds of disease given exposure compared to non-exposure.

The Eurhazcon study [13] is an example of a case–control study. This was set up to test the hypothesis that women in Europe living close to landfill sites containing hazardous waste were at higher risk of congenital malformations. All cases and controls were births to women living within 7 km of a site. Cases were births with congenital malformations and controls were normal births. Cases were found to be 33 per cent more likely to live close by (within 3 km of a site), even when socio-economic deprivation levels were taken into account.

Could this association have been due to confounding factors? Confounding occurs when the effects of two exposures have not been separated and it is incorrectly concluded that the effect is due to the one rather than the other. Age and sex are common confounding variables. There are several methods to overcome this. First, a case–control study can be designed to take account of known confounding variables by matching. This is particularly useful for countering the bias that would result if cases and controls differed much in age or gender. Second, during the analysis of a case–control study the data can be stratified into an increasing number of subsets based on exposure to an increasing number of potentially confounding variables, but there is a limit as to how many variables can be considered in a single analysis.

Could an association be caused by bias? The control group is intended to provide an estimate of the exposure rate that would be expected to occur in the cases if there were no association between the study disease and exposure. Controls must be free from disease and similar to the cases with regard to past potential for exposure during the time period of risk under consideration. The biggest threats to the validity of a case–control study are: selection bias resulting from non-participation of subjects; differential misclassification of exposure biasing effect estimates towards the null value; and recall bias resulting from differential recollection of past events for cases and controls. Exposures associated with differential surveillance, diagnosis, referral, or selection of individuals can also lead to biased estimates of risk. Interviewer bias may occur where the interviewer, perhaps subconsciously, encourages certain responses to questions posed. Furthermore, for specific diseases there is a critical time window (or 'aetiologically relevant exposure period') within which, occurrence of a particular exposure may be relevant to causation of the disease. If exposures outside that time window are included, it can lead to a misclassification of exposure and reduce the likelihood that a true association will be recognized.

Cohort studies

A cohort study begins by identifying groups of people exposed and not exposed to a particular factor, and then finding out the occurrence of disease in the two groups. Contrast this with a case–control study, which begins by identifying people with and without the disease and then retrospectively tries to identify exposures in the two groups. Cohort studies may be prospective, when the disease occurs after the study has begun and the population characteristics have been identified, or retrospective. The essence of the study is to discover which factors predict the development of disease within the cohort. In cohort studies the incidence of the disease in the exposed is compared to the incidence in the unexposed, and the ratio is called the relative risk. Prospective cohort studies have the particular strength in that exposure can be accurately measured before disease occurs. However, cohort studies are subject to similar biases as case–control studies.

The difficulty of making predictions on the basis of cohort studies is illustrated by the example of the antioxidants beta carotene and heart disease. Consistent findings in cohort studies showed that people with high carotene intake had lower risk of heart disease, but when people were given beta carotene in intervention studies the finding was not replicated and no protection was seen. According to Davey Smith and Ebrahim [14], the best explanation of such unexpected findings is residual confounding in the observation studies. When those who are exposed (to high dietary intake of beta carotene) are very different people from those not exposed, it is unlikely that simply matching or statistically adjusting for differences in measurable socio-economic factors will remove all of the extraneous effects. In randomized controlled trials, where the exposure is controlled as an intervention, the effect of these other unspecified causal factors tends to be equalized out by the random allocation to intervention and control arms of the trial. In observation studies the problem of residual confounding can be minimized but not eliminated. Ways of minimizing it include increased efforts to measure more and more specific potential confounders, performing more sensitivity analyses to explore the potential bias of measurement error, and testing the consistency between the findings of individual-based and population-based studies [14].

An exciting new approach to the problem of residual confounding has been suggested through the exploitation of Mendelian randomization [14]. Genetic polymorphisms exist that are functionally related to metabolic processes involved in disease prevention or causation. For example, higher homocysteine levels in blood have been associated with reduced folic acid intake and higher rates of cardiovascular disease. If the high homocysteine is truly a cause of cardiovascular disease, increasing folate, which reduces homocysteine, would be an important public health intervention. However, the association could in theory be the result of residual confounding, and a randomized controlled trial would be the obvious final proof. In the absence of the results of such a trial genetic studies can help. People with a genetically reduced ability to metabolize homocysteine have higher homocysteine levels and mimic the exposure state of low dietary folate. Inheritance of the polymorphism is random and therefore not likely to be biased by associations with other lifestyle factors. In observation studies 2.6 μm/l higher homocysteine levels were associated with an increased risk of cardiovascular disease of 1.13. The defective gene increases homocysteine levels by a similar amount to that observed in observational studies, and was also associated with a relative risk of 1.16. The two measures of risk related to homocysteine were therefore virtually the same, suggesting that the observational data were not biased by residual confounding, but that it was the homocysteine level that was closely linked to the cause of cardiovascular disease [14]. Where the risk related to the polymorphism is not close to the results of observational studies, residual confounding can be expected, and predictions of the likely effect in intervention studies should be made more cautiously. As more and more polymorphisms related to specific disease pathways are discovered a new method of validating observational epidemiological studies opens up.

Hierarchy of evidence

From the foregoing it can be seen that not all types of epidemiological evidence are considered to be of equal weight in understanding causation (see Chapter 7). A hierarchy of evidence has been developed within the evidence-based medicine movement. One such example is that used in Wales to evaluate the evidence for the influence of housing on health [15] (see Table 4.1). Systematic reviews, which compile the findings of all relevant studies and seek to avoid publication bias, by the use of predefined quality criteria, are given more weight than individual studies. Intervention studies, because they minimize bias, are given more weight than observational studies. Within observational studies, cohort studies are given more weight than case–control studies because of the problem of measurement bias in the latter. However, many health issues do not lend themselves to investigation by randomized controlled trial. There is a danger in valuing evidence from these trials more highly than that from observational studies. Interventions with limited effectiveness, for which there is randomized controlled trial evidence, might wrongly be judged more worthy than those for which there is only evidence from observational studies. Similarly interventions, whose effectiveness has been clearly proved by observational studies (making a randomized trial unethical), might be undervalued. Ideally, observational studies would be used to study the causes of disease. Randomized controlled trials should then be carried out whenever feasible to assess the effectiveness of interventions to change those causes. HIA could then confidently state that if these interventions were undertaken there would be a favourable impact.

Prediction

The final stage of the process, moving from demonstrating valid associations within a causal framework to predicting the experience of future cohorts, is extremely difficult. Epidemiological studies tend to focus on single causes of single diseases, and even then statisticians and epidemiologists

Table 4.1 Hierarchy of epidemiological evidence

Evidence type	STOX classification	Health Evidence Bulletin Wales classification	Description
Systematic reviews	S1	I	Comprehensive systematic review containing at least one randomized controlled trial
	S2	IV	Comprehensive systematic review
Trials	T1	II	Randomized controlled trial
	T2	II	Non-randomized controlled trial
	T3	III	Before and after interventional trial
Observational studies	O1	IV	Cohort study
	O2	IV	Case–control study
	O3	IV	Cross-sectional/longitudinal study (including statistical data)
	O4	IV	Study using qualitative methods only
	O5	IV	Case study (e.g., single housing estate)
Expression of opinion	X	V	Formal consensus or other professional opinion (this category includes literature reviews where there is no indication of a systematic approach and models based on reviews of the literature)

have a major reluctance to predict what would happen in future populations. Extrapolating correlations over time is notoriously dangerous. Variations in genetic make-up of populations and in the prevalence of risk factors can upset predictions.

In both case–control studies and cohort studies the first outcome is a measure of association—an odds ratio or a relative risk (RR). The odds ratio is in itself an estimate of relative risk. These measures are useful in causal inference, but do not in themselves give a good idea of the public health importance of the risk factors or provide a basis for HIA. Risk factor prevalence needs to be taken into account, and this can be done by calculating the population attributable risk. In a cohort study the attributable risk is RR-1, and is a measure of the proportion of a disease in exposed people that can be attributed to the exposure. Population attributable risk is the proportion of a disease in the risk population that is attributable to the exposure. In case–control studies attributable risk and population attributable risk can be calculated if the incidence of the disease is known and if the controls truly represent the exposure of the parent population. Attributable risk is the relevant figure for predicting the effect of changing exposure in HIA.

A further refinement proposed by Heller and Dobson is the disease impact number and the population impact number [16]. Disease impact number is defined as 'the number of those with the disease in question among whom one event will be prevented by an intervention'.

$$\text{DIN} = \frac{1}{(\text{AR}_{\text{exp}} - \text{AR}_{\text{non}}) \times d_{\text{exp}}}$$

Where DIN is the disease impact number, AR_{exp} and AR_{non} are event rates in exposed and non-exposed groups, and d_{exp} is proportion of people with disease who are exposed.

The population impact number is defined as the number of those in the whole population among whom one event will be prevented by an intervention.

$$\text{PIN} = \frac{1}{(\text{AR}_{\text{exp}} - \text{AR}_{\text{non}}) \times d_{\text{exp}} \times d_{\text{tot}}}$$

Where PIN is the population impact number and d_{tot} is proportion of total population with the disease.

This approach could also be applied to aetiological environmental exposures. For example, Heller [17] has taken the results of the SAHSU study of landfills and congenital malformations and translated the relative risk of 1.01 for all anomalies, and of 1.05 for neural tube defects, using the baseline population risk, into disease and population impact numbers. These relative risks translate into one extra anomaly in 5903 exposed women, and one extra neural tube defect amongst 35,714 exposed women. The number of total population who would have to be exposed for one extra anomaly to be produced by the presence of landfills is 10,733 for any anomaly, and 64,935 for neural tube defects.

For most complex diseases there are multiple causes, and a significant amount of work has taken place to try to model the effect of several variables simultaneously (see Chapter 6). These models attempt to predict the effects on heart disease, stroke, and other diseases taking into account the interactions between variables. Pursuit of better epidemiological models is desirable, not only because it tries to make the best use of available evidence, but also because the modelling process itself exposes the assumptions and gaps in evidence needed to make informed decisions. Scoring systems have been developed as a basis for predicting risk of disease for individuals and groups. A fundamental problem is regression towards the mean, which means that any model will fit better to the data from which it was generated than it does when applied to a new set of data. Even when sophisticated methods have been used to develop these models, they may not work well when applied to other populations. In one study two risk scores for cardiovascular disease, the Copenhagen risk score and the Framingham risk score, were applied to a Dutch population. Although the results of the two scoring systems correlated well they gave significantly different estimates for the average ten-year risk of coronary heart disease [18].

It has to be acknowledged, however, that the requirement in HIA to predict health and disease is more complicated even than this situation. Consider a local authority wishing to direct its resources so as to maximize the health benefits from spending on housing, environmental improvement, recreation, and transport, all of which are widely believed to influence health. Clearly there are good reasons for improving all these domains in themselves, but if health gain is one extra reason, and is put forward as the prioritizing criteria, how can the local authority decide? Spending on each domain may influence many aspects of health and disease. Each health state may be influenced by spending in several domains. These influences may not all be in the same direction. As a minimum the epidemiological approach to HIA has the potential to offer a view on the likely paths by which decisions could impact on health. It may even provide estimates to quantify the impacts, at least to the point of putting them in rank order.

References

1 **Rothman KJ and Greenland S**. Causation and Causal Inference. Detels R, McEwen J, Beaglehole R, Tanaka H (eds.), *Oxford Textbook of Public Health* (Vol 1). Oxford: Oxford University Press, 2002.

2 **Weed DL**. On the use of causal criteria. *International Journal Epidemiology* 1997; **26**:1137–1141.

3 **Fielder HMP, Poon-King CM, Palmer SR, Moss N, and Coleman G**. Assessment of impact on health of residence living near the Nant-y-Gwyddon landfill site; retrospective analysis. *British Medical Journal*, 2000; **320**:19–22.

4 **Elliott P, Wakefield J, Best N, and Briggs D**. *Spatial Epidemiology, Methods and Applications*. Oxford: Oxford University Press, 2001.

5 **Elliot P, Stamler J, Nichols R, Dyer AR, Stamler R, Kesteloot H,** *et al*. Intersalt revisited: further analysis of 24 hour sodium excretion and blood pressure within and across populations. *British Medical Journal*, 1996; **312**:1249–1253.

6 **McVernon J, Andrews N, Slack MP, and Pamsay ME**. Risk of vaccine failure after Haemophilus influenzae type b combination vaccines with acellular pertussis. *Lancet* 2003; **361**:1521–1523.

7 Elliott P, Briggs D, Morris S, de Hoogh C, Hurt C, Kold Jensen T, Maitland I, Richardson S, Wakefield J, and Jarup L. Risk of adverse birth outcomes in populations living near landfill sites. *British Medical Journal* 2001; **323**:363–368.

8 Dolk H, Elliott P, Shaddick G, Walls P, and Thakrar B. Cancer incidence near radio and television transmitters in Great Britain, 11. All high power transmitters. *American Journal Epidemiology* 1997; **145**:10–17.

9 Handbook on Air Pollution and Health. *Department of Health Committee on the Medical Effects of Air Pollutants.* London: The Stationery Office, 1997.

10 Kunzli N, Medina S, Kaiser R, Quenel P, Horak F Jr, and Studnicka M. Assessment of deaths attributable to air pollution: should we risk estimates bases on time series or on cohort studies? *American Journal Epidemiology* 2001; **153**(11):1050–1055.

11 Economic Appraisal of the Health Effects of Air Pollution Report. Department of Health. www.doh.gov.uk/airpollution/eareport.htm.

12 Clayton D and McKeigue RM. Epidemiological methods for studying genes and environmental factors in complex diseases. *Lancet* 2001; **358**:1356–1360.

13 Dolk H, Vrijheid M, Armstrong B, Abramsky L, Bianchi F, Garne E, Nelen V, Robert E, Scott JE, Stone D, and Tenconi R. Risk of congenital anomalies near hazardous-waste landfill sites in Europe: the EUROHAZCON study. *Lancet* 1998; **352**:423–427.

14 Davey Smith G and Ebrahim S. Mendelian randomization: can genetic epidemiology contribute to understanding environmental determinants of disease? *International Journal Epidemiology* 2003; **32**:1–22.

15 Weaver N, Williams JL, Weightman AL, Kitcher HN, Temple JMF, Jones P, and Palmer SR. Taking STOX: developing a cross disciplinary methodology for systematic reviews of research on the built environment and the health of the public. *Journal of Epidemiology and Community Health* 2002; **56**(1):48–55.

16 Heller RF and Dobson AJ. Disease impact number and population impact number: a population perspective to measures of risk and benefit. *British Medical Journal* 2000; **321**:950–952.

17 Heller D. Risks from landfill sites can be presented in alternative ways. *British Medical Journal* 2001; **323**:1365.

18 Visser CL, Bilo HJG, Thomsen TF, Groenier KH, and Meyboom-De Jong B. Prediction of coronary heart disease: a comparison between the Copenhagen risk score and the Framingham risk score applied to a Dutch population. *Journal of Internal Medicine* 2003; **253**(5):553–562.

Chapter 5

The contribution of the social sciences to HIA

Juhani Lehto

According to one international definition, the basic aim of health impact assessment (HIA) is to provide decision makers with better understanding of the anticipated consequences of their potential decisions on the health of a population. It is expected that better understanding may be conducive to adjusting decisions to maximize the positive and minimize the negative health impacts [1].

The hypothetical and real worlds of decision making

In a hypothetical world of rational decision-making, HIA is applied in decision-making processes where:

- the proposed decision is unambiguous,
- the implementation of the decision is unambiguous,
- it is possible to map and measure the relevant consequences of the implementation of the decision,
- it is possible to determine exactly enough the population that is exposed to the consequences of the decision, including the distribution of the consequences between the different subgroups of the affected population,
- the affected population cannot change or evade the consequences by reacting to them,
- there is enough reliable and valid evidence to predict the total impact on the health of the affected population, including relevant subpopulations, of the consequences of the potential decision,
- all the conditions listed here apply also for the potential relevant alternative decisions available for the decision makers.

In this hypothetical world HIA needs science to provide technical tools for pointing out the causal linkages between the proposed decision and the health of the exposed population as well as for guiding the calculation of the direction and magnitude of anticipated changes in health. When the causal linkages include relationships based on action and reaction by human beings in a social context, social sciences are expected to indicate them and to make them calculable.

Social science research on decisions on projects, policies, and programmes in the public sector as well as in the private sector has continuously pointed out that the world of decisions is actually much more complicated than any simple model of 'rational decision-making' assumes. For instance, there are many different and often even contradictory objectives, interests, and values involved in the decision-making processes. There is large uncertainty with regard to the consequences of the decision. There are both intended and unintended, and anticipated and unanticipated consequences of the decisions. Decisions are often far from unambiguous. Decisions may not be implemented at all, implementation may be partial or it may not accord with the content of the decision. When decisions impact on human beings, the reactions of these people interact with the direct consequences of the decision and, thus, the outcome of the decision is often quite different from the direct consequences of the decision. Quite often the actual content of the decision and the actual consequences of the decision are not the issues that matter. Rather, the perception of the proposed decision and anticipated consequences, by the relevant observers, is more important for the decision makers [2,3]. It is in this 'real world of decision making', where HIA should be applied and developed.

One particular aspect of the real world of decision making is the finding that most often decisions are based on earlier decisions and their implementation depends on later decisions. For instance, the decision to start building a new highway is based on earlier decisions about regional development plans, about regional traffic plans, about investments, and so on. And the implementation will be influenced by decisions about annual budgets, about specific building plans and contracts with the building companies, and the like. It has been claimed that the project phase may be too late to change much even if the environmental or health consequences would suggest major changes to the project. Particularly, often it is much more expensive to change the plans in the later phase than in the early planning phases. That is why it is suggested that impact assessment should concentrate on earlier, strategic phases of decision making [4]. This creates one of the paradoxes of HIA: it should be carried out before the decision to be assessed is exact enough to be accurately assessed.

Assess the proposal before assessing its impact

The foregoing underlines that the proposed decision should be analyzed before starting to assess its anticipated health impact. The significance of this aspect of HIA is often underestimated. In the models and descriptions of the HIA process, this aspect is often located to the fast preliminary steps of HIA, called screening and scoping [5]. This may reflect the inferiority of social and political science in the development of HIA, in comparison to the position of epidemiology and the sciences of biomedical pathologies.

The experience of policy evaluation studies is valuable for the development of appropriate approaches to the assessment of policy proposals. The first generation of evaluation studies were based on an expectation of rational decision-making and clear harmony between the ends and means of policy decisions [2]. The expectation was very much like the expectations built in some models of HIA. However, it was soon learned that the real world of policy decisions was different. Thus, younger generations of policy evaluation have focused much more on the analysis of the policy and not only on the attainment of its objectives. Figure 5.1 is one way of presenting schematically different tasks for policy evaluation related to the two major aspects of ambiguity that are often found in policy decisions.

Figure 5.1 may also be used to reflect the task of assessing a policy proposal as the first phase of HIA. If the policy proposal can be located into the upper left corner, the assessment of the policy proposal may be described as rapid screening. You just read the documentation of the policy proposal. Then you are ready to start establishing the causal relationships between the outcomes of the implementation of the policy and the health of the affected population.

		Information about implementation	
		Comprehensive	Partial
Policy objectives	Unambiguous	*Controlling the implementation* ensuring that decisions are implemented	*Social engineering* searching for the right means to obtain the objectives
	Ambiguous	*Concensus building* helping in the process of defining the objectives	*Exploring the opportunities* helping in the process of finding the need for action and the alternative ways of acting

Fig. 5.1 Policy objectives and information about implementation.

If the policy proposal is located in the upper right corner of the picture, the first task is to develop a better analysis of the implementation process. For instance, are there institutions capable of and people motivated to implementation and is there money available to finance the implementation? Is the implementation dependent on other decisions and what are the prospects that those other decisions will be made? Is the expected implementation dependent on particular conditions in the policy environment and what are the prospects that those particular conditions will be met? The scientific evidence for answering these kinds of questions, if it exists at all, is provided by social science, including research on public administration, on organizational behaviour, and on the particular sectors expected to implement the policy. The vast policy evaluation literature may be recommended.

If the policy proposal is located in the lower left corner of the picture, the first task is to assess whether the diffuse or even contradictory objectives will lead to later changes in the policy or its implementation. Evidence provided by policy evaluation research [2,6] indicates that changes may be quite probable. Thus, the assessment of the policy proposal may result in an anticipation of the future changes in the policy or alternative paths of the development of the policy. The scientific evidence for such assessments might be found in the research on politics, management, and administration of public policies.

The policy proposal may also be located in the lower right corner of the picture. In my country, the regional and sectoral strategy documents often are examples of this type of policy proposals. At the same time, and as said earlier in this chapter, HIA of these documents may be particularly important. An analysis of a proposed strategy should focus both on potential competing, changing or contradictory objectives and on dealing with the ambiguity with regard to the implementation process of the strategy.

Public health experts are not often best educated to carry out analysis of policy proposals of other sectors. Thus, it may be thought that this task should be given to the 'insiders' of the policy sector or even to the 'insiders' of the preparation of the given policy. Probably they know best about both those objectives and interests related to the policy proposal that may not be openly stated in the policy documents. Probably they also know better the potential ambiguities in the anticipated implementation process. However, leaving the assessment of the policy proposal to 'insiders' contains major risks. They may be blind to the weaknesses of the proposed policy or may have vested interests to give a specific view on the proposal. If such blindness or vested interests influence the first phase of HIA itself, a high quality of the later phases of HIA cannot compensate for the bias of the first phase. Thus, the author would not recommend leaving the assessment of the policy proposal to the 'insiders'.

The outcome of the assessment of the proposal may be called 'the anticipated actual policy implementation'. It creates a basis for analyzing the causal links between the policy proposal and the health of the affected population.

Causal relationships mediated by human action and reaction

There are many different potential links between the anticipated policy implementation and the health of the affected population. For instance, the quality of air or drinking water of a certain population may be affected without causing any significant reaction by those affected. In this case, there is not much need for the application of evidence or expertise from social science. It is enough to apply the biomedical and epidemiological evidence about the relationship between air or water quality and human health.

However, people may also react to the anticipated changes. If air and water quality drops, some people may move away from the affected area. Other people may reduce their open-air physical exercise. Some people may stop drinking the water from the water tap and change to bottled beverages. There are many potential reactions. Some of them may be significant from the perspective of the health of the affected population; some of them may not be really significant. Quite often, different population groups react differently to changes in their environment. A serious HIA cannot leave the potential reactions unnoticed, because the reactions may also directly or indirectly impact on the health of some affected people. The minimum is to assess whether such reactions are significant. And if they are, then they have to be taken into consideration in the further assessment process.

Sometimes, the reaction by the affected population is the most essential factor in the assessment of the impact. One example is the policy of reducing the level or the duration of social benefits for the unemployed. Without any reaction by the affected people such a policy would increase poverty, particularly of those who only receive the minimum benefit. It would be rather easy to anticipate negative health impact, due to the causal linkages between poverty and ill-health. The proponents of the policy argue, probably, that reductions in social benefits will create an incentive for the affected to seek a job more actively. If jobs are available the result may be a reduction in poverty through a reduction in unemployment. If suitable jobs are not available the result will be an increase in poverty. And if the incentive leads to an intensified competition for low-paid jobs, the result may also be a reduction in salaries in low-paid jobs. This may, in turn, improve the competitiveness of the local enterprises, which may either employ more people or make more profits to be used somewhere else, both of which will impact the health of some people. Thus, it may be anticipated that the policy starts a long chain of actions and reactions and each of these actions and reactions may be dependent on different conditions in the policy environment.

The nature of causality in social sciences and the compatibility of causality with the freedom of human beings to make their own choices are perhaps eternal topics of the philosophy of the social sciences. For the purposes of HIA, it may be enough to say that in any case the causality mediated through the action by human beings is different from causality mediated through processes not involving choices made by human beings. Quite often, 'different' here means also more complicated. It also means that the social science evidence available for indicating the anticipated causal chain may often be feasible only for particular policy environments or even offer contradictory conclusions. Thus, social scientists may provide HIA with deeper understanding of the potential relationships between the proposed policy and the determinants of health of the affected population. They may also provide HIA with a better picture of the affected population. At the same time, they may not be very helpful in providing tools for quantifiable predictions of the anticipated course of causal processes.

Health impact mediated through the social determinants of health

The third significant contribution of the social sciences to HIA is evidence on the social (socio-economic, cultural, and other) determinants of health. It is widely known that social conditions such as low income, low education, weak position at the labour market and in the workplace, low quality housing, the lack of supportive family or other close community are predictors of lower than average health status (e.g., Marmot and Wilkinson [7]). The evidence on the relationship between changes in the social conditions and the health of the population affected by these changes is weaker and all the mechanisms between low social position and low health status are not known. Thus, there is not much evidence to make quantitative predictions on the health impact of anticipated changes in the social determinants of health. Quite often, science can only help in anticipating whether the impact is negative or positive but not how many more people will die or how much the incidence of diseases will increase.

Understanding the relationships

It may be concluded from the observations made here that social and political sciences may provide significant contributions to HIA of policies, programmes, and large projects.

In the beginning of this chapter, an international definition of HIA was quoted. According to that definition, 'the basic aim of health impact assessment is to provide decision makers with better *understanding* of the anticipated consequences of their potential decisions on the health of a population' [1] (italics added). It is understanding of the policy proposal, understanding the causality mediated through human action and reaction, and understanding the role of social determinants in this mediation that social science is best in providing. At the same time social science may be making the relationships between a policy and its health impact look as complicated as it often actually is. Instead of offering tools for making simple and exact predictions about lives saved or lost it may be more valuable by offering tools for understanding.

References

1   Gothenburg consensus paper. *Health Impact Assessment. Main Concepts and Suggested Approach.* Brussels: European Centre for Health Policy, 1999.

2   Fischer F. *Evaluating Public Policy.* Chicago, IL: Nelson-Hall, 1997.

3   Bolman L and Deal T. *Reframing Organizations. Artistry, Choice, and Leadership.* San Francisco, CA: Jossey-Bass, 1993.

4   Boothroyd P. Policy Assessment. In Vanclay F, and Bronstein D. *Environmental and Social Impact Assessment.* New York, NY: John Wiley & Sons, 1995.

5   Lehto J and Ritsatakis A. Health impacts assessment as a tool for intersectoral health policy. In Diwan V, Douglas M, Karlberg I, Lehto J, Magnusson G, and Ritsatakis A (eds.), *Health Impact Assessment: From Theory to Practice.* Gothenburg: Nordic School of Public Health, 2000.

6   Uusikylä P. Politiikan ja hallinnon arviointi (The evaluation of policies and administration). In Eräsaari R, Lindqvist T, Mäntysaari M, and Rajavaara M (eds.), Arviointi ja asiantuntijuus. Helsinki: Gaudeamus, 1999.

7   Marmot M and Wilkinson R (eds.). *Social Determinants of Health.* Oxford: Oxford University Press, 1999.

Chapter 6

Quantitative approaches to HIA

Mark McCarthy and Martin Utley

Concepts of HIA

Concepts of HIA can be usefully classified according to three essential characteristics: whether the focus is on disease or health; whether it is a participative process or one based on expert opinion; and whether the approach is qualitative or quantitative. These approaches can often be complementary, and in any particular HIA each may have a part to play. It should however be recognized that they represent fundamentally different conceptions of evidence and, importantly, the role of evidence in HIA.

The WHO charter in 1948 described health as 'not the mere absence of disease but positive physical, psychological, and social well-being'. While this definition is heuristically interesting and reflects the complexity of human experience, there are considerable problems in implementing it. Far more research has been done on the causes of disease than on causes of health because disease is more tangible and measurable than health. Also, while disease is the province of health care services, the social and psychological determinants of health are often seen as outside health care. In practice, then, health impacts are usually described as diseases rather than states of well-being, particularly when attempts are made at quantifying impacts.

Participation of those likely to be affected by the development, policy, or intervention under investigation is often considered a necessary component of HIA. Indeed, for some practitioners, the process of participation is more important than the result, with HIA becoming a means of empowerment at an individual and community level and a means of advancing a political agenda. The attitudes of the affected community towards issues of disease, health, and risk will, to an extent, determine the impact of any proposal and a more complete picture of health impact is gained by drawing from a wide range of sources.

Yet communities themselves are not necessarily well informed about potential health impacts of phenomena they have not experienced, and they assess both proposals and evidence related to proposals from their own subjective viewpoint. This is the difference between approaches in HIA that aim to generate or collate expert opinion on health or disease impacts and approaches that react to, exploit, or ignore expert opinion depending on how it fits in with the existing beliefs and objectives of the community. For example, a website for a community group 'against' a proposed incinerator advised its members to downplay the health dimension because expert evidence indicated that there was very little health risk from modern incinerators.

The relationship between participative HIA and expert HIA is in this sense akin to that between a prosecuting or defence counsel and an expert witness in a court of law. It is important that the two processes are distinct and that expert HIA is independent and is seen to be independent of any interested parties. Indeed, one of the goals for a recent WHO programme has been 'to decouple epidemiological assessment of the magnitude of health problems from advocacy by interest groups of particular health policies or interventions' [1].

Health impact assessments are often qualitative. They scan a range of phenomena (of the policy or development) and indicate potential health (or more usually disease) effects. The imputation of cause in this process may be based on evidence or supposition; but it is an important issue that most diseases are multi-causal. While road accidents may be fairly directly associated with road traffic, the causes of diseases such as heart disease are multiple—thus smoking, diet, lack of exercise, and genetic disposition are only some of the contributing factors. While HIAs may draw freely on this epidemiological evidence, interpretation is contentious. Attributing causality depends both on the strength of evidence (e.g., supported by Bradford Hill principles) and that the expected effects depend on whether the change will actually happen (e.g., how many people will be exposed to what concentrations of pollution and for how long?).

In this chapter, HIA will be considered from the perspective of disease, experts, and quantification. Impacts of disease (including accidents and mental illness) are clearer than notions of 'health'; perceptions by an expert can be based on scientific evidence rather than the subjective judgements of the general public; and a statement that (e.g.) one death may occur is clearer than a statement simply that there is an 'appreciable' risk, or some other non-quantified statement. This approach is utilitarian: that is, it proposes information for the good of the public as a whole rather than taking a rights-based approach for an individual in a group.

Environmental impact assessment

Impact assessment started in the environmental field through the National Environmental Policy Act (1969) in the United States. The US Environment Protection Agency (USEPA) has developed a quantitative approach of risk assessment based on toxicological experimental and observational data. Risk assessment is usually divided into four steps:

♦ Hazard identification—identifying a potentially harmful substance;

♦ Dose–response assessment—assessing the level of dose and potential effects;

♦ Exposure assessment—identifying the group's exposure by number and duration;

♦ Risk characterization—describing the probability and severity of effects.

Thereafter, the task of *risk management* is to agree through a political process how much risk to accept and how it is to be distributed, while *risk communication* is encouraging informed debate with the public.

Comparative risk assessment has a similar function to environmental impact assessment. Comparative risk assessment evaluates and compares potential health, ecological, and quality-of-life impacts. It was first used in the United States in 1987 for regions, states, and cities, and was used internationally by USEPA in the 1990s to inform policy and public investment decisions. More recently it has been used within the WHO Burden of Disease project.

In the EU, a directive in 1985 required member states to introduce legislation for environmental impact assessment (EIA) to be undertaken. Initially these were limited to major economic developments, but the range of projects requiring EIA has subsequently been increased. EIA practice has created a relatively systematic approach, concerned with different environmental domains of ecology and physical characteristics. The evidence base for the EIA is an environmental statement, usually prepared by experts independent of the developer. The environmental statement presents quantified information in the public domain for discussion, drawing on a range of approaches including modelling.

The UK EIA Regulations (Town and Country Planning Regulations, 1999), recommend that information within environmental statements should include: 'A description of the aspects of the environment likely to be significantly affected by the development, including, in particular, population, fauna, flora, soil, water, air, climatic factors, material assets . . . an estimate, by type and quantity, of expected residues and emissions (water, air and soil pollution, noise, vibration, light, heat, radiation, etc.) resulting from the operation of the proposed development'. Thus, health is one dimension of the EIA. Environmental impact assessment, is discussed further in Chapter 12.

Health impact models

Models are representations of logical thinking. They may be conceptual, verbal, diagrammatic, or mathematical; and these may be linked: a mathematical model for impact analysis will be built on a conceptual model, and is often represented initially in a diagram. Equally, where there is a lack of numerical material, a verbal or diagrammatic model may still be useful. Four quantitative approaches to health impact include PREVENT, POHEM, Global Burden of Disease, and ARMADA.

PREVENT is a mathematical model developed in the 1980s which uses epidemiological data to predict the population effects of health promotion interventions. The model provides a basis for estimating impacts of changes in population due to heart disease risk factors such as hypertension, blood cholesterol, exercise, smoking. It has been used for comparisons of health policy between countries and also to identify impacts of potential policy changes within countries.

A useful example has been modelling the health impact of a policy to increased physical exercise in the United Kingdom [2]. Two strategies were modelled: the first was a 25 per cent increase in the proportion of adults who were moderately physically active; while the second was a similar increase in the proportion who were vigorously active. The two strategies would themselves be quite major changes in population behaviour. Interestingly, although the modelling indicated relatively small reductions in coronary heart disease death rates, the health impact was greater for the strategy of increasing moderate activity (0.15 per cent reduction over 25 years) compared with increasing vigorous activity (0.06 per cent reduction). The model also showed that targeting behaviour change on males over 45 of age who already took some exercise would provide the greatest population benefit.

POHEM is a longitudinal microsimulation model of health and disease developed by Statistics Canada [3]. The model simulates a representative population at the individual level in order to draw conclusions that apply to higher levels of aggregation. POHEM enables alternative health interventions to be compared while taking into account the effects of disease interactions. It includes data on risk factors, disease onset and progression, health care resource utilization, direct medical care costs, and health outcomes (http://www.statcan.ca/english/spsd/Pohem.htm).

POHEM currently models lung and breast cancer, coronary heart disease factors, arthritis, and dementia. As an example, it has been used to evaluate the impact of changing risk factors, diagnosis, and therapies for lung cancer, and can assess cost-effectiveness per life year gained [4]. A wider review of 36 microsimulation models in health and social care has been presented by Spielauer [5].

The Global Burden of Disease (GBD) has been developed by WHO over the past decade. 'The concept of disease burden refers to a systematic and internally consistent quantification of health problems of a defined population, preferably using a summary measure of population health that integrates mortality and morbidity information' [6]. The primary activity of the GBD has been the development of comparable, valid, and reliable epidemiological information on a wide range of diseases, injuries, and risk factors. In the phase of the project between 1999 and 2002, 25 risk factors were selected to obtain data on the prevalence of risk factor exposure and hazard size. From these data, population attributable fractions were estimated. The years lived with a disability were calculated using a model, DisMod, based on lifetable analysis.

Building up the model from attributable risk has implications for the effect estimates. Attributable risks for a disease may add up to over 100 per cent and the size of risk for a particular factor depends in part on the order of statistical analysis. Moreover, not all risks are avoidable; and avoidable risks continue to act on a population while the existing burden of attributable disease works its way out.

To apply the GBD approach to a specific country, De Hollander and colleagues made a provisional assessment of environmental disease burden in the Netherlands [7]. They identified 17 environmental exposures for which there were reasonable data relating to public health outcomes. Attributable fractions were combined with incidence in the Dutch population to calculate annual number of cases, and these were converted to disability-adjusted life years by estimates of severity and duration. The results (Table 6.1) provide a sense of relative impact for different environmental factors, both broad (e.g., accidents, damp houses) and specific (e.g., lead, benzene).

ARMADA is a mathematical model specifically developed for HIA [8,9]. The concept was initially broad: to create a model to compare baseline disease rates (mortality, morbidity) in a defined population against the effects due to implementation of an economic development. (The approach can equally be used to assess the expected beneficial impacts of a policy.)

In developing this quantitative model for HIA, it became clear that the range of epidemiological evidence available was limited. Twelve impact areas are usually described in the environmental statement (see Fig. 6.1). Some of

Table 6.1 Importance of environmental factors in burden of disease in the Netherlands

Up to 6%	Water supply and hygiene Food safety Indoor air pollution
Up to 2%	Occupational environment Road traffic accidents
Up to 0.5%	Pesticides Outdoor air pollution Lead Noise Electro-magnetic force Ionization Arboviruses Protozoans

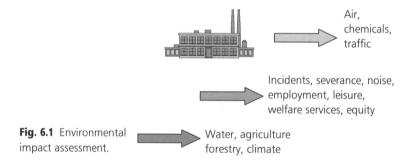

Fig. 6.1 Environmental impact assessment.

Air, chemicals, traffic

Incidents, severance, noise, employment, leisure, welfare services, equity

Water, agriculture forestry, climate

these are predominantly ecological (non-human) concerns; some have a human impact but the scientific relationship to health is not well demonstrated; only in three areas could epidemiology provide quantitative evidence for health impacts.

Air. The risk estimate of air pollution was assessed using epidemiological data from a prospective cohort study in the United States and applied to three European populations [10]. Although several environmental air pollutants are usually described in environmental statements, PM_{10s} are representative of levels of air pollutants from incinerators. The mechanism for the health effect of long-term exposure to particles is not fully understood: it appears mainly to affect deaths from cardio-respiratory disease.

Chemicals. These can have their effects both through absorption (respiratory and skin) and through the food chain. Risk estimates of chemicals are mainly derived from animal bioassay, although some data are also available from occupational exposure (both of these are levels higher than ordinary population exposures). The US Integrated Risk Information System describes oral reference doses, and inhalation reference concentrations, which are an estimate of a daily exposure to the human population (including sensitive subgroups) that is 'likely to be without appreciable risk of deleterious effects during a lifetime' [11]. Environmental risk estimates are traditionally constructed for an adult 'hypothetical maximum exposed individual', that is, a person exposed to the maximum permissible levels of the listed compounds for a lifetime (taken as 70 years). This assumes the person is always resident, breathes air at the point of maximum concentration of pollutants, and eats only locally grown food. Food chains may concentrate chemicals, with potentially greater effects on humans than respiratory inhalation, but the exposed population is much more widely distributed: a food chain model is needed to calculate a composite risk of cancer.

Road accidents. This relates to the number of vehicles and number of pedestrians exposed, although the relationship is complex. There is more traffic in towns than country roads, but town journeys are slower and shorter. Many factors contribute to road accidents, including age of the driver, driving behaviour (including speed, alcohols, and drugs), road engineering, climate, and time. For ARMADA we used an average risk based on UK national levels of trips and accidents (UK road deaths are the lowest per capita in Europe). This implies a 'no-threshold' approach similar to carcinogens, since one vehicle journey is capable of causing the death of one human (or more). The rate could be subdivided for various vehicle types (goods vehicles cause more deaths per trip than cars) or for population characteristics (road accidents occur at all ages, inversely by social class).

An environmental statement may also include other concerns impacting on human populations, but where epidemiological evidence of disease effects is lacking. For example, a relationship between the level of social contact between people in a street and the level of traffic is intuitively likely, as traffic on busy roads can form a barrier. Nevertheless, the direct impact on 'health' (or disease) has not been demonstrated, and the finding is also open to methodological questions. For example, do the people who find themselves, or choose to be, in the different roads have different social habits anyway? Perhaps houses on a main road are flats above shops where single people live without strong community contacts, while families with small children may move to streets with lower traffic. This 'confounding' is a frequent problem of epidemiological studies. Whatever the true relationship, it is at present impossible to provide a quantitative estimate of the effects of community separation in causing disease.

The ARMADA quantitative health impact model has been applied in two contrasting ways—for environmental development and for public policy.

Scenario 1. A new incinerator was planned in a town on the south coast of England. The incinerator was designed to take all the municipal waste for the city of 180,000 for a 30-year period. It was to be sited upwind and close to the seashore. The local resident population within the wind plume was about 10,000. Data on the expected levels of emissions and traffic were drawn from the environmental statement. Using the wind-plume population for air and chemical exposure, and the city population for traffic exposure, the extra disease for the incinerator was estimated to be about '0.15 of a person' in 30 years (Fig. 6.2). This contrasts with a total mortality over the same period of perhaps 60,000 people: not only is this a low extra level of death but it is also entirely beyond direct observational studies to detect it.

Scenario 2. An HIA was made of a government initiative to support research in the automotive industry. A panel of experts was convened to provide estimates of the expected degree by which car safety might be improved through research. Taking into account the projected trend in vehicle accident death rates, the phased introduction of the design measures was converted into an estimate of benefits for deaths and hospital admissions. Using a 'willingness-to-pay' method, reductions in costs to the NHS could also be calculated from these health benefits.

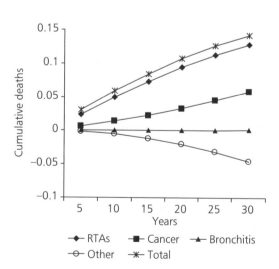

Fig. 6.2 Excess mortality over 30 years.

These quantitative assessments offer a measure for comparison, either for alternative proposals for investment, or for the health impacts of other hazards. Discussion of quantitative measures can contribute to the decision process.

Reflections

Some of the limitations to quantitative HIA models have been illustrated in this chapter. Satisfactory epidemiological evidence on health effects is available for only some of the areas covered by EIAs, and the evidence available uses different approaches for assessment of risk. There is little understanding of the implications of extrapolating risks between different populations—for example, the estimates of health impact of air pollution in ARMADA are based on US disease experience during the 1970s–1990s. Also, there is debate on how the impacts are distributed in different social groups. How does a quantitative estimate take into account social class effects, or migration in and out of a population over time?

There are further limitations to using quantitative HIA. The epidemiological evidence needs to be specified clearly and interpreted appropriately, since different assumptions may lead to different conclusions [12]. Moreover, uncertainties at different points may multiply during the assessment. And yet, providing a rough quantification is scientifically more appealing than making a qualitative, perhaps unfounded, statement that 'A will cause B'. A quantitative model provides a criterion to compare against other information for the decision (e.g., costs of the intervention) and describes impacts using standard public health measures of deaths or disease.

Quantitative models will be increasingly used in HIA, in the same way that models have been developed for other fields of environmental assessment. The challenge is to develop epidemiological knowledge for HIA in the social fields. When the social relationships of health are formally quantified, much less activity in HIA will be based on surmise and contention.

Acknowledgements

We are grateful to Professor Steve Gallivan for his substantial contribution to developing ARMADA, and to Dr Jane Biddulph and Mr Jake Ferguson for their contributions in applying ARMADA. We also thank the UK Department of Health for financial support.

References

1 WHO Geneva, (Ref: Global Programme on Evidence for Health Policy Discussion Paper No. 50, World Health Organization October 2002).

2 Naidoo B, Thorogood M, McPherson K, and Gunning-Schepers LJ. Modelling the effects of increased physical activity on coronary heart disease in England and Wales. *Journal of Epidemiology and Community Health* 1997; **51**:144–150.

3 Wolfson MC. POHEM a framework for understanding and modelling the health of human populations. *World Health Statistical Quarterly* 1994; **47**:157–176.

4 Houle C, Will B-P, Berthelot J-M, and Evans W-K. Use of POHEM to estimate direct medical costs of current practice and new treatments associated with lung cancer in Canada. Paper No. 99 Statistics Canada 1996.

5 Spielauer M. Dynamic mircrosimulation of health care demand, health care finance and the economic impact of health behaviour Part II: survey and review. Interim Report IR-02-036/May, International Institute of Applied Systems Analysis, Austria, 2002.

6 McMichael A, Pastides H, Pruss A, Corvalan C, and Kay D. Update on World Health Organisation's initiative to assess environmental burden of disease. *Epidemiology* 2001; **12**:277–279.

7 De Hollander AEM, Melse JM, Lebret E, and Kramers PGN. An aggregate public health indicator to represent the impact of multiple environmental exposures. *Epidemiology* 1999; **10**:606–617.

8 McCarthy M, Biddulph JP, Utley M, Ferguson J, and Gallivan S. A health impact model for environmental changes attributable to development projects. *Journal of Epidemiology and Community Health* 2002; **56**:611–616.

9 Utley M, Gallivan S, McCarthy M, Biddulph J, and Ferguson J. ARMADA: A computer model for evaluating the effect of environmental factors on health. *Health Care Management Science* 2003; **6**:137–146.

10 Kunzli N, Kaiser R, Medina S, Studick A, *et al*. Public-health impact of outdoor and traffic-related air pollution: a European assessment. *Lancet* 2000; **356**:795–801.

11 Environmental Protection Agency. http://www.epa.gov/iris/limits.htm

12 Nurminen M, Nurminen J, and Corvalan CF. Methodologic issues in epidemiological risk assessment. *Epidemiology* 1999; **10**:585–593.

Chapter 7

Evidence and HIA

Mark Petticrew, Sally Macintyre, and
Hilary Thomson

It is a general principle within the health sector that the benefits and costs of
interventions should be well established before they come into routine use,
and organizational apparatus exists to scrutinize the evidence that the benefits
of surgical, medical, and other types of intervention outweigh their harms.
However, despite the fact that major periods of an individual's life are spent
exposed to the outputs of the housing, education, transport, and other sectors,
the full health effects of non-health policies generally remain unknown.
Individual policies or programmes are routinely introduced without prospective assessment and often without consideration of their health impacts; health
has instead been assumed to be the province of the health sector.

Good research evidence on the positive and negative health impacts of policies is often absent because non-health sector policy and other interventions
are generally not implemented in such a way as to facilitate robust outcome
assessment. Governments tend not to think in terms of discrete interventions,
but tackle problems on a broad front using a range of strategies, which makes
disentangling their effects extremely difficult. For example, in the case of road
safety, we have better roads, speed cameras, seat belts, airbags, advertisements,
and other schemes, and yet it is difficult to say how much of the reduction in
traffic accidents is due to any of these individual interventions, or combinations of these, or to secular trends [1].

Meaningful health impact assessment (HIA) however clearly depends on the
availability of robust quantitative and qualitative evidence of such impacts. This
chapter considers what sort of evidence may be required, what may be available,
and some of the current gaps. First, however we cannot avoid rehearsing briefly
what evidence may mean to those conducting and using HIAs.

What does evidence 'mean' in HIA?

There are at least two broad evidential camps in HIA. Some HIA reports and
papers emphasize the need to collect evidence of the actual impacts of

policies, programmes, and projects, and in this context 'evidence' is a scientific term relating mainly to scientific research. This is similar to the *Chambers Dictionary* definition, which refers to the results of systematic investigations aimed at increasing the sum of knowledge [2]. Other types of evidence seen as admissible in HIA appear to be closer to the *Oxford English Dictionary* definition, in which 'evidence' refers to testimony, or to other information from which inferences can be drawn. In recent papers in health journals, HIA increasingly refers to the former: scientifically collected data on the effectiveness of policies or other interventions in improving health; that is, systematically collated information on the balance of benefits and harms [3,4].

As with any other sort of evaluation or assessment the sort of evidence that is required for HIA is of course dependent on the question that is being asked. The aim of HIA is often described in quite broad terms: for example, the purpose may be described as being simply to carry out an HIA of some policy or project, and this is expected to involve a quest for information on a range of health and other impacts. However HIA can and is often broken down into its component questions, and evidence of different types can be sought to answer these.

HIA incorporates a range of different types of information, some of which is research based and can help with predicting likely impacts. Research evidence is just one of many components shaping HIA; others suggested by the WHO Gothenburg Consensus Paper include issues of democracy, equity, sustainability, and ethics. However this chapter concentrates primarily on the role of research.

Evidence from primary studies

For predictive purposes HIA requires reliable data on the likely health effects of policies. What evidence then is needed to shed light on the health and related impacts of social interventions? For example, the effects of new road building on injuries, noise, pollution levels, general well-being, community severance, and other direct and indirect outcomes? The size of the actual impacts may be derived from quantitative studies (if available), and information on the existence and nature of those impacts can be derived from qualitative studies (though not exclusively). Although often referred to as an exploratory hypothesis-generating exercise, qualitative analysis of lay reports can be useful in evaluation [5]. By providing rich descriptors of relevant local area effects, and the influence of and interaction between lay perceptions, local context, and the intervention, mechanisms for health impact may be illuminated [6]. Information on unanticipated outcomes can also be derived from both quantitative and qualitative sources.

When we wish to know whether this road-building project is likely to be acceptable to the community, then the answers may come from qualitative data, and new qualitative research may be indicated (e.g., with focus groups or interviews), or answers may be sought from less formal consultations. An HIA incorporates a range of possible questions and different sorts of data are required to answer them fully. In clinical settings the hierarchy of evidence is used as a way of prioritizing the type of information required, but the application of such hierarchies to public health questions is contentious, and given the different types of question that need to be answered, no one single hierarchy is applicable. (The same can also be said of clinical decision-making.) For this reason, thinking in terms of matrices rather than hierarchies may be helpful when considering the strengths and weakness of different types of research evidence for answering different sorts of question. The example in Table 7.1 is adapted from health care decision making [7]. An example of its use for prioritizing evidence in child health is available elsewhere [8].

Evidence from systematic reviews and HIAs

Systematic literature reviews represent an important source of information on quantitative information on the outcomes of social interventions. The most well-known source of such reviews is the Cochrane Database of Systematic Reviews (CDSR) [9], but for many public health practitioners this source is often seen as inappropriate. This is because Cochrane is often thought to be concerned solely with clinical interventions, and with randomized controlled trials. These are often believed to have little of value to contribute to HIA, because HIA tends to focus primarily on projects where the health impacts are an indirect consequence of policies [10]. However it is worth noting that Cochrane is not solely a database of clinical reviews; the Cochrane collaboration includes a public health and health promotion field, which actively promotes the conduct and dissemination of reviews in these fields [11]. More specifically relevant reviews are also likely to come in time from the Campbell Collaboration Social Welfare, Education and Crime review groups [12]. Another important source of quality-assessed systematic reviews in United Kingdom is likely to be the Database of Abstracts of Reviews of Effectiveness at the University of York, which is being expanded to include reviews of public health interventions. It currently includes abstracts with commentaries for over 1300 reviews, many of which are of preventive and public health interventions (available free at york.ac.uk/nhscrd/dare.html). The UK Health Development Agency Evidence-Base is another important source of information of relevance to HIA, including systematic reviews (www.hda.org).

Table 7.1 An example of a typology of evidence

Research question	Qualitative research	Survey	Case–control studies	Cohort studies	RCTs	Quasi-experimental studies	Non-experimental evaluations	Systematic reviews
Effectiveness Does this work? Does doing this work better than doing that?				+	++	+		+++
Process How does it work?	++	+					+	+++
Salience Does it matter?	++	++						+++
Safety Will it do more good than harm?	+		+	+	++	+	+	+++
Acceptability Will the programme or project service be acceptable to the community?	++	+			+	+	+	+++
Cost effectiveness Is it worth paying for?					++			+++
Appropriateness For example, is this the right type of policy for this community?	++	++						++
Satisfaction Are users, providers, and other stakeholders satisfied?	++	++	+	+				+

Source: Adapted from Muir Gray (1997).

The widespread assumption that systematic reviews are overly concerned with randomized controlled trials (RCTs), which are of little relevance to public health and to HIA, is unfounded. While all Cochrane reviews seek to include RCTs, approximately a third of them also (and sometimes, only) include non-randomized controlled studies, or uncontrolled studies (such as interrupted time series, or quasi-experimental studies). In a sample of 960 reviews from the 2002 Issue 2 Cochrane database 32 per cent contained study designs other than RCTs. Non-Cochrane systematic reviews very commonly include other study designs—indeed systematic reviews and meta-analyses of observational studies are extremely common, though all too often methodologically unsound [13].

More generally, it is important that reviews of the evidence that are used in HIAs are systematic. Unsystematic trawls of the literature are no more accept-able in public health than in health care. This is not a new demand; the absence of systematic reviews of the evidence was one criticism levelled at the evidence presented to the public inquiry into the application to build a pressurized water reactor at Hinkley Point power station [14]. Without an explicitly evidence-based approach to the epidemiological evidence presented to HIAs, there is the risk that the conclusions will be based on poor evidence, recent studies as opposed to all relevant studies, or the best studies, and be subject to the partic-ular biases of experts. Moreover, the HIA may not be scientifically or indeed legally defensible, and the reputation of HIA itself will suffer [3].

However, it is still the case that systematic reviews of policies are relatively uncommon, and HIA may not often be based on these types of evidence. This is partly because few reviews have been produced, but also because the body of primary evidence is often too small to provide clear guidance to decision makers. Moreover, in the case of many interventions, the best available evi-dence of effectiveness (RCTs and other outcome evaluations) comes from evaluations that have been conducted in the United States, and the generaliz-ability of such evidence is still unclear. In HIA as in other fields the external validity as well as the internal validity of the studies needs to be assessed. There are few validated tools to aid in this task, though there are examples from health care that (with amendment) may be more widely applicable [7].

Absence of evidence in evaluations

The real problem remains that there is little sound evidence of any sort—either qualitative or quantitative—of the health impacts of social interventions to aid sound, scientifically-based HIAs. This is not simply a problem with the lack of RCTs, which has been highlighted elsewhere [15,16]. The problem is even wider; simply, there are relatively few prospective evaluations of any design of the wider health and social impacts of most policies. Housing is relatively well represented, for example, there is one ongoing RCT using a stepped-wedge design, 14 controlled studies of varying vintage (with seven more ongoing) and five uncontrolled prospective studies, though many of the completed studies are old [17]. In many other areas there are very few quantitative or qualitative outcome evaluations.

The gaps are particularly large in the case of transport. Transport policy is widely held to be one of the major social determinants of health, though sound research evidence as to how health may be promoted by this means is notably absent. It is particularly difficult to locate studies of the most effective means of bringing about modal shifts in transport choices. Furthermore, the best available evidence continues to be centred around more biomedical outcomes, rather than wider health impacts. In public health the term 'health' may well refer to an 'absence of illness', but as far as outcome evaluations of road building are concerned, the outcomes still mainly consist of injuries, noise, and pollution [18]. Health impact assessment questions about the effect of roads on general well-being, inequalities, social exclusion, and community severance are difficult to answer convincingly at present [19].

Designing studies to assess health impacts

One possible reason for this dearth of evaluative research relates to the potential difficulty of evaluating the secondary health impacts of policies and social interventions. Among these difficulties can be included the problems of obtaining pre- and post-intervention data, and the need to sample and analyze data at the community, rather than the individual level. At a methodological level, the use of interrupted time-series designs commonly used to assess the impacts of policies may involve a different set of problems; these include issues of internal validity (such as the difficulty of identifying appropriate controls), and of external validity (generalizability of the results to other areas, or populations) [19]. The lack of awareness among many academic researchers of the practicalities of public policy-making and implementation may be another problem [20]. The need for appropriate health indicators for assessing policy impacts has also been emphasized, for example, indicators need to be locally relevant, objective, valid, sensitive to change, and appropriate to the policy in question [21,22].

Qualitative empirical work too can be used to inform HIA of area-based change as well as contributing to academic understanding of health and place [23]. Although qualitative research is not able to produce quantifiable assessments of impacts, generalizations at a conceptual level may emerge [24], which are useful to inform the development of appropriate outcome measures in subsequent quantitative evaluations. Although not purporting to be qualitative research studies, community consultation exercises often have a central place in HIA [25] and the principles of rigorous qualitative research could usefully be applied to the process of sample selection, data collection, and interpretation [26].

What does HIA mean?

Earlier we suggested that the term 'evidence' may have multiple meanings. Like 'evidence', HIA is a term that may mean different things to different people, without the actors involved realizing that it is being used in different senses. For those within the HIA movement, it tends to be used to refer to scrutinizing proposed projects, programmes, or policies in advance for possible health effects (e.g., is this proposed new housing development likely to have any effects on human health? If so, are these potentially positive or negative effects? And, if negative, is there anything that could be done to mitigate against adverse effects?). However, for others with an interest in public health or evidence-based policy making, HIA may be taken to mean the empirical study, using some form of before-and-after design, of the actual health effects, positive or negative, of projects, programmes, or policies (e.g., this housing development improved the mental health of mothers with children, and reduced asthma admissions among children, but the increase in rents meant there was less income available for food or holidays, so the longer-term effects might need to be monitored or mitigated). For many in the public health and community development fields, this latter exercise is evaluation rather than HIA. (See Chapter 36 for further discussion of this point.)

That actors may not be aware of the meanings attributed to the term by others can be illustrated by two anecdotes. One involved discussions about the development of a research programme, between two researchers and a Director of Public Health (DPH). The researchers mentioned the possibility of examining the effects on local population health of an imminent factory closure; the DPH's response was that this was not HIA, but 'just an evaluation'.

The second concerns the Independent Enquiry into Inequalities in Health, whose first and overarching recommendation was: 'as part of Health Impact Assessment, all policies likely to have a direct or indirect effect on health should be evaluated in terms of their impact on health inequalities and should be formulated in such a way that by favouring the less well off they will, wherever possible, reduce such inequalities' [27]. Four years after the publication of the report, members of the enquiry were asked whether by this they had meant predictive screening of potential impacts on health inequalities, or monitoring of actual impacts on health inequalities. Members reported that there had not been any explicit discussion among them about which of these they had intended, but that probably they would advocate both. It should be noted that the wording of the recommendation (and the subsidiary ones relating to it) could be interpreted in either way, so those thinking of predictive estimates of likely effects could believe that is what the enquiry was recommending, while those thinking of monitoring actual effects could equally think that was what was being recommended. The problem of course is that without this latter monitoring and evaluation exercise we will continue to lack robust empirical information to feed into the former, predictive exercise. The other problem is that those outside the HIA field may erroneously assume that lots of monitoring and retrospective studies have taken place, and that therefore the predictive exercises are well grounded (e.g., lay people and many in public health and planning are extremely surprised when told how few empirical studies there are of the effects of housing developments or refurbishments on health, since they assume that many such studies will have been done). In the pressure to engage in predictive HIA, and in screening and scoping of potential impacts, we should not forget that there is a basic need for evaluation and monitoring of actual impacts.

Can HIA help promote the production of new evidence?

Health impact assessment may itself sometimes be in a position to promote the production of relevant evidence. That is, if the HIA is prospective, then there may be the possibility of collecting evidence of health impacts directly, rather than seeking to identify existing evidence from other sources (such as systematic reviews). This may of course not always be possible, as resource and time constraints on HIA practitioners would often preclude new evaluations, but this would at least ensure that the evidence that is collected is locally relevant, and over time an evidence base would be collected that would directly inform future HIAs. This approach depends on HIA being seen by practitioners not just as an opportunity to bring existing evidence to bear on decision making, but also as an opportunity to stimulate the production of new research on health impacts. In many cases this may of course not be feasible, but the possibility of collecting relevant evidence of actual impacts, as opposed to speculating on possible impacts, needs to be considered more often than at present.

Conclusion

There is a tendency in public health and in other fields to consider any information collected from any source as evidence, and the range of acceptable evidence is often very broad, ranging from research findings to anecdotes. This inclusive approach is sometimes justified on the grounds that the legal definition of evidence is similarly broad. HIA, as with any other decision-making task, does indeed require a range of different types of information, some of which is derived from research. However not all information sources are equally suitable or reliable in answering all sorts of questions. As is often the case, the type of question asked usually suggests the sort of evidence that is needed to answer it.

To use an often-repeated phrase from the HIA literature, the 'stage is set' for HIA [28]. Unfortunately, it sometimes seems that many of the most important 'props' are currently missing. In particular, we have few evaluations of actual impacts to feed into the HIA process, and relatively few systematic reviews of the health effects of social interventions and non-health policies. We also need to make better use of the existing evidence, including that of variable quality. Most importantly, if HIA is to fulfil its potential in improving the public health and reducing health inequalities, we need more evaluations of the actual effects of policies and other interventions.

Acknowledgements and funding

MP and HT are funded by the Chief Scientist Office of the Scottish Executive Health Department. MP and SM are part of the ESRC Evidence Network.

References

1 Parish R. Consumer protection and the ideology of consumer protectionists. In Duggan A. and Darvall L. (eds.), *Consumer Protection Law and Theory*. Sydney: Law Book Company, pp. 229–243, 1980.

2 Davies H, Nutley S, and Smith P. *What Works: Evidence-Based Policy and Practice in Public Services*. Bristol: The Policy Press, 2000.

3 Joffe M and Mindell J. A framework for the evidence base to support Health Impact Assessment. *Journal of Epidemiology and Community Health* 2002; **56**:132–138.

4 Morrison D, Petticrew M, and Thomson H. Health impact assessment. *Journal of Epidemiology and Community Health* 2001; **55**:219–220.

5 Shaw I. *Qualitative Evaluation*. London: Sage, 1999.

6 Popay J and Williams G. Public health research and lay knowledge. *Social Science and Medicine* 1996; **42**:759–768.

7 Muir Gray JA. *Evidence-Based Healthcare*. London: Churchill Livingstone, 1997.

8 Petticrew M and Roberts H. Evidence, hierarchies and typologies: horses for courses. *Journal of Epidemiology and Community Health* 2003; **57**:527–529.

9 The Cochrane Library [database on disk and CDROM]. The Cochrane Collaboration. Oxford: Update Software.

10 Birley MH. *The Health Impact Assessment of Development Projects*. London: HMSO, 1995.

11 **Waters E and Doyle J.** Evidence-based public health practice: improving the quality and quantity of the evidence. *Journal Public Health Medicine* 2002; **24**:227–229.

12 Campbell Collaboration website.

13 **Egger M, Schneider M, and Davey Smith G.** Spurious precision? Meta-analysis of observational studies. *British Medical Journal* 1998; **316**:140–144.

14 **Sutcliffe J.** Environmental impact assessment: a healthy outcome? *Project Appraisal* 1995; **10**:113–124.

15 **Oakley A.** *Experiments in Knowing: Gender and Method in the Social Sciences.* Cambridge: Polity Press, 2000.

16 **Macintyre S.** Evidence based policy making. *British Medical Jorunal* 2003; **326**:5–6.

17 **Thomson H, Petticrew M, and Morrison D.** Housing interventions and health—a systematic review. *British Medical Jorunal* 2001; **323**:187–190.

18 **Egan M, Petticrew M, Hamilton V, and Ogilvie D.** Health impacts of new roads: a systematic review. *American Journal of Public Health* 2003; **93**:1463–1471.

19 **Britt C, Kleck G, and Bordua D.** A reassessment of the DC gun law: some cautionary notes. *Law & Society Review* 1996; **30**:361–397.

20 **Brownson *et al.*** Policy research for disease prevention: challenges and practical recommendations. *American Journal Public Health* 1997; **87**:735–739.

21 **Joffe M and Mindell J.** A framework for the evidence base to support Health Impact Assessment. *Journal of Epidemiology and Community Health* 2002; **56**:132–138.

22 **Wills J and Briggs D.** Developing indicators for environment and health. *World Health Statistics Q* 1995; **48**:155–163.

23 **Thomson H, Kearns A, and Petticrew M.** Assessing the health impact of local amenities: a qualitative study of contrasting experiences of local swimming pool and leisure provision in two areas of Glasgow. *Journal of Epidemiology and Community Health* 2003; **57**:663–677.

24 **Baum F.** Researching public health behind the qualitative-quantitative methodological debate. *Social Science and Medicine* 1995; **40**:459–468.

25 **McIntyre L and Petticrew M.** Health Impact Assessment: A Literature Review: MRC SPHSU Occasional Paper No. 2 (http://www.msoc-mrc.gla.ac.uk/Reports/PDFs/Occasional-Papers/OP-002.pdf), 1999.

26 **Mays N and Pope C.** Assessing quality in qualitative research. *British Medical Journal* 2000; **320**:50–52

27 **Acheson D.** *Independent Inquiry into Inequalities in Health.* London: Department of Health, 1998.

28 **Ratner P, Green L, Frankish C, Chomik T, and Larsen C.** Setting the stage for Health Impact Assessment. *Journal of Public Health Policy* 1996; **18**:67–79.

Chapter 8

The role of lay knowledge in HIA

Eva Elliott, Gareth Williams, and Ben Rolfe

The people are excluded from forming judgement on
various matters of public interest on the ground that
expert knowledge is required, and that of course the
people cannot possess . . . The debunking of the expert
is an important stage in the history of democratic
communities because democracy involves the assertion
of the common against the special interest.
(*From a speech made by Aneurin Bevan in 1938, quoted
in [1].*)

Introduction

In recent years evidence about the determinants of health has accumulated
significantly [2]. Emerging sources of longitudinal data have enabled much
more sophisticated analysis of the interrelationships between different risk
factors than was previously the case [3,4]. The new pluralism implied
by these perspectives can be seen in the formulation of the Dahlgren and
Whitehead [5] model of the 'social determinants of health'. The porosity of
the membranes between the different 'layers of influence', and the complex
interactions it suggests between economic conditions, social structure, social
relationships and networks, individual behaviour, and psychosocial factors,
stands in sharp contrast to the Black Report's categorical explanations for
health inequalities [6].

It is not, therefore, simply the availability of new or maturing sources of
longitudinal data that have enabled more complex forms of analysis to take
place. There is also a new willingness to think creatively about purportedly
explanatory concepts and categories. Understanding the determinants of
health means understanding what sociologists call the 'recursive' relationships
between social structure, context, and human agency [7]. What this means is
that we need to understand the interacting effects of underlying structures

and processes, the immediate contexts in which these appear in different forms, and the interpretation and response of people to those risk factors. Such a demand requires us to be very flexible indeed about data and its interpretation.

Health impact assessment (HIA) concerns itself with 'the determinants of the determinants' of health [8] as these are explored in particular local contexts, with the involvement of multiple stakeholder perspectives and sources of evidence. If HIA is something like: '. . . a combination of procedures, methods and tools by which a policy, programme or project may be judged as to its potential effects on the health of a population, and the distribution of those effects in the population' [9], then it is involved in a very practical way in understanding and working with the recursive relationships between underlying structures and processes and their interpretation by different groups of people in particular settings or contexts.

Knowledgeable opinion

Concern has been expressed about the dangers of relying on opinion and unreliable consultation processes in HIA, and the consequent need to 'decouple' the technical aspects of HIA from both community development and more political aspects of the process [8,10]. We would argue that if an understandable concern with technical rigour and robustness in HIA fails to engage with the new pluralism of our understanding of the determinants of the determinants it will be irrelevant to real-life situations. In our view it is wrong to see the lay views expressed in consultations simply as 'opinion', or to see the consultation process as something separate from the more technical knowledge-generating aspects of HIA. Indeed, without 'lay knowledge', it is impossible to focus in a satisfactory way on underlying causes or the determinants of the determinants [11]. In relation to the role of lay perspectives in epidemiology it has been suggested that 'popular epidemiology' is:

> . . . the process whereby lay persons gather scientific data and other information, and also direct and marshal the knowledge and resources of experts in order to understand the epidemiology of disease. In some of its actions popular epidemiology parallels scientific epidemiology . . . Yet popular epidemiology is more than public participation in traditional epidemiology, since it emphasises social structural factors as part of the causal disease chain. [12]

Although Brown is concerned with the very specific context of toxic waste, this kind of lay monitoring and analysis of the data of everyday life can be seen to be operating in more generic contexts. One respondent in a study of lay perceptions of health risks illustrates very well the way in which 'opinion' can in fact contain just that contextualization of underlying causes that is essential to the practice of HIA:

> Smoking and drinking and drug-taking. I put it down to one thing . . . until money is spent on these areas . . . there doesn't seem to be much point in trying to stop people smoking and what else. As long as the environment is going down the pan the people will do down with it. [13]

The watchword for all respectable commentaries on the conduct of HIA is 'robustness'. However, effective HIA also has to involve many stakeholders, holding different interests, and speaking in different voices. This is not just a problem of how to set up and use appropriate methods such as focus groups, stakeholder seminars, or citizens' juries. The argument of this chapter is that the way forward is not to separate the technical from the community or political aspects of HIA, but rather to recognize in the processes of consultation and opinion gathering, not just ethically important opportunities for participation and capacity building, but the social conditions for the transformation of knowledge and understanding about the technicalities of HIA.

HIA in the Garw Valley, South Wales

The political context

The Welsh Assembly Government has an established commitment to developing HIA throughout Wales [14] (see also Chapter 18). As well as a means of informing decisions, a key objective of this commitment is to develop a better understanding of how HIAs operate in real settings. The HIA referred to here came about as a response to the concern of Bridgend County Borough Council's cabinet member for social services and housing. Local authority housing built in the 1970s was in a noticeable state of decay and residents suffered the humiliations associated with economic decay such as lack of employment, low income, increased vandalism, and ill-health. Concern for the close connections between housing and health, and the need to take the well-being of residents into account in any decisions about options, led to the decision by the Welsh Assembly Government to commission an HIA in a small village in the Garw Valley north of Bridgend in South Wales [15]. The rationale was that it would provide council officers with an evidence-based resource to inform their decisions, and the Welsh Assembly Government with a detailed local case study to inform a better understanding of the interconnections between housing actions and their potential consequences for population health.

A particular case of general decline

Exploration of the socio-economic circumstances of the area highlighted both the particular character of the village itself and the experiences it shared with other deindustrialized areas in the South Wales coalfields. A complex set of factors appeared to have shaped the village's current housing state, which in turn, impacted on the physical and psychological well-being of its remaining residents. Former coalfield communities are stark exemplars of how the determinants of the determinants interrelate and shape population health. Like other communities in the former coalfield area, the decline had not just been economic, but social and cultural [16].

In this particular village there was the added problem of a large number of poorly designed houses. However the dilapidation of housing was not simply due to bad design but also a declining population, creating a large number of void properties, and a lack of appropriate employment and recreation, particularly for children and young people. In effect, the vacant housing stock and the empty spaces within the housing development became the exclusive recreational playgrounds for a small number of young residents and the dumping ground for unwanted household items and waste.

The process

Against the background of this decline, people in the village were in the process of trying to revive civic activity in an attempt to improve local facilities and community life. The residents were actively trying to rekindle a sense of pride in an area of considerable natural beauty, and as the HIA project developed it was important to acknowledge this pre-existing local activity. A steering group was set up, which included a number of key representatives (the Council's Principal Housing Officer, a representative of the Local Health Board, a county borough councillor, a community councillor, a member of a local community development organization, a representative of a housing association working in the area, two representatives from the National Assembly for Wales (public health strategy and housing), the researcher and the project manager representing Cardiff University and two local people). The two local residents involved were invited onto the steering group by the researcher on the basis of contacts made through the local community development organization. Each had recently become actively involved in community-based activities. A third local person attended one steering group meeting but although her involvement was most valuable, other commitments meant that she could play no further direct part in the steering group. The community councillor was unable to attend the steering group due to illness. All meetings took place in the village's community centre.

Lay knowledge in HIA provides a better understanding of how the determinants of health interrelate and impinge in the real and meaningful conditions in which people find themselves. In other words analysis of cases of local involvement contributes to the cogency of our theoretical reasoning about HIA in general.

The value of involving local people in HIAs is not simply social and political. It also contributes to a knowledge and understanding of people's complex responses to social and economic structures and as such provides the evidence for what is possible. In terms of seeking solutions to problems that may arise from implementing policies, programmes, and projects lay knowledge brings a historical perspective that is critical to the appropriateness and effectiveness of contemporary decisions [20]. The importance of lay knowledge in a multi-agency and multidisciplinary approach to healthy decision-making is not that it provides a separate form of evidence, but that it creates the conditions for thinking in new ways. Lay voices, in dialogue with other professionals and academic experts, provide the foundations for a 'civic intelligence', which is grounded in a better understanding of the human condition in different contexts. It would be a mistake to allow the postmodern possibilities of a pluralist approach to HIA to be straitjacketed by a modernist methodological framework tied to traditional hierarchies of evidence and authority.

References

1 Smith D. *Aneurin Bevan and the World of South Wales.* Cardiff: University of Wales Press, 1993.

2 Marmot M and Wilkinson R. (eds.) *The Social Determinants of Health.* Oxford: Oxford University Press, 1999.

3 Power C, Matthews S, and Manor O. Inequalities in self-rated health in the 1958 birth cohort—lifetime social circumstances or social mobility. *British Medical Journal* 1996; **313**:449–453.

4 Wadsworth M and Kuh D. Childhood influences on adult health: a review of recent work from the British 1946 national birth cohort study, the MRC National Survey of Health and Development. *Paediatric and Perinatal Epidemiology* 1997; **11**:2–20.

5 Dahlgren G and Whitehead M. *Policies and Strategies to Produce Social Equity in Health.* Stockholm: Institute for Futures Studies, 1991.

6 Department of Health and Social Security. *Inequalities in Health: the Report of a Working Group (the Black Report)*, London: DHSS, 1980.

7 Williams GH. The determinants of health: structure context and agency. *Sociology of Health and Illness* 2003; **25**:131–154.

8 Joffe M and Mindell J. A framework for the evidence base to support health impact assessment. *Journal of Epidemiology and Community Health* 2002; **56**:132–138.

9 European Centre for Health Policy. *Health Impact Assessment: Main Concepts and Suggested Approach.* 1999.

10 **Parry J and Stevens A.** Prospective health impact assessment: pitfalls, problems, and possible ways forward. *British Medical Journal* 2001; **323**:1177–1182.

11 **Popay J, Williams GH, Thomas C, and Gatrell A.** Theorising inequalities in health: the place of lay knowledge. In Bartley M, Blane D, and Davey-Smith G (eds.), *The Sociology of Health Inequalities*. Oxford: Blackwell, 1998.

12 **Brown P.** Popular epidemiology and toxic waste contamination: lay and professional ways of knowing. *Journal of Health and Social Behaviour* 1992; **33**:267–281.

13 **Williams GH, Popay J, and Bissell P.** Public risks in the material world: barriers to social movements in health. In Gabe J (ed.), *Medicine, Health and Risk* (*Sociology of Health and Illness* Monograph Series). Oxford: Blackwell, 1995.

14 National Assembly for Wales. *Developing Health Impact Assessment in Wales*. Cardiff: National Assembly for Wales, 2000.

15 **Elliott E and Williams GH.** *Housing, Health and Well-Being In Llangeinor, Garw Valley: A Health Impact Assessment*. Cardiff: Welsh Assembly Government, 2002. http://www.hiagateway.org.uk/Resources/showrecord.asp?resourceid=45

16 **Bennett K, Beynon H, and Hudson R.** *Coalfields Regeneration: the Consequences of Industrial Decline*. Bristol: Joseph Rowntree/The Policy Press, 2000.

17 **Allen T.** Housing Renewal—Doesn't it make you sick? *Housing Studies*, 2000; **15**:443–461.

18 **Woolcock M.** Social capital and development: concepts, evidence and applications. *Presentation at a workshop on understanding and building social capital in Croatia, Zagreb, 6 March 2002.*

19 **Mitchell J Clyde.** Case and situation analysis. *Sociological Review* 1983; **31**:187–211.

20 **Mallinson S, Popay J, Elliott E,** *et al.* Historical data for health inequalities research: a research note. *Sociology* 2003; **37**.

Chapter 9

Planning an HIA

Jennifer Mindell, Mike Joffe, and Erica Ison

Review of different 'levels' of HIA

The phrase health impact assessment (HIA) is used to signify many different activities, of varying complexity, and requiring a wide range of resources. 'Desktop' appraisal is generally undertaken by officers in an organization over a few person-hours to gain a snapshot of the health impacts to inform proposal direction. Similarly, a 'mini' HIA involves the use of existing information and evidence and seeks no community or stakeholder participation [1]. Rapid appraisal is also characterized by the use of information and evidence that is already available or easily accessible but also involves in most cases a half-day 'rapid' stakeholder appraisal workshop. Although termed 'rapid', the preparation for, and report writing generated by, the workshop is labour-intensive, albeit for a short timespan; the attendance of sufficient individuals also represents a substantial commitment, if their time is costed [2]. Rapid appraisal is described in greater detail in Chapter 11.

Comprehensive appraisal (also called maxi HIA [1]) involves the collection of new data. This might include a survey of local residents, a detailed literature review, and/or a primary study of health effects of the same proposal elsewhere or, for a concurrent HIA, of the proposal as it is implemented. It usually requires a prolonged and substantial time-commitment from a number of people and is resource-intensive [3].

Which 'level' of HIA is undertaken will depend on:

- the timescale of the proposal, since an HIA report will be unable to affect decisions taken before the report is written;

- the resources available for the HIA (time, staff, expertise, community development);

- the importance of the proposal or the potential health effects.

For example, where there is a good quality, up-to-date systematic review available of all the relevant scientific evidence on a subject, a rapid or mini

HIA may be easily undertaken or, in a more comprehensive appraisal, a greater proportion of resources can be directed towards other components of the HIA process, such as community participation.

Models of HIA

Initially, two basic models of health were used by HIA practitioners: the biomedical, and the social or socio-economic. In the biomedical model, the focus is on disease and ill-health, and the causal mechanisms through which they arise. This model has most commonly been used in environmental health impact assessment (EHIA). The social or socio-economic model of health focuses on the broader determinants of health, such as employment, housing, education, and social support. The biomedical model of HIA has been referred to as 'tight' perspective HIA, and the social model as 'broad' perspective HIA [4]. The biomedical model is founded in the disciplines of epidemiology and toxicology, and the social model in the social sciences. In the former, quantitative evidence and modelling are used predominantly. In the latter, qualitative evidence and stakeholder knowledge are given greater prominence. These two approaches are not mutually exclusive: most HIA practitioners employ elements of both, with a variable balance between the two

A large number of different models of HIA now exist with guidelines, toolkits, and reviews available (Table 9.1). Guidance for topic-specific HIA, such as regeneration[24,25], and for the use of HIA in healthy public policy (HPP) [26,27] have also been published. The WHO [28] has recently produced a technical briefing on HIA as a tool to include health on the agenda of other sectors.

Reviews of HIA methodologies have been published by the Liverpool Public Health Observatory [13], the University of Northumberland [29], the Welsh Assembly [4], the WHO [30], London Regional Office of the NHS (National Health Service) Executive [31], the Institute of Public Health in Ireland [32], and the Northern and Yorkshire Public Health Observatory [33]. The English Health Development Agency has also reviewed systematic reviews of HIA [34]. Others have considered methodological issues relevant to the conduct of HIA [1,35].

Stages in HIA

The HIA process comprises six main stages: screening, scoping, appraisal or risk assessment, preparation and submission of report and recommendations to decision-makers, and monitoring and evaluation [36].

Screening is the first stage and its primary function is to filter out those proposals that do not require HIA because the proposal has a neutral or negligible

Table 9.1 HIA models and guidelines

Model of HIA	Examples and references	Main focus	Identification of health impacts
Policy analysis	British Columbia [5]	Possible impacts of public policy on determinants of health	Checklist
Based on EIA	Australia[6], New Zealand [7,8], Liverpool Health Impact Programme [9,10], Bielefeld EHIA [11], EHIA [12], Prospective HIA (Manchester) [13], Merseyside [14,15], British Medical Association [16], Lerer [17], Australia [18], Canada [19]	Protecting and improving public health by anticipating adverse health effects to incorporate mitigation at the planning stage Quantitative assessment of environmental factors	Checklist + local concerns + risk assessment
Economic appraisal	English Department of Health [20]	Monetary values	Experts from a range of disciplines
Elements of EIA/ British Columbia HIA model/ democracy	Swedish County Councils [21], Scotland [22,23]	Determinants of health	The Swedish model assumes an extensive understanding of impacts on influences on health. The other two are based on the Swedish model but while the Scottish model uses a systematic comprehensive framework to identify all relevant impacts, including reviewing the literature, 'expert' informants, focus group discussion, interviews, and routine data, the latter uses a checklist

impact on health, and/or it is possible to amend the proposal easily due to the existence of a robust evidence base both for impacts and effective interventions. Screening should be conducted systematically using a set of criteria (sometimes listed in a 'screening tool') against which proposals are judged. Screening enables scarce resources to be targeted towards those proposals that will benefit from further assessment. For proposals that are selected, the results of screening will provide a basis for the conduct of subsequent stages in the process. 'Desktop appraisal' is similar to screening but does not have the function of selection. The Netherlands has considerable experience in screening of policies for health impacts: Putters describes administrative barriers and other issues [37].

Scoping is the second stage of HIA and its function is to set the boundaries for the HIA (sometimes described as setting the HIA's terms of reference). The following items are delineated during scoping: the elements or aspects of the proposal to be assessed; the proposal's non-negotiable aspects; aims and objectives of the HIA; values underpinning the HIA; the populations or communities affected by proposal implementation; the geographical area covered; any vulnerable, marginalized, or disadvantaged groups within the affected population/community; stakeholders for the HIA and the nature of their involvement; potential health impacts of concern; background information for the HIA (evidence base, HIAs of similar proposals, baseline profile of the population/community, and specific local conditions affecting proposal implementation); methods to be used during appraisal/risk assessment; timescale for the HIA; management arrangements; work programme; resources (human, financial, and material) available and those required; decision-making forum(s) to influence; and arrangements for monitoring and evaluation of the HIA and its outcomes. The baseline profile will describe the demographics and health status of the affected population(s) and include identification of existing inequalities and of excluded or vulnerable groups, who may be inherently so or at increased risk from the proposal. For example, people with severe pre-existing cardiopulmonary disease are at greatest risk of further exacerbations and early death from pollution [38,39].

Appraisal or risk assessment is where potential positive and negative impacts on health of a proposal are identified (quantitatively or qualitatively). Many different methods can be used during this stage, for example, from modelling to stakeholder workshops. The choice of method is determined partly by the model of HIA being used but is also constrained by factors such as timescale or resources available. Risk assessment can neglect the effects a proposal may have on the determinants of health, so a useful model for HIA is the Policy/Risk Assessment Model (PRAM) [40]. Appraisal/risk assessment is the stage of HIA that defines the length of the process (e.g., 'rapid' or 'comprehensive').

The fourth and fifth stages of the HIA process involve the preparation of the report and recommendations and then submission of the report and recommendations to decision makers. It is essential that this occurs within the decision makers' timeframe, meeting deadlines for consultation period or scheduled meetings, and in an appropriate format for the target audiences.

Involvement of decision makers in the HIA process through membership of the steering committee and/or participation in stakeholder workshops increases the likelihood of the findings of the HIA being considered relevant to the decision-making process, a prerequisite to acceptance of recommendations. In some HIAs, the submission of the report is accompanied by a presentation at the decision-making forum.

Although the primary audience for the report and recommendations is the decision makers, it is also important to communicate the main contents of the report and recommendations to all stakeholders, especially those who have participated in the process. The content, format (e.g., web, printed newsletter, poster), and presentation of the communication should be designed according to the needs of stakeholders and their preferred way of accessing information. Access to the full report and recommendations should always be provided if different types of summaries have been produced as the main form of dissemination to stakeholders.

Monitoring and evaluation is the sixth stage of HIA. There are several components to monitoring and evaluation: evaluation of the process of HIA, monitoring the acceptance and implementation of recommendations (impact evaluation), and monitoring and evaluation of indicators and health outcomes after the proposal has been implemented. In the New Zealand and EHIA models, monitoring is performed to ensure compliance of a project with the conditions attached to the consent [7] but most guidance refers to monitoring of health determinants, outcomes, or indicators.

Process evaluation of HIA is important as a source of learning, as part of the drive towards quality improvement, and as a mechanism of quality assurance. Mechanisms for monitoring the acceptance and subsequent implementation of recommendations are vital: to assume that because a recommendation has been accepted it will be implemented is misguided. However, even when decision makers are committed to an assessment of potential health impacts, the results of an HIA are only one of many sets of factors that will influence their decision.

Outcome evaluation is fraught with difficulties. A successful HIA may persuade decision makers not to implement a proposal or to make substantial modifications, so the anticipated effects of the initial proposal cannot be monitored to examine the accuracy of the predictions. Many predicted health effects cannot be monitored using only routine data; even when data are available, as only a small proportion of any relevant health outcome can usually be attributed to a change resulting from a project, programme, or policy, random fluctuations will generally mask achievable changes in health outcomes.

Selection of outcomes

There are a number of issues to consider when choosing outcomes to examine in an HIA: what is important may not be measurable and that which is measured routinely or can be measured may be unimportant [41]. For example, community severance is a recognized adverse effect of traffic, dividing communities, limiting access to goods, services, and social networks, and impeding independent mobility for many disadvantaged groups [42]. As there is no simple or routine measure of such severance, no quantified assessment of severance can be made at baseline nor estimates made of the effects of proposals. Those who favour quantified assessments, necessary for economic appraisal or for other explicit trade-offs, may give more weight to those outcomes that have been measured (such as pollutant levels or estimates of deaths caused by pollution) than to a qualitative statement ('community severance will be exacerbated').

Even if the burden of disease [43] (attributable risk) from a given cause is high, the reduction attributable to a *fall* in exposure to that cause *consequent on a proposal*, that is, the *achievable* reduction, may be very small [3]. Thus, if the link between cause and effect is well established, it may be better to use a proxy measure, such as educational attainment, rather than the eventual health outcomes (death rate before the age of 75) when measuring changes in small populations.

Assessing impacts on equity

Assessing effects on inequalities in health is integral to most models of HIA. A possible gradient of effects or susceptibility across the whole population (by educational level or occupational social class, e.g.) should be considered as well as the more common approach of considering impacts on specific excluded or vulnerable groups. In most cases, it will be possible to identify vulnerable groups, defined by age, disease, ethnicity, deprivation, or other disadvantage. Such groups are characterized (e.g., the number of people and their location) during the profiling stage.

Assessing which inequalities represent inequity is a matter of judgement. Whereas 'inequality' refers to differences (being unequal or 'variations'), equity is generally used to convey unfair or unjust differences [44]. The former can often be measured but the latter is harder to assess. For example, educational attainment may differ between individuals. The unequal results can be monitored objectively but whether or not it represents inequity includes elements of judgement and viewpoint. Differences in opportunity, such as parental support or quality of teaching, indicating inequity, are much harder to determine; if the only difference is due to the students' own effort, opinions may vary as to whether this represents inequity. There are also situations in which inequalities are advocated (such as providing more resources per capita in areas of higher need) in order to reduce inequity.

Proposals may impact on equity in four ways:

No differential effects (e.g., the same percentage change in mortality is anticipated among affluent and deprived groups): existing inequalities may be perpetuated and health status may diverge further.

Differential susceptibility because of greater susceptibility of some people, for example, the greater risk from pollution for people with severe cardiorespiratory disease. For quantifiable effects, different exposure-effect estimates may be available [39].

Differential susceptibility because of a greater number of susceptible individuals within some sections of the population: both cardiovascular and respiratory disease are more common in less affluent and less educated groups in the United Kingdom.

Differential exposure: various groups may be exposed to different changes in exposure. Air pollution in London is correlated with deprivation and the predicted falls are greatest in the most deprived areas[45]. As those most susceptible to exposure also have the highest baseline exposure, changes in exposure may have marked effects on inequalities in outcome.

The process of estimating effects on inequalities and vulnerable or excluded groups is the same as for the whole population but the effects are described or calculated after stratifying the population into such subgroups and uses the relevant prevalence of risk factors, effects of exposure, and changes in exposure for each of the groups. The effects are then compared with the effect on the population as a whole or with other groups.

Literature searching and involving technical experts

Health impact assessment differs in a number of ways from other evidence-based practice for which syntheses of the evidence regarding a policy's effects are compiled: assumptions of reversibility of adverse factors; a diverse evidence base utilizing studies from different disciplines and a range of designs; involvement of a range of individuals from different backgrounds and with varying priorities, concerns, and prior beliefs; the need to make recommendations to decision makers regardless of the quality of the evidence; and tight timescales as the norm [46]. Much useful information is available only as reports not published in scientific journals ('grey' literature). This presents three problems: identification, since few such reports are indexed on nationally or internationally accessible catalogues or databases; obtaining copies; and assessing the rigour of the work.

Where topics are frequent subjects of HIA (e.g., regeneration and transport) readily available 'off-the-shelf' reviews conducted proactively by technical experts could enable local expertise to focus on local concerns and on engaging the community [40]. Such 'off-the-shelf' reviews of evidence, made available via the Internet [47], can expedite robust local HIA [48] and aid separation between the technical work of HIA and the political processes of policy development and decision making [40]. Further discussions of the nature and role of 'evidence' in HIA are presented in Chapters 4 and 7.

People involved in HIA

The stakeholders in an HIA are the people involved in or affected by proposal development and implementation. Stakeholders will be drawn from public, private, and voluntary sectors, and include communities or groups affected. In any HIA, it is important to identify all stakeholders, and include as many as possible to ensure ownership of the process [49]. However, stakeholder participation can be constrained by the timescale and resources available for an HIA. Key informants are stakeholders whose roles, and/or standing in a community, mean that they have experience, knowledge, or information of relevance to the proposal.

Assessors are the practitioners who undertake primarily the appraisal or risk assessment, and the preparation of the report and recommendations. They may have much experience of conducting HIA but are often local authority or health service public health staff with minimal HIA experience. A steering and/or management group, comprising representatives from key stakeholder organizations and ideally representatives from the affected community, is often appointed to oversee the process and outputs of an HIA. Decision makers are the people who receive the HIA report and recommendations following appraisal/risk assessment, and decide which to accept in relation to the proposal. Decision makers may be involved in the process of HIA, and in some instances they may be the same group as the steering group, but in many cases they are not. In large or complex proposals, there may be more than one group of decision makers.

Community involvement as full and active stakeholders is one of the values underlying HIA (see Chapter 8). There are various ways in which the community can be involved, and, for some HIAs, practitioners will 'piggy-back' onto forms of consultation that are already established in the community, such as citizens' juries, to avoid consultation 'fatigue'. Community involvement can be difficult to achieve, particularly when trying to ensure representativeness of views (especially from 'hard-to-reach' groups), but it is important to obtain the perspectives of at least some of the community affected. However, some HIAs have been, and continue to be, undertaken without community involvement. In instances of proposal development, this is valid when officers are at an early stage, or when public consultation has occurred and the results are included as components of the appraisal/risk assessment. In some cases, HIAs have been led by a community.

The process of writing the report and formulating recommendations requires integration of the views of technical experts with those of the communities. Involvement of the community from the earliest stage is likely to ensure that issues of concern to the community are considered as part of the technical review as well as aiding acceptance of technical evidence by the community. The evidence should be summarized in a form that is usable by all stakeholders (see Chapters 11 and 20) to ensure discussion can draw on both published evidence and stakeholders' experiences. When recommendations are written, a brief statement of the supporting evidence should be included in the report.

References

1 **Parry J and Stevens A.** Prospective health impact assessment: pitfalls, problems, and possible ways forward. *British Medical Journal* 2001; **323**:1177–1182.

2 **Ison E.** *Rapid Appraisal Tool for Health Impact Assessment in the Context of Participatory Stakeholder Workshops. Commissioned by the Directors of Public Health of Berkshire, Buckinghamshire, Northamptonshire and Oxfordshire.* London: Faculty of Public Health Medicine, 2002.

3 **Mindell J and Joffe M.** Predicted health impacts of urban air quality management. *Journal of Epidemiology Community Health* accepted for publication.

4 **Kemm JR.** *Developing Health Impact Assessment in Wales. Better Health, Better Wales.* Cardiff: Health Promotion Division, National Assembly for Wales, 1999.

5 Ministry of Health and Ministry Responsible for Seniors. *Health Impact Assessment Tool Kit: A Resource for Government Analysts.* British Columbia, Canada: Population Health Resource Branch, Ministry of Health, 1994.

6 National Health and Medical Research Council. *National Framework for Environmental and Health Impact Assessment.* Canberra: Australian Government Publishing Service, 1994.

7 Public health commission. *A Guide to Health Impact Assessment.* Wellington: Public Health Commission, 1995.

8 New Zealand Ministry of Health. *A guide to health impact Assessment.* Wellington: Ministry of Health, 1998. Access Date: 27.02.03. http://www.moh.govt.nz/moh.nsf/

9 **Birley M and Peralta GL.** Health impact assessment of development projects. In Vanclay F, Bronstein DA (eds.), *Environmental and Social Impact Assessment.* Chichester: John Wiley & Sons, 1995.

10 Birley M. Global perspectives on health impact assessment. 1st UK HIA conference, 16 Nov 1998, Liverpool, London: Health Education Authority, 1999.

11 Fehr R. Environmental health impact assessment: Evaluation of a ten-step model. *Epidemiology* 1999; **10**:618–625.

12 Clark BD. *Integration of health impact assessment into the procedures of the Convention. Ad hoc working group workshop on strengthening the health component of the Convention on Environmental Impact Assessment in a Transboundary context, 26–28.02.1997.* Bilthoven: WHO-ECEH, 1997.

13 Winters L. *Health Impact Assessment—a Literature Review.* Liverpool: Liverpool Public Health Observatory, 1997.

14 Winters L and Scott-Samuel A. *Health Impact Assessment of the Community Safety Projects Huyton SRB area.* Liverpool: Liverpool Public Health Observatory, 1997.

15 Scott-Samuel A, Birley M, and Ardern K. *The Merseyside Guidelines for Health Impact Assessment.* Liverpool: Merseyside Health Impact Assessment Steering Group, 1998. Access Date: 27.022003. http://www.liv.ac.uk/~mhb/publicat/merseygui/index.html

16 Birley M, Boland A, Davies L et al. *Health and Environmental Impact Assessment. An Integrated Approach.* London: Earthscan Publications Ltd, 1998.

17 Lerer LB. Health impact assessment. *Health Policy & Planning* 1999; **14**:198–203.

18 enHealth Council. *Health Impact Assessment Guidelines.* Canberra: Commonwealth Department of Health and Aged Care, 2001. Access Date: 27.02.2003. http://www.health.gov.au/pubhlth/publicat/document/env_impact.pdf

19 Kwiatkowski RE and Gosselin P. Promoting health impact assessment within the environmental impact assessment process: Canada's work in progress. *IUHPE—Promotion and Education* 2001; **8**:17–20.

20 Department of Health. *Policy appraisal and health.* London: HMSO, 1995.

21 Federation of Swedish County Councils. *Focusing on Health.* Stockholm, Sweden: Landstingsförbundet, 1998. Access Date: 27.02.2003. http://www.lf.se/hkb/engelskversion/enghkb.htm

22 Conway L, Douglas M, Gavin S, Gorman D, and Laughlin S. *HIA Piloting the Process in Scotland.* Glasgow: SNAP, 2000. Access Date: 27.02.2003. http://www.gla.ac.uk/external/ophis/PDF/hia.pdf

23 Douglas MJ, Conway L, Gorman D, Gavin S, and Hanlon P. Developing principles for health impact assessment. *Journal of Public Health Medicine.* 2001; **23**:148–154.

24 Cave B and Curtis S. Developing a practical guide to assess the potential health impact of urban regeneration schemes. *IUHPE—Promotion and Education* 2001; **8**:12–16.

25 Cave B and Curtis S. *A Practical Guide. Health Impact Assessment for Regeneration Projects.* London: East London and The City Health Action Zone, 2001. Access Date: 13.01.2003. http://www.geog.qmul.ac.uk/health/Vol1.pdf

26 Taylor L and Blair-Stevens C (eds.). *Introducing Health Impact Assessment (HIA): Informing the Decision-Making Process.* London: Health Development Agency, 2002. Access Date: 27.02.2003.
http://www.hiagateway.org.uk/search/showrecord.asp?resourceid=228

27 Mahoney M and Durham G. *Health impact assessment: a tool for policy development in Australia.* Deakin, Australia: Faculty of Health and Behavioural Sciences, Deakin University, 2002. Access Date: 27.02.2003. http://www.hbs.deakin.edu.au/HealthSci/Research/HIA/

28 *A technical briefing on Health Impact Assessment (HIA). Background paper EUR/RC52/BD/3—WHO Regional Committee for Europe, Copenhagen, 16–19 September 2002.* Copenhagen: WHO Regional Office for Europe, 2002. Access Date: 27.02.2003. http://www.euro.who.int/document/rc52/ebd3.pdf

29 **Milner S and Marples G.** *Policy Appraisal and Health Project. Newcastle Health Partnership, Phase 1—a Literature Review.* Newcastle: University of Northumberland, 1997.

30 **Lehto J and Ritsatakis A.** *Health Impact Assessment as a Tool for Intersectoral Health Policy.* Brussels: WHO Europe Centre for Health Policy, WHO Regional Office for Europe, 1999.

31 **Ison E.** *Resource for Health Impact Assessment.* London: NHS Executive London, 2000. Access Date: 27.02.2003. http://www.londonshealth.gov.uk/allpubs.htm#Top

32 **Elliott I.** *Health Impact Assessment: An Introductory Paper.* Dublin: The Institute of Public Health in Ireland, 2001. Access Date: 20.05.2003. http://www.phel.gov.uk/hiadocs/hia_introductory_paper_ireland.pdf

33 Northern and Yorkshire Public Health Observatory. *An Overview of Health Impact Assessment.* Durham: N&YPHO, 2001. Access Date: 20.05.2003. http://www.nypho.org.uk/ reference/occ_paper/OC01.pdf

34 **Taylor L and Quigley R.** Health impact assessment—a review of reviews. 2002. 29-10-2002. Access Date: 27.02.2003. http://www.phel.gov.uk/hiadocs/HIA_evidence_brief_FINAL.pdf

35 **Mahoney M and Morgan RK.** Health impact assessment in Australia and New Zealand: an exploration of methodological concerns. *IUHPE—Promotion and Education* 2001; **VIII**:8–11.

36 **Cameron M.** *A short guide to health impact assessment.* London: NHS Executive London, 2000. Access Date: 27.02.2003. http://www.londonshealth.gov.uk/pdf/hiaguide.pdf

37 **Putters K.** *Health Impact Screening.* Rijswijk: Ministry of Health, Welfare and Sport, 1997.

38 **Thurston GD.** A critical review of PM10-mortality time-series studies. *Journal of Exposure Analysis and Environmental Epidemiology* 1996; **6**:3–21.

39 **Goldberg MS, Bailar JC, III, Burnett RT,** *et al. Identifying Subgroups of the General Population that may be Susceptible to Short-term Increases in Particulate Air Pollution: A Time-series study in Montreal, Quebec. 97.* Cambridge, MA: Health Effects Institute, 2000.

40 **Joffe M and Mindell J.** A framework for the evidence base to support health impact assessment. *Journal of Epidemiology and Community Health* 2002; **56**:132–138.

41 **Mindell J, Hansell A, Morrison D, Douglas M, and Joffe M.** What do we need for robust and quantitative health impact assessment? *Journal of Public Health Medicine* 2001; **23**:173–178.

42 **Watkiss P, Brand C, Hurley F, Pilkington A, Mindell J, Joffe M and Anderson R.** *Informing Transport Health Impact Assessment in London.* London: NHS Executive London, 2000. Access Date: 27.02.2003. http://www.londonshealth.gov.uk/pdf/transhia.pdf

43 **Murray CJL and Lopez A.** *The Global Burden of Disease.* Boston, MA: Harvard University Press, 1996.

44 **Kawachi I, Subramanian SV, and Almeida-Filho N.** A glossary for health inequalities. *Journal of Epidemiology and Community Health* 2002; **56**:647–652.

45 **King K and Stedman J.** *Analysis of Air Pollution and Social Deprivation.* A report produced for the Department of the Environment, Transport and the Regions, The Scottish executive, The National Assembly for Wales and Department of Environment for Northern Ireland. London: ONS, 2000.

46. **Mindell J, Boaz A, Joffe M, Curtis S, and Birley M.** Enhancing the evidence base for HIA. *Journal of Epidemiology and Community Health.* Accepted for publication.

47. Health Development Agency. HIA gateway. Access Date: 27.02.2003. http://www.hiagateway.org.uk

48. **McIntyre L and Petticrew M.** *Methods of Health Impact Assessment: A Review.* Glasgow: MRC Social and Public Health Sciences Unit, 1999. http://www.msoc-mrc.gla.ac.uk/Reports/PDFs/Occasional-Papers/OP-002.pdf

49. **Mindell J, Ison E, and Joffe M.** A glossary for health impact assessment. *Journal of Epidemiology and Community Health* 2003; **57**:647–651.

Chapter 10

HIA: a practitioner's view

Kate Ardern

This chapter discusses the experience of practising health impact assessment (HIA), some of the problems encountered, and some of the unexpected benefits enjoyed.

Why do an HIA?

In order to avoid being diverted into the investigation of issues, which will not influence the decisions about the project, programme, or strategy under review, one must remain focused on why the HIA is being done. The purpose of conducting an HIA is to assist an organization ensure that they actively contribute to the improvement of health and the reduction of inequalities or, at the very least, do not inadvertently make these things worse. The HIA process achieves this by offering a framework for the systematic assessment of how a proposal will affect population health, and how these effects will be distributed between the different population groups.

Who owns the HIA?

Ownership of the HIA is critical to its success or failure especially in organizations that are new to HIA. At the start of any HIA it is important to make sure that the work has political and high-level management support to ensure that the recommendations proposed by the study will be given due consideration. In a recent example, involving the Strategic Housing Statement for Liverpool, there was a clear political and strategic context for an HIA. A very critical Audit Commission report had been published on the city's housing strategy pointing out the need to strengthen links between housing and health [1]. The HIA was able to provide a timely response to this and the citywide Strategic Housing Partnership took ownership of the work [2]. The Strategic Housing Partnership has a wide-ranging and influential membership including registered social landlords, the North West Regional Government Office, Liverpool City Council and the Housing Corporation, and the three Liverpool Primary Care (Health) Trusts.

The steering group

The underpinning values of HIA are sustainability and democracy. These are particularly important when considering issues of accountability for the conduct of an HIA, for the reporting of its recommendations, and for monitoring outcomes. A steering group is advisable to oversee the HIA process. Although it is difficult to be prescriptive about membership, it should include individuals and representatives of organizations and groups, who have an interest in, or are affected by, the proposal.

The steering group should devise a work programme encompassing all the required components of the HIA; for example, in the Strategic Housing Statement HIA referred to here, each member of the steering group was required to lead on a particular area of research thereby ensuring ownership by the group of the HIA. It is also helpful to have an external reference group made up of those who have knowledge and experience of the subject but are less closely involved with the HIA. This group can support and advise the steering group as necessary.

In addition to ensuring the overall progress of the HIA, the steering group is responsible for identifying the people and organizations that need to act on the HIAs findings. The steering group must ensure that there are mechanisms for liasing and communicating with these political and strategic bodies. The steering group should have administrative support to formally record proceedings and to 'chase up' people when necessary. This is a crucial role but often forgotten in the planning process. In the Strategic Housing Statement HIA, an officer of Liverpool City Council undertook this role and ensured that all steering group members were clear about the actions expected of them and completed them on time. Keeping to deadlines is especially important when the HIA forms part of a submission to a planning inquiry, planning application, or other tightly regulated process such as a submission of an outline business case or Community Plan.

Communication

Communicating with key stakeholders is critical to the success or otherwise of an HIA. Organizations such as local authority committees, Primary Care (Health) Trust boards and local strategic partnerships (combined health and local authority committees) often require formal feedback. When an HIA is submitted to a planning inquiry, a nominated senior officer may be required to present the evidence. The Directors of Public Health, whose health authorities had submitted an HIA, were called to give evidence at the planning enquiries into Alconbury Airfield [3] and Finningley Airport (Chapter 26). In such cases it is wise to seek legal advice from counsel.

In high-profile HIAs it is advisable to seek advice on dealing with the media from a communications or press officer. HIAs of big projects or policies may require a formal launch and the steering group need to consider the cost of this in planning. It is also courteous and important to provide feedback to all who have participated in an HIA, especially members of the public who have given of their time to contribute. For the Strategic Housing Strategy HIA [2] a simple but informative newsletter, which summarized the issues raised and the progress made, proved an excellent feedback tool. Other innovative ways of communicating the results to different stakeholders could include the Internet or the use of drama, film, art, photography, or music.

What sort of HIA?

The type and complexity of an HIA depends on several key issues, for example, resources (time, staff, and funds) available and local 'politics'. Practitioners of HIA need to be very pragmatic in their approach but as rigorous in their work as is possible within inevitable constraints—the important thing is to recognize and document the limitations of the HIA.

Tools for measuring health outcomes together with sustainable financial and social gains are key to enabling partners to invest wisely for health. Public health professionals can provide guidance on the health status of the population, the major factors that contribute to it, and the actions by which ill-health may be prevented. However, although public health professionals have a clear and important role to play in HIA, there are many other people in sectors outside the health service, whose policy decisions impact on health. It is necessary to equip these people with the means of assessing the health dimensions of their decisions so that all sectors can contribute to health improvement. However for many organizations there is an apparent conflict between the desire do an HIA and the resources available for the task. Furthermore many organizations are under pressure to undertake impact assessments focused on other issues besides health, for example, sustainability, social inclusion, and value-for-money. The prospect of adding yet another assessment to this list may seem overwhelming. Yet in many cases there is considerable overlap between HIA and these other impact assessments although duplication may not be recognized given that for many individuals, the term 'health' is still synonymous with 'disease' and 'healthcare'. These difficulties have led some organizations to look to the development of other ways of doing HIA. In the North West region of England the Regional

Development Agency, whose planners were risking 'impact overload', has explored integrating the various impact assessment tools.

The North West Rapid HIA Tool

Work elsewhere on integrated tools coincided with work undertaken by the author to develop a more flexible tool based on the Merseyside guidelines [4] but drawing on the World Bank wealth accounting system developed in 1995 [5]. This new tool is termed the North West Rapid HIA Tool [6].

The World Bank accounting system recognizes that the wealth of a society depends on more than monetary worth. It identifies four measures of wealth or capital, natural (or environmental), produced assets (economic), human resources, and social capital. For a sustainable society to function the four types of capital must exist. In the North West Rapid HIA Tool this has been expanded to five types of capital by splitting natural resources and environmental protection into separate capital types (Fig. 10.1).

The tool has adapted features taken from the Merseyside Guidelines [4], Best Value Sustainability Appraisals (an impact appraisal system used by English Local Authorities) [7,8], and Social Inclusion toolkits [9,10]. Assessment against the capital system involves a simple three-stage process consisting of screening, evaluation of impacts, and recommendations for further action (Table 10.1).

Screening

HIA should assist and not delay the decision-making process. In order to make the most efficient use of available expert resources, one must be selective about what work is undertaken. Screening is the procedure whereby projects, programmes, or policies are selected for HIA. Candidate projects for HIA, programmes, or policies are rapidly assessed in relation to the five issues listed in Table 10.2. The procedure is necessarily crude, but is essential if HIA is to be effectively deployed.

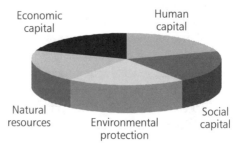

Fig. 10.1 Capital system for rapid HIA.

Table 10.1 The three stages of the North West Rapid HIA process

Screening of policy, programme, or project for health impacts	It is essential that areas with the greatest likely impacts on the health of the most vulnerable groups in the population be given priority for assessment within the constraints of budget and timescales
Evaluation of impacts and potential impacts on factors that determine health	These influences can be considered in a matrix comprising a series of questions under the headings set out in the Capital System. The relative importance of these health impacts, in the short, medium, and long term, and the risks to health versus health gain should be considered
Recommendations for further action	If the risks to health of a proposal outweigh the potential health gains, for instance, if the proposal will exclude the poorest members of the community, further action to address these negative health impacts should be considered

Table 10.2 Matters to be considered in screening

Economic issues	The size of the project and of the population(s) affected The costs of the project, and their distribution
Outcome issues	The nature of potential health impacts of the project (crudely estimated) The likely nature and extent of disruption caused to communities by the project The existence of potentially cumulative impacts
Epidemiological issues	The degree of certainty (risk) of health impacts The likely frequency (incidence/prevalence rates) of potential health impacts The size of any probable health service impacts The likely consistency of 'expert' and 'community' perceptions of probability (i.e., risk), frequency, and severity of important impacts
Strategic issues	The need to give greater priority to policies than to programmes, and to programmes than to projects, all other things being equal (this results from the broader scope—and hence potential impact—of policies as compared to programmes and to projects) Timeliness (ensuring that HIA is prospective wherever possible and Planning Regulations and other statutory frameworks) Whether the project requires an Environmental Impact Assessment Relevance to local decision-making

The North West Rapid HIA tool's evaluation matrix

If screening suggests it is worth going further, the next stage of the North West Rapid HIA process is evaluation of impacts on capital (this is similar to the step that most authors refer to as risk assessment). This involves completion of the matrix, shown in Fig. 10.2, by answering questions on potential grouped under five capital types. The impacts have to be considered separately for the development, operational, and life span phases.

1. Impacts on Human Capital:

Which groups of people within your local population do you think the project/programme/proposal or service will directly affect or encourage other to affect? (*mark positive impacts with a tick and negative ones with a cross*)	Development/ building phase	Operational phase	Life span
The population of "x" as a whole			
Persons aged 65 years and over			
Children aged 0–14 years			
Persons aged 15–25 years			
Disabled adults			
Long term unemployed people aged 26 to 65 years			
Black and ethnic minority populations			
Low income rural households			
Low income urban households			
Increases equal opportunities			
Low pay workers			
Sub total (no. of +ve ticks)			
Sub total (no. of −ve crosses)			

2. Impacts on Natural Resources Capital:

Do you think the project/programme/proposal or service will directly affect or encourage other to affect? (*mark positive impacts with a tick and negative ones with a cross*)	Development/ building phase	Operational phase	Life span
Reduce the production of greenhouse gases			
Encourage use of renewable forms of energy			
Increase the provision of affordable, energy efficient housing			
Increase use of brownfield sites			
Protect/conserve urban green space/amenities			
Promote sustainable use of agricultural land			
Promote efficient use of water supplies			
Protect drinking water quality and availability			
Increase % of reusing, recycling or composting of waste			
Increase % of energy recovered from waste			
Increase % of use of recycled materials			
Sub total (no. of +ve ticks)			
Sub-total (no. of −ve crosses)			

Fig. 10.2 (continued)

3. Impacts on Environmental Protection Capital:

Do you think the project/programme/proposal or service will directly affect or encourage other to affect? *(mark positive impacts with a tick and negative ones with a cross)*	Development/ building phase	Operational phase	Life span
Improve the quality of public transport			
Increase the number of Green Travel Plans			
Reduce the use of private transport			
Reduce deaths and serious injury from road traffic accidents in persons aged 0 to 25 years			
Increase % of journeys made by walking and cycling			
Improve access to public services for non car-users			
Reduce the number of poor air quality days			
Reduce number of hazardous pollution incidents			
Protects habitats and species			
Maintain or increase woodland and tree cover			
Reduce urban noise levels towards WHO safe limits			
Protects /conserves existing historic buildings			
Minimises need for new infrastructure			
Improves the quality /quantity of housing stock			
Reduce the proportion of housing deemed unfit for habitation			
Reduce the number of food poisoning cases in "x"			
Sub total (no. of +ve ticks)			
Sub-total (no. of −ve crosses)			

4. Impacts on Social Capital/Progress:

Do you think the project/programme/proposal or service will directly affect or encourage other to affect? *(mark positive impacts with a tick and negative ones with a cross)*	Development/ building phase	Operational phase	Life span
Enhance human health and safety			
Increase average life expectancy towards the UK average (All cause SMR for people aged 74 years and under to improve)			
Improve access to healthcare for low income households			
Reduce the number of people who misuse drugs or alcohol			
Improve the quality of local healthcare services			
Increase opportunities for physical activity and sport			
Support community networks			
Reduce community severance			
Reduce the rate of domestic burglaries			
Reduce the rate of domestic violence			
Reduce the rate of assaults			
Improve community safety			
Reduce fear of crime			
Reduce the rate of racially motivated crime			
Improve access to healthy foods for low income households and non car users			
Support regeneration of existing areas of decline			
Increase public participation in democracy and local decision-making (% voting in elections)			
Sub total (no. of +ve ticks)			
Sub total (no. of −ve crosses)			

Fig. 10.2 (continued)

5. Impacts on Economic Capital:

Do you think the project/programme/proposal or service will directly affect or encourage other to affect? (mark positive impacts with a tick and negative ones with a cross)	Development/ building phase	Operational phase	Life span
Increase life-long educational opportunities			
Increase access to higher education for 16–18 year olds from low income households			
Improve the rates of GCSE passes in economically deprived areas			
Increase use of local goods and services by business			
Support development of community-based businesses			
Increase provision and quality of ILMs			
Create sustainable employment opportunities for local people			
Improve occupational health provision			
Increase local income levels			
Increase economic diversity			
Reduce the unemployment rate in Pathways areas			
Reduce the rural unemployment rate			
Increase opportunities for developing sustainable business practice			
Increase sustainable local, regional, national and international communications links			
Increase the use of e-based technology			
Improve accommodation for SMEs and start-up businesses			
Increase opportunities for "Family Friendly" employment policies			
Sub total (no. of +ve ticks)			
Sub total (no. of −ve crosses)			

Fig. 10.2 The North West Rapid HIA tool's evaluation matrix.

In completing each question on the matrix one has to consider the potential impact on individuals, groups, and organizations paying particular regard to disadvantaged groups. For each affected group, both positive and negative potential impacts and the likelihood of their occurring in the short, medium, or long term must be thought of. The response to each question should be supported by documented evidence (quantitative and/or qualitative) so that there is consistency in the way health issues are interpreted and analyzed for each question.

Where impacts are identified one has to consider if the proposal can be modified to maximize positive impacts on health or minimize negative impacts. Where modifications are indicated their practicality and the time needed must be thought of. If further research is indicated, one has to think how it will be done and by whom. Alternative options or modified proposals may be presented at this stage and these may require further research or investment. It may be necessary to suspend a decision while further work is done.

When the matrix has been completed the number of positive ticks for each section in each phase is counted. Proposals that score less than six positive ticks in any section, in any phase, or less than 26 ticks in total for each phase will require further action and/or developmental work to ensure that they contribute to health improvement. Ideally there should be a balance in each of the five 'capitals' (Fig. 10.1).

Who pays for HIA?

One of the major difficulties in conducting HIAs is identifying staff and resources. Some areas in need of regeneration in England were designated Health Action Zones (HAZ). These zones were given significant budgets and in Liverpool the senior officers of the HAZ partners agreed to support a programme of HIAs. An HIA of the Local Transport Plan [11] was one of the HIAs funded under this programme (costs shown in Table 10.3).

Before the HAZ was established, the four former Merseyside Health Authorities included support for HIA as part of their contract with the Liverpool Public Health Observatory. Other authorities include a requirement to do HIA work in the job plans and contracts of specific staff.

However for many HIAs, funding to support the assessment will need to be sought on a 'one-off' basis. Merseyside Guidelines for Health Impact Assessment recommend that a budget for an HIA is identified from the project finance and suggest a figure of 0.1 per cent of the total budget. This is consistent with the current funding arrangements for environmental impact statements.

Implementing HIA recommendations

If HIA is going to be value-for-money, then it must produce worthwhile and useful recommendations. This means that the recommendations must:

◆ Be practical.

◆ Aim to maximize health gain and minimize health loss.

◆ Be socially acceptable (a degree of pragmatism may be inevitable).

◆ Consider the cost of implementation.

Table 10.3 Costs of Local Transport Plan HIA study

Elements of cost	In pounds
Systematic review of attitudinal surveys/data	1500
Air quality study	4000
Stationary, travel	1000
Focus group work	2000
Researcher's time (CPHM)	1500[a]
Total revenue	11,000

[a] In addition to incorporation of HIA into current work objectives.

- Consider the opportunity cost.
- Include preventative as well as curative measures.
- Be prioritized in terms of short-, medium-, and long-term objectives.
- Identify a lead agency or individual.
- Identify the drivers and barriers to change
- Be agreeable to that lead agency.
- Be capable of being monitored and evaluated

The list given here is, of course, not definitive and as HIA develops other criteria will be added. Birley [12] has observed that recommendations should be specific rather than general so that their implementation can be monitored. He also notes that if the multidisciplinary teamwork is poor and the recommendations only reflect one person's viewpoint then the logistics of implementation may not be appreciated and key agencies may disown the recommendations.

Sharing the learning

Finally, when undertaking an HIA it is important to be highly self-critical throughout the study and seek to identify lessons learnt (both good and bad) and to document them. It is a valuable, if painful, process and helps not only with the design of future HIAs but also with the monitoring and the implementation of recommendations. The list that follows identifies some of the strengths and weakness of the HIA of the Strategic Housing Statement for Liverpool undertaken by the author [2] and referred to earlier in the chapter.

Weaknesses

- There was little community involvement in the project steering group, it was a very officer-led process.
- There was no representation from the private rented/private sector involvement despite our best efforts.
- Steering group members were undertaking project work within already very busy working environments and a number of our original timescales were too ambitious.
- Use was not made of the Young People's Parliament.

Strengths

- This study has highlighted the need to direct adequate resources to accessing hard to reach groups, and has pointed to some possible ways of doing this.

- This was the first experience of the HIA process for the majority of people involved in the study, yet all participated in the study with enthusiasm and embraced the holistic model of health that is at the core of the HIA process.

- The allocation of responsibility for project management to members of the steering group engendered a sense of ownership of the study by the steering group members and ensured that political processes both within and across organizations were handled sensitively.

- The role of the chair was very important in maintaining the momentum of the study and ensuring the project was well resourced. This HIA illustrates well that you do not have to have an HIA 'expert' as the chair of the steering group to ensure a successful outcome. The chair was also able to ensure that the project had effective and efficient administrative support, which was invaluable.

- Within the HIA we had examples of organizations working together for the first time, for example, Liverpool City Council's Housing Policy Team and the Childrens and Young Peoples Bureau. The importance of this cross-agency working cannot be underestimated. As well as new relationships being forged, we also saw the cementing of others, for example, South Liverpool PCT and CDS Housing looking for innovative sustainable solutions to meet the housing needs of key healthcare workers.

References

1   Audit Commission. Housing Best Value Inspection: Liverpool's Housing Strategy, July 2001.

2   Webster J and Ardern KD. Health Impact Assessment of Liverpool City Council's Housing Strategy Statement, Liverpool City Council, Strategic Housing Partnership of Liverpool and North, Central and South Liverpool Primary Care Trusts, April 2003.

3   Jewell T. *Alconbury Health Impact Assessment Report.* Cambridge, Cambridgeshire Health Authority, 2000.

4   Scott-Samuel A, Birley M, and Ardern K. *The Merseyside Guidelines for Health Impact Assessment. The Merseyside Health Impact Assessment Steering Group.* Liverpool, University of Liverpool, 1998.

5   Munasinghe M and Shearer W. *Defining and Measuring Sustainability: The Biogeophysical Foundations* (Vol 1). World Bank Publications, January 1995.

6   Arden K, Kilfoyle M, and Hussey. Chapter 5. In Kilfoyle M (éd.), *Health—A Regional Development Agenda: A Special Report to the North West Regional Development Agency by the Health Task Group North West.* Liverpool Health Authority and Liverpool John Moores University, May 1999. www.nwda.co.uk/publications.

7   Office of the Deputy Prime Minister. Best Value and Local Councils Report, 2002. http://www.local-regions.odpm.gov.uk/bestvalue/legislation/byparish/index.htm

8 Office of the Deputy Prime Minister. Sustainability Appraisal of Regional Planning Guidance: Final Report, 2001. http://www.planning.odpm.gov.uk/prp/sustrpg/index.htm

9 Tools to Support Participatory Urban Decision Making, United Nations Centre for Human Settlements Habitat Programme, September 2001.

10 Bridging the Divided City: Report of the Dialogue on Social Inclusion in Cities 18th meeting of the Commission on Human Settlements, United Nations Centre for Human Settlements Habitat programme, 14 February 2001.

11 **Ardern KD, Watson A, Stewart L, Hall D, and Felstead T.** Health Impact Assessment of the Merseyside Local Transport Plan, Merseytravel and the Merseyside Health Action Zone, June 2000.

12 **Birley MH, Boland A, Davies L, Edwards RT, Glanville H, Ison E, Millstone E, Osborn D, Scott-Samuel A, and Treweek J.** *Health and Environmental Impact Assessment: An Integrated Approach.* London, Earthscan/British Medical Association, 1998.

Chapter 11

Rapid appraisal techniques

Erica Ison

Since the introduction of health impact assessment (HIA) in the United Kingdom by a group of practitioners working in Merseyside, England, the methodology has evolved and been applied in two main ways:

- as the 'classic' form in which there is a six-step process—screening, scoping, appraisal, report-writing, influencing decision making, and monitoring and evaluation;

- as an approach in which only certain steps in the HIA process are used to inform the work of organizations in the public, private, or voluntary and community sectors—that is, individual steps are 'embedded' into normal working practice. The two steps most frequently 'embedded' are screening and appraisal.

The focus of this chapter is appraisal using rapid appraisal techniques. It should be noted that appraisal can also be referred to as risk assessment, especially in models of HIA derived from environmental impact assessment (EIA), which are based on measurement and mathematical modelling of environmental, toxicological, and epidemiological data (referred to as tight perspective HIA by Kemm [1]). Given the need to assess accuracy, validity, and consistency of quantitative data, and the lack of computer programs for risk assessment in HIA at present, it is unlikely that risk assessment will be undertaken during rapid appraisal. Rapid appraisal can be used in two types of situation:

- during the third step of the classic HIA process; and

- embedded during normal working practice where an HIA approach is used to explore the potential for health protection and improvement in a proposal.

What is rapid appraisal?

There are two types of rapid appraisal techniques:

- participatory, in which as many as possible of the stakeholders for an HIA are involved;

- non-participatory, in which only a small number (one or more) of stakeholders are included.

Participatory rapid appraisal techniques

Participatory techniques tend to be those used most frequently when rapid appraisal is part of the classic process of HIA.

The first rapid appraisal technique used in HIA in the United Kingdom was adapted from that developed by the World Health Organization (WHO) as an effective way of consulting a relatively large number of widely dispersed health workers in the field. This adapted form is known as a participatory stakeholder workshop in HIA.

Other participatory techniques described in this chapter include:

◆ 'open' events; and

◆ business forums.

Using participatory rapid appraisal techniques as part of the classic HIA process is the way in which most organizations will embark upon the use of HIA as a methodology.

Non-participatory rapid appraisal techniques

Non-participatory techniques tend to be those that are most frequently embedded into the normal working practice of an organization, for example, desktop HIA [2]. In general, they are used by officers to ensure that health impacts are taken into account during the design, planning, and development of a proposal, or during the selection of bids in strategic procurement or contracting processes.

Using non-participatory techniques is unavoidable if the concept of assessing any proposal's impacts on health is to become routine at an early stage of planning.

Defining characteristic of rapid appraisal techniques

Although many people define rapid appraisal techniques by the length of time it takes to complete them, for example, The Merseyside Guidelines [3], this definition is simplistic, and leads to confusion about what constitutes rapid appraisal. The defining characteristic of rapid appraisal is that no new data are collected. As the collection of new data takes time, it is true that an appraisal which meets this definition will take a relatively short amount of time to complete. Thus, the information used during rapid appraisal is that which is readily available, irrespective of whether it is:

◆ evidence from the literature;

◆ the results of already-completed HIAs in the grey literature;

◆ local data (both routine and non-routine) relating to the community involved, or their circumstances.

There is also confusion about the distinction between screening and rapid appraisal. Some practitioners (especially in Netherlands) refer to rapid appraisal techniques—participatory or non-participatory—as screening or as if they are equivalent to screening. However, within the classic HIA process, screening and rapid appraisal are steps with different functions. Embedded or non-participatory rapid appraisal techniques have different functions to both of these. Table 11.1 shows the distinctions between 'embedded' rapid appraisal, screening in 'classic' HIA and rapid appraisal in 'classic' HIA.

The appeal of rapid appraisal

The appeal of rapid appraisal is the relatively short amount of time in which potentially useful outputs can be generated. There is also the perception that rapid appraisal is less intensive in its use of resources, particularly financial. Thus, the image of rapid appraisal is that it is a practicable, feasible, affordable, and manageable aspect of the HIA process or approach.

The reality of rapid appraisal

Participatory rapid appraisal

The reality of using participatory rapid appraisal techniques is that it can be intensive in the use of resources. To obtain good-quality outputs that will be helpful to decision makers and other stakeholders, it is necessary:

◆ to prepare appropriately;

◆ to report the results effectively.

Both of these tasks are important, and require time, expertise (in HIA or some other types of assessment), and proper execution. They also require the participation of many people, which is intensive in terms of human resources from more than one organization.

Moreover, as appraisal is only one step in the process of classic HIA, the use of participatory rapid appraisal techniques saves time mainly in relation to this step. The time taken for the first two steps—screening, and scoping—will be the same if performed properly, irrespective of the type of appraisal selected. Rapid appraisal might reduce the time taken for report writing because the volume of material generated can be less than that generated using non-rapid techniques. However, the time taken to influence decision makers may be similar irrespective of the type of appraisal because similar tasks are involved. It is unlikely that the time taken for the various modes of monitoring and evaluation, particularly of health outcomes, will be different simply because rapid appraisal was used.

Table 11.1 Features of embedded rapid appraisal techniques, and of screening and rapid appraisal as steps in the classic HIA process

	Embedded rapid appraisal techniques	Screening in classic HIA process	Rapid appraisal in classic HIA process
Function	To support processes of proposal planning, design, and development, or of strategic procurement and contracting	To select proposals to which the classic process of HIA needs to be applied; one of the criteria for selection will include potential impacts on health	To identify potential impacts on health, and ways of addressing those impacts
Accountability	Through usual line and performance management channels	Mainly through usual line and performance management channels, but may be under the direction of a group of key stakeholders overseeing an HIA programme	Through a steering group who have an agreed terms of reference for a specific HIA
Protocol	Involves a tool: for the design of proposals, a 'general' tool; for strategic procurement and contracting process, a tool designed specifically according to a general model	Involves either a general screening tool or set of criteria against which it has been agreed proposals will be assessed	Guided by a scoping document for a specific HIA agreed by steering group, containing the parameters or 'boundaries' for the HIA

People involved	Individuals in an organization or partnership	One or more personnel from an organization or partnership, who may submit results to other stakeholders	Stakeholders for a specific HIA under the guidance of the 'assessors' appointed by the steering group
Distribution of results	Results kept as an internal record	Results may be reported to a group of key stakeholders overseeing an HIA programme, and should be accessible especially for use later on in classic HIA process	Results are reported to the steering group in full, before wider dissemination to all stakeholders
Community consultation	None	Unlikely; a community representative could be involved if there is a well-developed HIA programme underway	Members of the community are stakeholders, and should be invited to take part[a]

a A situation in which community members may not be invited to take part in participatory rapid appraisal might be the use of a stakeholder workshop to train professionals in HIA, which may have a dual purpose of generating useful outputs about the health impacts of a particular proposal, that is, learning by doing.

Non-participatory rapid appraisal

For non-participatory rapid appraisal techniques to be effective, it is important to develop appropriate tools. If tools are to be fit for the purpose, and usable, it is important to develop them in conjunction with people who will be using them. Once a prototype has been developed, it is highly advisable to pilot it. Thus, the development of useful tools is intensive of human resources, and also requires time and expertise.

Over the last 18 months, the author has been involved in developing both general and specific non-participatory rapid appraisal tools that can be used to embed HIA as an approach into normal working practice.

The strengths of rapid appraisal techniques

Participatory rapid appraisal

The greatest strength of participatory rapid appraisal is the structured participation of many stakeholders from different sectors and disciplines in the assessment of a proposal using a systematic framework. All other strengths of participatory rapid appraisal stem from this, for example:

- the bringing together of different people who have a common interest, that is, in the development and implementation of a proposal;
- the rapid interrogation of the knowledge base and experience of people who live in and work for/with a community—known as tacit knowledge;
- the framing of recommendations to modify a proposal that reflects not only its multifaceted impacts but also the collaborative response of different stakeholders.

Non-participatory rapid appraisal

The strength of non-participatory rapid appraisal techniques is the introduction of health and well-being as an important consideration at an early point in the design of proposals, whether as part of an internal process of proposal development or for the assessment of external bids from different potential contractors during strategic procurement.

Criticisms of rapid appraisal

Criticisms of rapid appraisal techniques cover both outputs and process. The major criticism made about the outputs from participatory and non-participatory rapid appraisal is the predominantly qualitative nature of the results and the lack or paucity of evidence applied.

Participatory rapid appraisal

Participatory rapid appraisal techniques may be criticized because their outputs rely on opinion, or speculation, expressed by stakeholders at consultation event(s). The major criticism of the process is that the representation of stakeholders may be inadequate. This can arise for several reasons:

- in the generally short time available, it may not be possible to identify all stakeholders;
- even if all relevant stakeholders can be identified, it may not be possible to find a suitable contact or representative in an organization, and their correct details, to invite them to the consultation event(s);
- some stakeholders may choose not to participate for various reasons.

Stakeholders may choose not to participate because they do not understand what HIA is, and feel unable, or lack the confidence, to contribute—this applies to community members and to some professionals. The consultation events may not be arranged at a convenient time, and/or venue to allow stakeholders to attend. This is particularly relevant when working with the community, and small- and medium-sized businesses. Stakeholders, from the community and voluntary sector and from small businesses may not have the resources to contribute, unless remunerated or subsidized. Those invited may not see the relevance of the HIA or may be sceptical about the power of the technique, or those leading the HIA, to make a difference. Some organizations, particularly in the public sector, may feel it compromises their neutrality with respect to the proposal. As a result, the representation of stakeholders from all sectors may be partial. This is critical when dealing with the community, and in particular hard-to-reach groups, such as black and minority ethnic (BME) communities, older people, young people, and refugees and asylum seekers.

Non-participatory rapid appraisal

The output of non-participatory rapid appraisal techniques may be criticized because they reflect only the perspectives and knowledge base of the single person, or the relatively few stakeholders, involved.

Benefits of rapid appraisal techniques
Participatory rapid appraisal

The benefits of participatory rapid appraisal techniques are:

- they increase understanding among stakeholders;

- health is introduced as an important issue to be addressed by both health-related and non-health agencies;
- they improve partnership working in general, and on specific proposals.

Non-participatory rapid appraisal

The major benefit of using non-participatory rapid appraisal techniques is that a consideration of health, and ways of achieving health gain, becomes routine during normal working practice of organizations and partnerships whose functions affect the public health.

When to use rapid appraisal

Rapid appraisal techniques can be used in a variety of situations and for many different reasons. However, it is important to distinguish between those situations when it is advantageous, those when it is appropriate, and those when it is merely the best available technique given the circumstances and constraints. Table 11.2 compares the circumstances for use of participatory and non-participatory techniques.

Organising participatory rapid appraisal in classic HIA

When using rapid appraisal techniques in a classic HIA, the appraisal step is preceded by screening and scoping steps just as for HIAs in which non-rapid techniques are used. The quality of outputs from any type of appraisal, and the steps that follow, is likely to be improved if the outputs of each step are used to inform the following steps.

The first step in the classic process of HIA is screening, primarily used to select proposals on which it is appropriate to undertake HIA. Screening is most effectively done using a screening tool or a set of criteria to aid selection decisions. Screening should provide a systematic and transparent framework against which the proposal can be assessed. Screening will generate outputs to inform scoping, the second step in the HIA process.

Preparation for scoping involves establishing a steering group for the HIA comprising key stakeholders, who can be identified during the screening step if a screening tool is used. Once this has been done a first meeting of the steering group, including inviting potential members and preparing the agenda needs to be arranged. Prior to the meeting draft terms of reference for the steering group and a draft scope for the HIA can be prepared using information from screening ready to be discussed and if agreed ratified at the first meeting of the steering group.

Table 11.2 Situations in which the use of participatory and non-participatory rapid appraisal are advantageous, appropriate, or merely the best available

Effective use	Appropriate use	Best available technique
Participatory rapid appraisal		
◆ The need to bring stakeholders *together* to explore different facets of a proposal, increase understanding of its implications, and co-ordinate responses	◆ After scoping, near the beginning of a classic HIA process that may be lengthy, or is being undertaken on a complex proposal/set of proposals,[a] to identify impacts that need to be investigated further	◆ Limited resources (financial and human in particular) ◆ Limited time in which to generate outputs to influence decision makers
◆ The need to explore knowledge base and experience of people who live in and work for the community	◆ As a complement to non-rapid techniques, for example, surveys, in which the conjunction of stakeholders can illuminate/expand upon responses	
◆ The need to encourage interest and *engagement* in the HIA process	◆ As a complement to non-rapid techniques to map sources of information relevant to the HIA, and to identify gaps in the evidence base, and/or routine and non-routine local data	
Non-Participatory rapid appraisal		
◆ Initial planning and design of a proposal ◆ Guiding an internal process of proposal development	◆ Initial planning and design of a proposal ◆ Guiding an internal process of proposal development ◆ In the assessment and selection of bids in strategic procurement and contracting processes	◆ Lack of time to consult other stakeholders even using participatory rapid appraisal techniques ◆ Lack of resources to consult other stakeholders even using participatory rapid appraisal techniques

[a] This can be referred to as screening—however, the use of rapid appraisal at this point is to identify the potential impacts on health that need to be assessed indepth within the HIA process as opposed to selecting a proposal for entry into the HIA process.

Scoping builds on the outputs of screening. It is the point in the classic HIA process when the boundaries for a specific HIA are identified and recorded, including:

◆ elements of the proposal to be assessed;

◆ the community or population affected—including any vulnerable groups— and any other stakeholders involved;

- relevant geographical boundaries, local conditions, and circumstances;
- parameters for the HIA—aims, objectives, values, and methodology(ies) to be used;
- information requirements for the HIA;
- management arrangements for the HIA;
- decision makers and other stakeholders it is important to influence;
- parameters for various modes of monitoring and evaluation.

Thus, the scope as a document is the 'blueprint' for an HIA. It informs not only the preparation for appraisal, but also all the later steps in the classic HIA process.

Before moving onto the appraisal step the following information needs to be prepared

For all appraisals:

- document containing the elements of the proposal to be assessed;
- baseline profile of the community/population affected;
- summary of local data relevant to the proposal (from routine and non-routine sources)—this can be combined with the baseline profile of the community;
- review of evidence in the literature;
- digest of relevant points from previous HIAs on similar proposals or on the same community/population.

For appraisals using participatory techniques:

- a brief introduction to the methodology for stakeholders who have little or no experience of HIA;
- A summary of other HIAs in progress, and those that have been completed, in the local area may also be useful.

Participatory rapid appraisal

Three techniques of participatory raid appraisal techniques will be outlined. Which technique or techniques is most appropriate in a specific HIA will depend on the stakeholders involved. The three techniques are:

- stakeholder workshops for service providers, and community organizations and community representatives; the format of a stakeholder workshop has been described elsewhere by the author [4];
- open events for the community/members of the public—using techniques from PLANNING FOR REAL exercises [5];
- business forums—aimed at the private sector adapted from the format for stakeholder workshops.

'Open' events and business forums were developed by the author as adjuncts to participatory stakeholder workshops. These formats have the advantage of increased engagement of communities and businesses in the HIA process and they may be more attractive to stakeholders who are reluctant to participate in HIA for reasons discussed earlier in this chapter.

The 'open' event

The advantages of an 'open' event for the community affected by a proposal are that they are convenient, they offer choice, and appear relevant and user-friendly. People do not have to commit to the long period of time involved in attending and participating in a workshop (3–4.5 h if travelling time is included). Open events can be held on a Saturday or the evenings when most people are available. People are able to choose from several different ways of giving their input and have a relatively unlimited opportunity to ask questions freely in relation to their particular concerns about a proposal. To a certain extent, open events avoid imposing a professional paradigm on members of the public.

An open event should be held in a large space, such as a community centre hall, near to the community affected. The event should be advertised locally using newspapers, leaflets, and posters in community settings. The event should be open for 3–4 h, allowing people to drop in when they like and for as long as they wish. Saturday afternoons are a good time for such events. There should be displays about the proposal, the HIA, the evidence, and relevant local data. These should include graphic elements so they will be helpful to people with poor literacy skills or whose first language is not English. A range of ways for people to make their input to the HIA should be provided such as:

◆ filling in a questionnaire;

◆ talking to a facilitator;

◆ using a map to identify and mark locations where issues/problems are associated with a proposal, for example, where people feel unsafe on an estate;

◆ adding to graffiti boards.

Handouts should be provided for people to take away as they leave the open event.

Business forums

The main advantage of a business forum is convenience. The events are usually held in the early evening, when most business proprietors or employees are able to attend, and the length of time involved is about half that for a stakeholder workshop (business forums are typically 2 h long).

A business forum should be held in a relatively large room near the business community affected. The businesses to be invited are identified by liasing with officers in the business development section of the council. The forum is also advertised in local newspapers. A typical forum might consist of:

- talk about the proposal (15 min);

- explanation about HIA, description of the HIA of the proposal (15 min);

- exploration of concerns about the proposal (15 min);

- exploration of positive expectations of the proposal (15 min);

- identification of impacts of proposal from business sector's point of view (25 min);

- identification of ways of managing those impacts (25 min);

- discussion of further involvement of the business sector in the process (10 min).

As with open events, displays about the proposal, the HIA, the evidence, or relevant local data, should be put up in the room. The displays should include graphic elements. Handouts should be provided for people to take away.

Preparation for participatory events

The tasks involved in preparing for the three participatory techniques are shown in Table 11.3. In particular the following points have to be borne in mind

- Ascertain a convenient date and ensure the availability of assessor or lead facilitator and other facilitators. For open events and business forums, be aware of any cultural and religious festivals that should be avoided to ensure attendance.

- Book a venue. For all events, ensure it has disabled access and for open events and business forums, ensure it is located within easy reach of the community being consulted.

- Prepare a contact list of stakeholders: for open events, include community organizations, for example, tenants and residents associations.

- Book catering: for all events, be aware of dietary needs of participants for religious reasons, for example, halal.

- Send out invitations: for stakeholder workshops and business forums, try to send out the programme with the invites; for business forums, the material may need to be translated into the language(s) used by some of the business community.

- Book interpreters: for open events and business forums, ensure languages of the main BME groups are represented, including British Sign Language

Table 11.3 Tasks involved in the preparation of participatory techniques

	Stakeholder workshop	Open event	Business forum
Admin/secretarial support	√	√	√
Preparation of contact list of stakeholders	√	√	√
Preparation of invitation	√	√	√
Preparation of programme	√		√
Preparation of materials for responding to HIA, for example, questionnaires		√	
Preparation of information for participants ◆ handouts ◆ visual displays	√	√	√
Advertising the event: preparation of newspaper ads and leaflets/posters		√	√
Appointment of assessor(s)	√	√	√
Use of facilitators	√	√	√
Use of interpreters	?	√	√
Venue	√	√	√
Catering	√	√	√
Crèche (if relevant)	?	√	√
Preparation of venue: handouts available and displays of material	√	√	√
Dissemination of summary	√	√	√
Dissemination of full report	?		

(BSL) if relevant; BSL may be necessary for a stakeholder workshop—assess this following receipt of replies to invitation.

◆ Advertise the events: for open events and business forums, place advertisements in local newspaper(s) two weeks running before the event, and for open events distribute leaflets and posters at venues in community, for example, schools, religious facilities, community centres—leaflets and posters may need to be translated into the language(s) used by BME groups in the community.

◆ Prepare information for participants: for stakeholder workshop, if possible, send documents relevant to the appraisal to participants in advance but ensure they are available on the day.

◆ Prepare the venue: for open events, ensure that different ways of inputting to the HIA can take place in specified areas, including a separate area for interpreters to work with relevant BME community members.

It is also advisable to develop a communications strategy for the HIA if the proposal is, or has the potential to be, contentious or a source of conflict among certain stakeholders.

Reporting an HIA that includes rapid appraisal

As for any HIA, the results of rapid appraisal need to recorded and presented to inform decision makers and other stakeholders. The report of an HIA should incorporate the following information (see Fig. 11.1):

- an outline of the proposal being assessed;
- an introduction to the HIA of that proposal;
- a baseline profile of the community or population affected by the proposal;
- a summary of local conditions or circumstances relevant to the proposal (from sources of local routine and non-routine data);
- evidence from the published literature;
- information from HIAs that have been conducted on similar proposals and/or on the same community or population;
- results of the appraisal, including impacts on health and interventions to address those impacts.

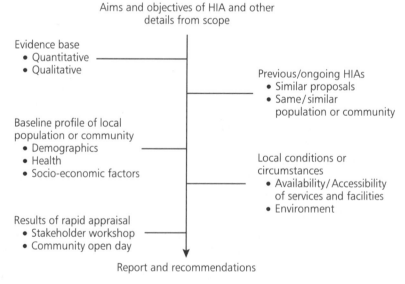

Fig. 11.1 Writing the report of an HIA.

The amount of time required to collate and combine this information should not be underestimated, irrespective of whether the results of the appraisal have been generated using rapid appraisal techniques.

For details of reporting the results of a participatory stakeholder workshop, see Ison [4].

Depending on the techniques used during an open event, it is important to produce an accurate transcript of responses—for example, input to the graffiti board(s), input to any visual features such as maps, and responses expressed verbally to a facilitator. Once accurate transcripts have been produced, it is necessary to analyze the responses, both those in the transcripts and those to the questionnaires. The results can then be incorporated into the report as for those of a stakeholder workshop.

Reporting the results of a business forum is similar to reporting the results of a stakeholder workshop, as a similar but shortened format is used. Again, it is important to produce an accurate transcript of the event, before responses are analyzed and incorporated into the report.

The future

The use of rapid appraisal techniques in the future will be strengthened by:

- research that eliminates gaps in the evidence base about the wider determinants of health;
- the development and structuring of the evidence base for use in HIA;
- the improvement of routine sources of local data relevant to HIA;
- the development of robust tools, particularly for non-participatory techniques;
- the development of participatory techniques, by learning from other disciplines, for example, community development.

Acknowledgements

Many colleagues have been supportive during the development of various rapid appraisal techniques: Kukuwa Abba, Fiona Adshead, Ben Cave, John Chapman, Anthea Cooke, Iris Elliott, Brid Farrell, Janice Gibbons, Sian Griffiths, Jan Hart, Sophie Heighway, Kim Julicher, Karen Lock, Angela Mawle, Sheila Paul, Roger Seddon, Anita Tandon, Alison Talbot-Smith, and Jenny Wright; it is also important to acknowledge all the people, from whatever sector, who have come to an event and taught me how to improve participatory rapid appraisal in HIA.

References

1 **Kemm J.** *Developing Health Impact Assessment in Wales.* Cardiff: Health Promotion Division, National Assembly for Wales, 1999.

2 **Parry JM and Stevens AJ.** Prospective Health Impact Assessment: problems, pitfalls, and possible ways forward. *British Medical Journal* 2001; **323**:1177–1182.

3 **Scott-Samuel A, Birley M, and Ardern K.** The Merseyside Guidelines for health impact assessment. Liverpool: Merseyside health impact assessment Steering Group, Liverpool Public Health Observatory, 1998.

4 **Ison E.** Rapid Appraisal Tool for Health Impact Assessment. A task-based approach. Eleventh iteration. 2002. http://www.hiagateway.org/ under Resources

5 Planning for Real www.nifonline.org/NIP30301PlanningForReal.asp

Chapter 12

Lessons from EIA

Alan Bond

Environmental Impact Assessment (EIA) is a process in which there is now more than three decades of experience. Over these three decades, it has been implemented as a legal procedure in many countries, and has undergone review at international [1,2], regional [3,4], and national levels [5–7]. In addition, the techniques used for assessment of various types of impacts have themselves improved and EIA has evolved to become a more effective and worthwhile process. Lessons have already been learned which can benefit the development of HIA and perhaps enable a more rapid maturation into an internationally accepted and used assessment process.

This chapter reviews the development of EIA to set the context against which lessons can be drawn. The review will briefly consider the rationale for EIA and will indicate the extent of its spread across the globe as a legal procedure. The chapter will go on to explain how EIA works. This will encompass a consideration of the procedural nature of EIA, which is a crucial point when it comes to the possibility for legal action. It will also consider the role of stakeholders, the potential for conflicts of interest to arise as well as public consultation. Finally it considers the funding of an EIA, which is a practical point of great significance when considering the impartiality of the assessment.

The chapter will end by drawing on the experience with respect to HIA. First, the existing consideration of health in EIA will be considered, and this will be followed by an indication of some of the lessons that HIA can learn from the EIA experience developed over these three decades. Finally, short consideration will be given to the potential for integration of HIA within EIA—although this is presented in terms of feasibility rather than as a recommendation.

Rationale for EIA

The roots of EIA as a legal process lie in increasing concern in the 1960s that development was proceeding unchecked, and that foreseeable environmental damage was occurring. In particular, books like Rachel Carson's *Silent spring*

[8] raised awareness of man-made environmental problems and, by the end of the decade in which *Silent spring* was published, EIA was a legal process in the United States.

The rationale behind EIA is simply that if decision makers are aware of the potential impacts a proposed development may have, they will be better informed and make better decisions. However, EIA can be far more effective than this, and lead to more environmentally friendly design as modifications can be made to proposals, based on the EIA, before any document is passed to decision makers.

The need for EIA is well acknowledged [9], and is primarily an academic consideration now that the need is stipulated in legislation in many countries. Despite this, it has always been difficult to prove its value although progress is being made in demonstrating its utility and cost effectiveness [10]. The need for EIA is often illustrated through the use of case studies to demonstrate the implications of development where consideration was not given in advance to the likely impacts. Examples frequently include dam developments because these are large projects and have clearly identifiable impacts, many of which are health impacts (like increase in disease through provisions of suitable habitat for disease vectors).

Expansion of EIA across the globe

Whilst academics might disagree on what the first examples might be of processes that can be considered to be EIA (e.g., roots in the United States around the time of the Second World War [11]; early example in 1548 in England to examine effects of iron mills and furnaces of the Weald [12]), there is no disagreement that the world's first EIA legislation was the National Environmental Policy Act of 1969 in the United States. This development spurred on other countries to develop their own systems, some based very much on the NEPA model, others developing along very different lines. By the year 2000, 112 countries were known to have EIA systems in place [13], although their relative effectiveness is likely to be highly variable.

What an EIA involves

In general, EIA is a process that leads to the production of a document, often called an Environmental Impact Statement (EIS). This provides information to decision makers to help them make appropriate decisions about development applications. The EIA involves a number of iterative stages, which are normally defined by legislation or administrative guideline. A generic EIA process is described here, although it is important to realize that not all of the stages will feature in every procedure. Also, even where stages are included in procedures, there may be a difference between the legal requirement and actual practice.

Procedural basis for EIA

EIA is a procedure usually defined in regulations of some form. Such procedures require that a certain number of steps are carried out, or that documents are produced that consider specific information. Such a procedural emphasis can mean that an EIA can be carried out fully compliant with procedure (as each of the steps has been carried out), but which is actually of little value to the decision makers.

Figure 12.1 details how an EIA procedure might look, although it is likely that legal procedures will include some of the stages included in Fig. 12.1 rather than all of them. Also, different procedures may duplicate stages (e.g., requiring draft EISs to be produced followed by final versions after a round of consultations), or they may require independent organizations to be involved in some of the stages (e.g., scoping and review).

At the concept stage of any development, a number of alternative ways of proceeding can be considered. The alternatives can be design alternatives or location alternatives. Once the alternative choices have been considered, the actual design itself can take place. This involves specific choices regarding the siting of various components of the development—such as car parks, paths, and buildings.

If a development is likely to have impacts on the environment, then a decision is made on whether a formal EIA will have to take place to take into account these impacts (positive and negative)—such a decision is known as the screening decision. The EIA process is, it must be stressed, iterative. This is demonstrated at this early stage of screening where the requirement for a formal EIA and its associated cost implications can lead the developer to reassess the project design with a view to reducing the significant impacts to a level where EIA is not legally required.

Where it is decided that a formal EIA is required, the next stage is to define the issues that need to be addressed, that is, those impacts that might have a significant effect on the environment. This is known as scoping and is essential for focusing the available resources on the relevant issues. Production of an EIS then involves a number of steps that are themselves iterative. The first of these is, for the scoped issues, to gather all the required baseline information. This is information on the current status of the environment likely to be affected by the development, and must take account of current trends so that the status of the environment in the future, in the absence of the development, could be predicted.

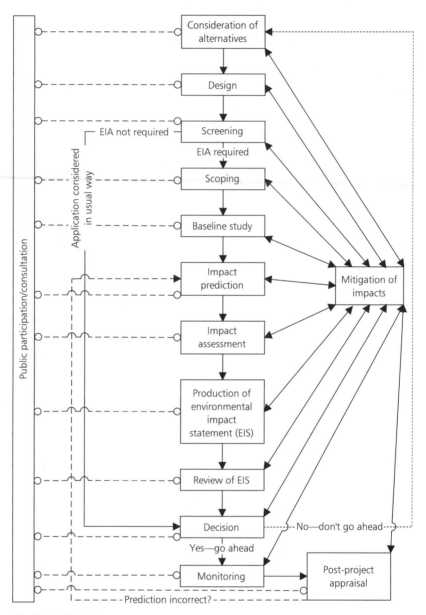

Fig. 12.1 The EIA process.

Impact prediction is forecasting the change in the environmental parameter in question because of the presence of the development. Impact assessment then involves an analysis of the predicted impacts to determine both their importance and significance. The outcome of an EIA is usually a document, known as an EIS.

Once the EIA is complete, the EIS is submitted to the decision-making authority who need to review the document to determine whether the EIA has been adequate, whether the information is correct, whether it is unbiased, and whether further information is required.

Ideally, monitoring should be undertaken to determine whether the impact predictions were accurate and ensure that no unexpected effects are occurring and, once the development is in place, a post-project appraisal should be carried out to help to improve the standard of predictions in the future.

Frequently, the assessment of impacts will reveal damaging effects upon the environment. These can be alleviated by mitigation measures. Mitigation involves taking measures to reduce or remove environmental impacts.

The public should be extensively involved in EIA and, if successful, public participation can help to ensure that development is acceptable and that it is not environmentally unacceptable.

Stakeholders and questions of conflict of interest

There is a general consensus that EIA can only be effective if adequate consultation takes place with all the stakeholders. The key stakeholders are the developer, the decision authority, the local resident, and the local environmental groups, and these can all have different objectives for any consultation in EIA [14]. Whilst these are the key stakeholders, there are many other groups or individuals who need to be consulted, and a basic checklist of stakeholders is reproduced in Table 12.1 [15].

The involvement of so many different stakeholders, each potentially having different aspirations about any proposed development, can lead to conflict.

Table 12.1 Checklist for identification of stakeholders [15]

- Local residents
- Trades unions
- Media
- Religious groups
- Local community groups
- Commercial associations
- Educational institutions
- Non-governmental organizations
- Independent experts
- National and local government agencies with responsibility for management of natural resources and welfare of people likely to be affected by the project

A typical situation where conflicts arise is where decisions have been made before the EIA process takes place whereby the EIS becomes 'less like a document for discussion than a focus for disagreement—ammunition or justification, whichever side of the debate you happen to be on' [9]. EIA, then, can fuel conflict unless those responsible for the assessment enter into consultation with an open mind and a willingness to change their designs in some way or another.

Public consultation

All EIA systems require some form of public consultation, although this can vary between mere information provision, to a high level of involvement in the process and even in the decision making. Consultation needs to take place early in the EIA process, and this is commonly interpreted as meaning considerable involvement at the scoping stage to determine the key issues to be examined. There are a variety of techniques that can be used, which are adequately described by others, for example [14,16–19]. It is not possible to generalize about the types of techniques that are most appropriate as it tends to be case and context specific.

In a health context, consultation at the scoping stage should lead to the identification of health issues perceived to be important by stakeholders and should be an efficient way of determining some of the health issues to be examined.

Funding of EIA

Typically, EIA has been established as a system that tries to avoid placing undue costs on decision makers. The norm in most systems is therefore for the developers to fund the EIA and produce the EIS for consideration by the decision maker [20]. This situation does lead to concerns over bias in systems where there are no independent bodies (as exist in the Dutch EIA system, e.g.) setting terms of reference and checking the final EIS against these prior to their being accepted by decision makers.

In terms of actual cost, a study funded by the European Commission found that costs typically amounted to less than 0.5 per cent of capital costs, and costs above 1 per cent of this value were the exception and occurred only in relation to controversial projects, projects sited in sensitive areas, or where good EIA practice was not followed [21].

Coverage of health in EIA

Most EIA procedures require the assessment of the effects of proposed developments on, amongst other things, human beings. It is the interpretation of 'human beings' that determines the extent to which health is considered in EIA and, indeed, if it is at all. Where such decisions are left to the developer, the approach taken is often to meet legal compliance with minimum expense— research in the United Kingdom has shown that this can lead to poor coverage of health. A review of 39 EISs covering a variety of development sectors indicated that 72 per cent did not list in their tables of contents sections or chapters dealing with human health. Forty-nine per cent of the EISs did not contain any analysis of potential impacts on human health at all, and 67 per cent did not even provide sufficient information to allow an estimation of the population size that might be affected by particular impacts. Human health impacts were thought to be adequately assessed in 28 per cent of the sample [22]. A recent handbook on EIA [23] lays the blame on this poor consideration of health on the text of the EIA Directive of the European Union [24], which, it says, in requiring an assessment of potential impacts on human beings is referring to 'demographic changes rather than socio-economic or health effects'.

Poor coverage of health in UK EISs was also found by Russell and Gallagher [25], and poor coverage in US EISs was identified by Arquiaga *et al.* [26]. The International Study of the Effectiveness of Environmental Assessment concluded that a key area requiring more work was 'the closer integration of environmental, with social, health and other impacts' [2]. However, the picture is not so bleak everywhere: the Dutch Commission for EIA recognized the importance of health as a component in EIA as far back as 1994 and include such issues in their terms of reference when necessary [27]. Thus, the level of inclusion of health in EIA across the world is clearly variable [28–31], although the inclusion of health in EIAs in practice around the world is far less common than the legislation would have us believe [32].

Canter [33], writing predominantly about the US EIA system, suggests that a health impact focus should be included in an EIA where the answer to any of the following questions is yes:

1. Does the nature of the proposed project (or activity) involve the handling of or emissions to the environment of materials such that their physical, chemical, radiological, or biological nature may be harmful to human health?

2. Is the location of the proposed project, together with its nature, likely to give rise to conditions that would alter the occurrence of natural hazards in the study area?

3. Could the implementation of the proposed action eventually give rise to conditions that would reduce or increase the number of adverse health-impact causing factors?

Lessons for HIA from EIA experience

A number of studies have identified ways in which the EIA process could progress, based on problems identified with current practice [2, 34–38], and even with systems regarded as functioning effectively, improvements have been suggested [39]. Based on these studies, a number of lessons for HIA, based on the experience of EIA, are categorized below:

1. Capacity building
   - Capacity needs to be built in terms of understanding of health impacts, especially within decision-making bodies. At the very least, this requires regular training.
   - Guidelines need to be developed for each context where HIA may be applied.
   - There needs to be a publicly accessible repository for HIA experience, preferably on the Internet, to allow current experience to be shared; this can be facilitated further with a requirement that copies of HIA are made accessible.

2. Decision making
   - Health needs to be a clear criterion in decision making.
   - The legal standing of HIA in a decision-making context needs to be clarified (and it needs to be made a consideration).
   - Decision making needs to be open and accountable.

3. Quality control
   - HIA documents need to undergo quality review to ensure that they present adequate information and consider appropriate mitigation measures.
   - Should a development proceed, monitoring needs to be carried out to verify the findings of the HIA and to provide feedback for future assessments.

4. Communication
   - Health professionals and decision makers need to work closely together.
   - Stakeholders should be involved particularly at the scoping stage and throughout an HIA, and should help to identify the health issues to be investigated and/or the health determinants.
   - All stakeholders should focus on accentuating positive health outcomes as well as aiming to avoid detrimental health impacts.

5. Procedural

- ◆ Screening needs to be strengthened such that health issues are given an appropriate level of consideration depending on the likelihood and significance of impacts, that is, there needs to be a hierarchy of studies, with full-blown HIA only required for proposed developments with potentially serious health impacts.

- ◆ HIA needs to start very early in the project planning cycle if it is to be beneficial for both developers and stakeholders

- ◆ HIA needs to be carried out at strategic levels to determine the potential health implications of policies, plans, and programmes. This allows better consideration of cumulative impacts.

Potential for integration of HIA and EIA

There is clearly potential for integration of HIA and EIA, and an international study has recommended this as a route towards better effectiveness of environmental assessment [2]. To this end, guidance has been produced in the United Kingdom on how this might be achieved [22], and progress has already been made in the United States [26]. The perception of the meaning of health in the EIA community is changing from 'illness' to 'wellness' [40], and it is accepted that better integration is needed if EIA is to be a tool for sustainable development.

There have also been examples of integration of health into EIA, which, whilst identifying some scope for improvement, do demonstrate that it is possible [28,41,42]. Caution does need to be exercised over this potential for integration however. The screening stage of EIA can often be based on strictly environmental considerations, and developments may be excluded from the need for EIA despite having potentially significant health effects. Even if EIA is carried out, where control of the content of an EIS is left to the discretion of the project proponent, cost considerations can ensure that minimum compliance with EIA procedures takes places at the expense of proper consideration of health issues.

Simple integration of health issues within an existing EIA procedure can lead to a number of problems. In the first instance, it may be the case that EIAs are dealt with by decision makers who have no real competence of dealing with health, and that health professionals may be marginalized in the decision-making process—there is evidence that this does happen in the United Kingdom [25].

Along similar lines to integrating EIA and HIA, authors have indicated the value of integrating HIA and social impact assessment (SIA) [43], but caution that such a step presents disciplinary, institutional, organizational, resources/capacity, and conceptual challenges.

In conclusion, integration of HIA with EIA is generally regarded as being a good move and has been shown to be possible and beneficial. However, a good deal of caution needs to be exercised as it will not work without considerable effort to get various organizations/departments working together, and will only facilitate the consideration of health in decision making in those cases where EIA is currently required—and this doesn't necessarily coincide with all those cases where significant health impacts may arise.

References

1 **Sadler B.** Taking stock of Environmental Assessment. *Environmental Assessment* 1996; **4**:125.

2 **Sadler B.** *Environmental Assessment in a Changing World: Evaluating Practice to Improve Performance.* International Study of the Effectiveness of Environmental Assessment. Page 248 Ottawa: Minister of Supply and Services Canada. 1996.

3 Commission of the European Communities. *Report from the Commission of the Implementation of Directive 85/337/EEC on the Assessment of the Effects of Certain Public and Private Projects on the Environment.* Luxembourg: Office for Official Publications of the European Communities, 1993.

4 **Lee N.** Environmental assessment in the European Union: a tenth anniversary. Project Appraisal, 1995; **10**:77–90.

5 **Barker A and Wood C.** An evaluation of EIA system performances in eight EU countries. *Environmental Impact Assessment Review* 1999; **19**:387–404.

6 **Wood C and Coppell L.** An evaluation of the Hong Kong environmental impact assessment system. *Impact Assessment and Project Appraisal* 1999; **17**:21–31.

7 **Wood C.** *Environmental Impact Assessment: A Comparative Review.* 2nd edn. Edinburgh: Prentice Hall, 2003.

8 **Carson R.** *Silent Spring.* London: Hamish Hamilton, 1963.

9 **Weston J** (ed.). *Planning and Environmental Impact Assessment in Practice.* Longman: Harlow, 1997.

10 **Tanvig L and Nielson A.** *A successful EIA for Billund Airport.* Commission of the European Communities, 2002.

11 **Wiesner D.** *EIA The Environmental Impact Assessment process—What it is and How to do One.* Bridport, Dorset: Prism Press, 1995.

12 **Barrow CJ.** *Environmental and Social Impact Assessment: An Introduction.* edn. London: Arnold, 1997.

13 **Bond AJ.** *Environmental Impact Assessment in the UK: Background, Basics, Context and Procedure.* Oxford: Chandos, 2000.

14 **Petts J.** Public participation and environmental impact assessment. In Petts J (ed.), *Handbook of Environmental Impact Assessment—Vol.1 Environmental Impact Assessment: Process, Methods and Potential.* Oxford: Blackwell Science, pp. 145–177, 1999.

15 Bond A, Bussell M, O'Sullivan P, and Palerm JR. Environmental Impact Assessment and the Decommissioning of Nuclear Power Plants—a review and suggestion for a best practicable approach. *Environmental Impact Assessment Review* 2003; **23**:197–217.

16 Department of the Environment Transport and the Regions. *Environmental Assessment: Public Participation*. London; Department of the Environment Transport and the Regions, 1998.

17 EIA Centre. *Consultation and Public Participation within EIA*. EIA Centre, Manchester: Leaflet Series **10**: Manchester; EIA Centre, 1995.

18 Kakonge JO. Problems with public participation in EIA process: examples from sub-Saharan Africa. Impact Assessment, 1996; **14**:309–320.

19 Palerm JR. Public participation in environmental impact assessment in Spain: three case studies evaluating national, Catalan, and Balearic legislation. *Impact Assessment and Project Appraisal* 1999; **17**:259–271.

20 Harrop DO and Nixon JA. *Environmental Assessment in Practice*. London: Routledge, 1999.

21 Land Use Consultants, Eureco, and Enviplan. *Environmental Impact Assessment: a Study on Costs and Benefits*. Brussels: European Commission, Directorate General Environment, Nuclear safety and Civil Protection, 1996.

22 British Medical Association. *Health and Environmental Impact Assessment: an Integrated Approach*. London: Earthscan publications, 1998.

23 Carroll B and Turpin T. *Environmental Impact Assessment Handbook*. London: Thomas Telford, 2002.

24 Council of the European Union. Council Directive 97/11/EC of 3 March 1997 amending Directive 85/337/EEC on the assessment of the effects of certain public and private projects on the environment. *Official Journal of the European Communities* 1997; **40**:5–14.

25 Russell SC and Gallagher E. *Health Issues in Environmental Impact Assessment in the UK*. 12th workshop: Reykjavik, Iceland, 14–18 May 1997 12th report: pp. 73–86.

26 Arquiaga MC, Canter LW, and Nelson DI. Integration of health impact considerations in environmental impact studies. *Impact Assessment* 1994; **12**:175–197.

27 Tersteeg V. *Integration of Health Assessment in EIA*. 12th workshop: Reykjavik, Iceland, 14–18 May 1997., 12th report: pp. 63–72.

28 Ahmad B, Hassan A, Birley MH, Fakhro K, and Alkuwari Z. Integrating health and environmental impact assessment in bahrain: opportunities and challenges. In *Assessing the Impact of Impact Assessment—Impact Assessment for Informed Decision Making. 22nd Annual Conference of the International Association for Impact Assessment. 15–21 June 2002*. The Hague, Netherlands, 2002.

29 Irvine J. *Health in Environmental Impact Assessment*. 12th workshop: Reykjavik, Iceland, 14–18 May 1997., 12th report: pp. 54–56.

30 Radnai A. *Consideration of Health Impacts in the Hungarian EIA Regulation*. 12th workshop: Reykjavik, Iceland, 14–18 May 1997., 12th report: pp. 92–99.

31 Pavlickova K and Banska H. *Health Impact Assessment as an integral part of environmental impact assessment in the Slovak Republic*. 12th workshop:Reykjavik, Iceland, 14–18 May 1997. 12th report: pp. 87–91.

32 Banken R. Public health in environmental assessments. In Porter AL and Fittipald JJ, (ed.), *Environmental Methods Review: Retooling Impact Assessment for the New Century*. Fargo: International Association for Impact Assessment, pp. 247–253, 1998.

33 Canter LW. *Environmental Impact Assessment.* 2nd edn. New York, NY: McGraw-Hill, 1996.

34 EIA Centre. *Evaluation of the Performance of the EIA Process.* Leaflet Series, **17**: Manchester EIA Centre, 1996.

35 Lawrence DP. EIA—Do we know where we are going? *Impact Assessment* 1997; **15**:3–13.

36 Fuller K and Binks M. The effectiveness of environmental Assessment. *Environmental Assessment* 1995; **3**:88–89.

37 Hickie D and Wade M. Development guidelines for improving the effectiveness of environmental assessment. *Environmental Impact Assessment Review* 1998; **18**:267–287.

38 Petts J. Introduction to environmental impact assessment in practice: fulfilled potential or wasted opportunity?. In Petts J (ed.), *Handbook of Environmental Impact Assessment—Vol. 1 Environmental Impact Assessment: Process, Methods and Potential,* Oxford: Oxford; Blackwell Science, pp. 3–9, 1999.

39 ten Heuvelhof E and Nauta C. The effects of environmental impact assessment in the Netherlands. *Project Appraisal* 1997; **12**:25–30.

40 Vemuri SR and Kelly P. Holistic health impact assessments. In *Assessing the Impact of Impact Assessment—Impact Assessment for Informed Decision Making. 22nd Annual Conference of the International Association for Impact Assessment. 15–21 June 2002.* The Hague, Netherlands, 2002.

41 Chung Y. Health risk assessment (HRA) in EIA. In *Assessing the Impact of Impact Assessment—Impact Assessment for Informed Decision Making. 22nd Annual Conference of the International Association for Impact Assessment. 15–21 June 2002.* The Hague, Netherlands: International Association for Impact Assessment, 2002.

42 Franssen EAM, Staatsen BAM, and Lebret E. Assessing health consequences in an environmental impact assessment—the case of Amsterdam Airport Schipol. *Environmental Impact Assessment Review* 2002; **22**:633–653.

43 Rattle R and Kwiatkowski RE. The way forward: should health and social impact assessments be integrated? In *Assessing the Impact of Impact Assessment—Impact Assessment for Informed Decision Making. 22nd Annual Conference of the International Association for Impact Assessment. 15–21 June 2002.* The Hague, Netherlands, 2002.

Chapter 13

Community development: the role of HIA

Maurice B Mittelmark, Doris E Gillis, and Bridget Hsu-Hage

Health impact assessment for the purpose of community development is not a new idea. Many communities in the Healthy Cities movement, for example, have used HIAs to examine how local environmental conditions influence community health and functioning [1]. Findings from HIAs of this type are useful both in planning and in citizen advocacy for better community living conditions. This type of HIA (referred to hereafter as HIA-CD) is different from usual HIA practice, in which pre-specified policies, programmes, or projects are the objects of assessment. When HIA-CD is conducted, environmental conditions in general are scanned, seeking to identify the key determinants of community health and functioning. The professional arena within which the idea of HIA-CD has emerged is health promotion, not impact assessment. The aim of this chapter is to sketch the rationale for HIA-CD and provide some illustrative examples.

Health promotion

To understand HIA-CD is to understand its building blocks—health promotion and community development. Health promotion is a broad concept. Under its rubric, one finds health behaviour change programmes as well as advocacy campaigns aimed at changing upstream health determinants. It is in the pursuit of the latter that HIA-CD is usually conducted. Further, health is conceived not merely as the absence of disease, but as a constellation of intra-personal and social resources for functioning well in society. To be healthy in this sense is to have an adequate reservoir of physical, mental, and social resources to enable a person to function in her various roles, and gain satisfaction from doing so. Simply put by Illich [2], health in this sense means autonomy. This does not discount other conceptions of health, focussed on risk factors and diseases, but rather illuminates an additional dimension that is of special relevance to HIA-CD.

Consistent with the foregoing, a key objective of health promotion is the stimulation of active, participating communities. The methods for achieving this can be described in volumes, or with elegant simplicity [3]:

$$\text{health promotion} = \text{health education} \times \text{healthy policy}$$

HIA-CD is tied to the far right of the equation, healthy policy. Its relevance to healthy policy lies in its ability to provide evidence about how policies and practices across community sectors, not just in the health care sector, influence a community's health, whatever aspect of health might be of concern [4]. 'Community' as the term is used here has two main characteristics. First, it is a geographical area often defined by formal administrative boundaries (city, town, village) or settings within (neighbourhoods, schools, workplaces). Second, a community comprises people who have a shared sense of identity and belonging, and shared norms, values, and helping relationships. This definition emphasizes community as system; not every person in a community has the same relationship to every other person in the community. Individual people of a community may be remotely connected, and unaware of one another's existence. Yet all people in the community are linked by the systems they create (or tolerate) that assist the collective to function.

As stated earlier, health promotion as a process seeks to stimulate active, participating communities. This implies, first, that there are dimensions of action and participation along which communities may be distributed, and second, that communities can move on these dimensions. 'Community development' describes purposive movement along these dimensions, in the directions of more action and more participation.

The process of community development often starts with identifying and prioritizing needs or objectives [5]. When promoting health is among the objectives, HIA-CD can be used as a tool to identify objectives for action. The community proceeds by developing the confidence, will, and resources needed to act on the objectives. The community then takes action, and through this process, develops its ability to collaborate to solve problems and create opportunities for community improvement.

There are in practice many forms that community development may take, depending on the size, structure, and resources that a community has. It may be guided by sophisticated health promotion planning models, such as Green and Kreuter's PRECEDE–PROCEED model [6], or by no model at all, relying on the experience and common sense of citizens acting together for their common good. Whatever its form, community development always has citizen involvement at the core, and very often the leadership comes from the citizenry, assisted by development professionals [7]. In concert with that, HIA-CD is not conceived of as a highly technical process managed by impact analysis professionals (as is the case for much of impact assessment). It is, rather, a relatively low technology process that can be managed by the lay community.

In this sense, HIA-CD is distinct from mainline HIA, which may often by necessity be a rather complex undertaking requiring tight professional control. Any trend to more complex approaches to HIA-CD would threaten to exclude average citizens from participation, and more critically, from the leadership functions. The critical challenge for the health promotion arena is to develop and disseminate effective HIA-CD processes that any person or group without formal professional training can master, with some study and practice.

We move now to descriptions of two community development initiatives in which health issues are the focus and in which HIA-CD has played an important role. The first is a brief case from Tasmania, Australia [8], which had its starting point in a specific community health concern, widespread respiratory illnesses. It illustrates HIA-CD as part of a comprehensive community movement to improve the public's health and regain the beauty of the natural environment. The second is a more extensive case, from Nova Scotia, Canada, which had its starting point in the need to engage citizens in health planning in the context of health care reform [9,10]. It illustrates how HIA-CD can help set a community's priorities when no specific health concern sets the agenda. These cases illustrate possibilities; they are not offered as prescriptions of how HIA-CD should be managed. The academic literature on HIA-CD is minuscule, and there is much work to be done before one can move from description to prescription, if ever. The term HIA-CD has been used in this chapter for the sake of convenience, and there is no widely accepted term for the activities and processes described here. In the case presentations, HIA-CD is recognizable not by label, but by the fundamental approach that emphasizes broad environmental scans seeking determinants of health, and considerable citizen involvement in, and leadership of, the process.

Case 1: Tasmania, Australia

In Tasmania, Australia, members of the neighbouring rural townships of Launceston and Upper Tamar Valley (referred to hereafter as the community) expressed concern over increased respiratory illnesses in the winter months. An investigation into air pollution, environmental health, and respiratory diseases commenced in 1991. It concluded that the main cause of the pollution was the use of wood-fired heating in winter, exacerbated by unfavourable topographical and meteorological conditions. Other factors were forest fires, poor waste incineration practices in the timber industry, and rural and domestic outdoor burning. The increase in the use of domestic wood heaters followed the surge of world oil prices during the 1970s. The improper use of wood heaters played a significant role in the high level of air pollution.

The community and government agencies worked hand-in-hand to identify the causes of pollution and develop strategies to reduce pollution levels. A community action group stimulated involvement of the media and school and academic leaders, to publicize the issue and educate the community about the effect of wood burning on the atmosphere and of the need to improve techniques of fuel combustion.

Technical reports were prepared that advised more stringent emission stipulations for new wood heaters, subsidies for upgrading of heaters, quality controls on firewood, continuing community education, and encouragement of homeowners to properly insulate their houses. The local government worked in partnership with the Australian Solid Fuel Heating Association Inc., in a proactive way to educate the community by offering a free advisory service to any domestic consumer who had a problem with smoke from a wood heater. The local Council improved a local law that controlled the construction and use of incinerators, restricted the operation of domestic incinerators to two days a month, and banned on-the-ground burning.

This case study illustrates the feasibility of changing local government policy and ordinances in response to community advocacy about a health issue raised by citizens. Critically, the advocacy initiative was armed with evidence from an impact assessment that identified the determinants of the health problem. The community demonstrated its willingness and ability to tackle a complex problem in partnership with government and industry. The three-year process undoubtedly strengthened the community's confidence and ability to take concerted action on a wide range of issues that might arise in the future. Thus, while the impetus for action was a problem with respiratory illnesses, the process for problem solving was community development supported by the use of HIA-CD.

One can better appreciate the relevance of this case to the theme of this chapter by considering a hypothetical, but highly plausible, alternative outcome to this case. Stimulated by citizen complaints, local government authorities could have immediately turned to a 'medical' perspective on the problem, and accepted top–down expert driven solutions. The diagnosis could have been that the immediate problem was that vulnerable people were foolishly exposing themselves to polluted air. Children and old people might have been advised to remain indoors during periods of high air pollution. Citizens in general might have been advised to wear face masks during times of highest risk. Cigarette smokers might have been advised of the extra hazard of tobacco use in combination with air pollution. Newspapers and electronic media could have responded by publishing and broadcasting daily air pollution alerts, always helpful on a slow news day. Local politicians could have engaged the Solid Fuel Heating Association in a vicious cycle of blame casting. Citizens could have expressed their frustration by fruitless grumbling to each other. The lesson to all could have been that forces beyond local control rule, others are in charge, we are victims, we are powerless. But that is not what happened.

Case 2: Nova Scotia, Canada

The People Assessing Their Health (PATH) project is an example of a community-based initiative designed to engage citizens in developing the process and tools necessary for undertaking HIA at the community level [9,10]. Its purpose was to enable citizens to play an active role in health planning within an emerging regional health system.

The PATH project began in 1996 in eastern Nova Scotia, Canada, at a time of significant restructuring in the provincial health system—part of the wave of health reform moving across the country in the 1990s. In 1994, *Nova Scotia's Blueprint for Health System Reform* [11] called for new governance structures to secure greater citizen involvement in health planning, by shifting decision making from the provincial to the regional and community levels. Four regional health boards were created to plan and manage health service delivery. It was also planned that community health boards in each health region would develop community health plans for primary health services based on local priorities. Health services were broadly defined to include the many social and environmental factors that determine health, not just the formal health care services offered in medical facilities.

It was within this context that the PATH project was designed. Its goal was to provide a means for people in three selected communities to identify, define, and assess all aspects of health in their communities in order to become effective participants in the new health system.

A new organization was not needed to launch PATH. Three existing organizations, all familiar with the health challenges of people living in eastern Nova Scotia, guided the PATH project in a partnership. The partners were a local women's resource centre, a university extension department, and a public health services unit. Representatives from these organizations, along with community development leaders from throughout the region, formed an advisory committee. Their perspectives on community development and health promotion shaped the process and tools for community HIA that emerged.

The PATH project, and its approach to community HIA, evolved from two principle convictions. The first is that community people possess valuable knowledge about what influences their health and what is needed to keep their communities healthy. The second is that citizens must be involved in community health planning and in making decisions about policies, programmes, and services that affect their health, if today's multifaceted population health problems are to be addressed successfully.

Members of the PATH project advisory committee recognized that, for decades, many people in eastern Nova Scotia had experienced significant barriers limiting their capacity to enjoy good health. Chief among the roadblocks were economic hardships due to limited employment opportunities, socio-economic stress due to limited education, inadequate income and social exclusion, and geographical isolation from the affluent centres of decision making. Not surprisingly, inequalities in health outcomes paralleled these social environmental health determinants, with prevalence rates of obesity, diabetes, and cardiovascular diseases higher in this remote region than in other parts of Canada [12].

Through the PATH project, people from many walks of life became engaged in broadening their understanding of what determines the health of the community. Adult education strategies were used to involve community members in discussing factors influencing their health, based on their own experiences [10]. In particular, story telling, using structure dialogue, enabled people to identify what it takes to make and keep their communities healthy.

In each community, a steering committee was responsible for guiding the process and a local person was hired by the committee and trained as a facilitator. The facilitator worked with committee members in creating a community HIA tool (CHIAT) for use in generating data from the community. The CHIAT was pilot tested in community workshops, revised, and distributed to community leaders. The CHIAT was designed to assist citizens in assessing the potential impact of policies and programmes on community health. Although the CHIATs were custom designed by each of the three communities, the three tools included the following common components:

- Vision statement of a healthy community
- Summary of identified determinants of health in the community
- Other factors considered key in building and sustaining a healthy community
- Values and principles guiding community members as they work together
- HIA worksheet
- Planning for action worksheet

- Illustrations communicating the uniqueness of the community
- Description of the process
- Acknowledgements

The advisory committee viewed community HIA as a strategy to promote population health by enabling citizens to think about health in a new way—from the perspective of the health of the community as a whole, rather than just individuals' illnesses. This reflected their understanding that health is influenced by a broad spectrum of factors beyond individual citizen's control, especially social and economic determinants of health.

Although the PATH project introduced community HIA at a critical time in the health reform process, the promised creation of an organizational structure to enable greater citizen involvement in health planning was delayed by the government's moratorium on the formation of community health boards. It was not until the next wave of health reform in 1999 that community health boards were mandated, thereby providing the means for significant citizen participation.

After the PATH project was completed in 1997, members of the advisory committees and other community development leaders formed the PATH Network as a means to continue discussing health and exchange ideas and resources related to building healthy communities. A number of public education sessions were held and follow-up projects proposed. In 2001, the PATH Network decided to evaluate the degree to which the PATH project continued after the formal phase closed. An external evaluator interviewed 33 key informants and reviewed documentation describing follow-up activities since the completion of the project.

Results of the evaluation revealed that following the close of the PATH project, two of the three communities continued to use the processes and tools they had earlier developed. However none had fully implemented community HIA in assessing the impact of programs or policies on their communities. Nevertheless, the PATH project was seen to have made a number of significant contributions. Key informants considered that it had increased people's understanding of the determinants of health and of the need to involve people from many sectors in building capacity for community health. Several participants in the PATH project went on to become members of community health boards, bringing their expanded vision of health to the decision-making process. Key informants also reported that there was a growing recognition among decision makers that evidence from many sources—including community generated knowledge as well as epidemiological findings—was needed to guide decision making at the community level.

Many lessons have been learned from the PATH project about the application of HIA at the community level. Through the evaluation process, key informants identified a number of factors they considered important to the adoption of the tools and process of community HIA. These included:

- the extent to which use of the CHIAT was integrated into the ongoing work of the community prior to and after the project;
- the presence of local champions who had been involved in the PATH project;
- how the community health boards evolved in the three PATH communities;
- the extent of support provided by public health field staff;
- the participation of people from the PATH communities in educational activities sponsored by the PATH Network, which were held after the project was completed.

Respondents also identified a number of barriers to applying community HIA, including:

- the sporadic unfolding of the health reform process and subsequent delay in development of infrastructure to support citizen involvement in health planning;
- lack of leadership in addressing the broad determinants of health;
- bureaucratic obstacles inhibiting the inter-sector collaboration required to take action on issues related to the broad determinants of health;
- lack of resources to support training and facilitation in order to familiarize members of community health boards with the purpose, value, and use of community HIA.

The PATH experience shows that implementing and sustaining community HIA requires adequate human and material resources and a supportive organizational structure. The experience of those involved in the PATH project suggests that capacity needs to be built at the individual, group, and system level if community HIA is to be successfully implemented. For example, individuals need to value community generated knowledge, be familiar with participatory means of decision making, and have leaders and facilitators supportive of community HIA.

Groups such as community health boards need to be able to work towards consensus based on a shared vision of health, and use effective approaches for engaging community members with diverse viewpoints. They also need access to credible research findings about their priority health issues as well as best practices appropriate for addressing their concerns. Systems need to support citizens involved in health planning by encouraging the use of community

HIA, offering orientation and training in the use of the process, and providing guidelines for how insights can be used to influence public policy.

Conclusion

A sceptic might well wonder if HIA has really been a part of the story of these cases, and even if this chapter has anything to do with HIA. A restricted definition of HIA might well exclude the approach described here, in which the starting point is not a specific policy, programme, or project whose health impact will be assessed. The starting point, instead, is citizen concern with a health issue, which stimulates a community development process, an essential part of which is a scan for environmental factors responsible for health problems. This last part has been termed HIA-CD, and it is suggested that narrower visions of HIA should be expanded to include this HIA-CD. The clear advantage is that the concept of HIA is becoming widely appreciated (feared) by politicians and bureaucrats, and citizen use of HIA-CD can increase the possibility that decision makers will listen when citizens speak. The possible disadvantage is that the term HIA will be watered down due to imprecise usage. This is a matter for discussion and debate, perhaps stimulated by this chapter.

In the authors' opinion, the clear advantage outweighs the possible disadvantage, and our task has been to describe how HIA has been used in processes of community development and to provide cases illustrating the main principles. The overview material and the Australian and Canadian cases are intended to accomplish three ends. First, we have described community development processes in which health issues are in focus. Second, we have illustrated how and where in the community development process HIA-CD fits. Finally, we have emphasized how citizens can be in the forefront of solving community problems, by taking control of health promotion processes.

References

1 World Health Organisation. Urban health topics: health impact assessment. 2002. Access Date: 10 February, 2003 http://www.who.dk/healthy-cities/How2MakeCities/20020114_2.

2 Illich I. *The Limits of Medicine—Medical Nemesis: the Expropriation of Health.* London: Marion Boyars, 1976.

3 Tones K and Tilford S. *Health Promotion; Effectiveness, Efficiency and Equity.* Cheltenham: Nelson Thornes, 2001.

4 Mittelmark MB. Promoting social responsibility for health: health impact assessment and healthy public policy at the community level. *Health Promotion International* 2001; **16**:269–274.

5   Ross M. *Community Organisation: Theory and Principles*. New York, NY: Harper and Row, 1955.

6   Green LW and Kreuter MW. *Health Promotion Planning: An Educational and Ecological Approach*. Mountain View: Mayfield Publishing Co., 1999.

7   Mittelmark MB. Health promotion at the community-wide level: lessons from diverse perspectives. In Bracht N (ed.), *Health Promotion at the Community Level*, 2nd edn. Newbury Park: Sage Publications, 1999.

8   National Public Health Partnership (Australia). The Role of Local Government in Public Health Regulation. National Public Health Partnership. 2002. Access Date: 10 February, 2003. http://www.dhs.vic.gov.au/nphp/legtools/localgov/localgov.pdf.

9   Gillis DE. The 'People assessing their health' (PATH) project: tools for community health impact assessment. *Canadian Journal of Public Health* 1999; **90** (suppl I): S53–S56.

10 Gillis DE and English LM. Extension and health promotion: An adult learning approach. *Journal of Extension* 2001; 39, Number 3, Access Date: 10 February, 2003. http://www.joe.org/joe/2001june/a4.html

11 Nova Scotia Department of Health. Nova Scotia's Blueprint for Health System Reform. Halifax, NS: Department of Health, 1994.

12 Colman R. The Cost of Chronic Disease in Nova Scotia. Report of the Genuine Progress Index for Atlantic Canada (GPI Atlantic) prepared for the Unit for Chronic Disease Prevention and Population Health of Dalhousie University, Halifax, NS and the Population and Public Health Branch of Health Canada, Atlantic Region, 2002.

Chapter 14

HIA at the international policy-making level

Anna Ritsatakis

The WHO and intersectoral action for health

Influenced by the Lalonde report released in Canada in 1974 [1], the need to promote and protect health through action in non-health sectors has been discussed in international circles for at least 30 years. With some exceptions, there has been more talk than action [2]. In recent years however, health impact assessment (HIA) has created interest as one way of putting intersectoral action for health into practice.

The WHO has been perhaps the strongest and most persistent voice calling for recognition of the role of health in development and the impact of socio-economic development on health. In 1977, through the Health for All (HFA) resolution, the World Health Assembly formally confirmed the need for an intersectoral approach to health development. The WHO European Region acted quickly adopting a HFA strategy in 1982 [3], which whilst recognizing possible conflicting interests of social and economic sectors stated unequivocally that 'health development both contributes to and results from wider socio-economic development'. In 1986 the Ottawa Charter called for 'healthy public policy' to direct policy makers in all sectors 'to be aware of the health consequences of their decisions and to accept their responsibilities for health' [4]. Thirty years since Lalonde, it is clear that health plays a central role in the world economy. Cross-sectoral action is still a key element in WHO's corporate strategy [5], and the WHO in partnership with other organizations set up a Commission on Macroeconomics and Health [6] in 2000 to recommend concrete health measures to maximize poverty reduction and economic development.

In the WHO European Region, intersectoral action for health has been promoted with increasing emphasis through a continuously revised HFA policy for Europe [7,8], and by supporting countries to implement their own HFA-type policies [2]. It is the cornerstone of programmes such as the Regions for Health Network [9], Healthy Cities [10], and the work of the recently established WHO centre for investment in health [11]. The current European HFA policy [12] explicitly states that by 2020 'Member States should have established mechanisms for health impact assessment and ensured that all sectors become accountable for their policies and actions on health'. To this end, members of the Regional Committee were briefed in 2002 on current experience and emerging issues in HIA in Europe [13].

European Community

Recognition of the importance for health of action in other sectors is given in Article 152 of the Amsterdam Treaty which states 'a high level of human health protection shall be ensured in the definition and implementation of all Community policies and activities' [14]. This is also reflected in the Community's strategy for public health, which aims to tackle the underlying causes of ill-health [15] and in documents informing the European Parliament's discussion on how health policy at the European level can best deliver added value [16].

As the policy framework was developing, the Commission was already investigating possible ways of integrating health in other policies. For example, an EU-funded project was led by Finland [17] to map the current state-of-the-art and provide analysis and guidance for further action in the context of European Community policies. Staff of the Commission were also actively involved in the HIA activities of the WHO European Region. Apart from the pressure coming as a result of decisions taken within the Community, the Commission was probably spurred on by the rapidity of HIA developments in some of their members states, particularly Sweden, the United Kingdom, the Netherlands, Finland, and Ireland. A policy report from the National Institute of Public Health in Sweden on the health impact of the EU Common Agricultural Policy [18], created considerable discussion and a call for reform of the CAP leading to healthy farming and food policy [19].

As a result the Commission is now piloting HIA as part of an integrated approach to impact assessment [20].

Other intergovernmental organizations

The World Bank's 1993 World Development Report [21] showed the importance of health in development. More recently, the Bank has taken a strong stand on attacking poverty [22], highlighting the need to ensure a more active voice for poor people and countries in global fora that affect their life, and their health [23]. Negotiations in the World Trade Organization, for example, are a continuous process taking place in Geneva where most of the least developed countries do not have adequate staff. The discussion around the Trade-Related Aspects of Intellectual Property Rights (TRIPS) particularly as this affects the distribution of medicine, has brought the potential impact on health of such internationally agreed policies to the forefront of public interest.

Such health-related activities are part of a wider movement throughout the UN organization, through which decision makers have recognized the urgent need to close the health and development gaps. Agenda 21 adopted more than a decade ago in the Rio conference indicated clearly that sound development is not possible without a healthy population, and that this can only be achieved if action is taken to reduce poverty, to consider new patterns of consumption, to improve education and housing, and to minimize the hazards of environmental pollution. The Beijing conference in 1995 highlighted the particular needs and potential contribution of women in such efforts. The 1997 Jakarta Declaration [24] on health promotion stated that both the public and private sectors should include equity-focused HIAs as an integral part of their policy development.

At the start of a new century, many of these aspirations were made more concrete in the form of UN Millenium Development Goals [25] to be achieved by 2015. In calling for a clear focus for the World Summit on Sustainable Development in Johannesburg in 2002, the Secretary-General highlighted five areas for immediate attention: water, energy, health, agriculture, and biodiversity (WEHAB). In relation to health he points out that we need to know better how and where to act and that research and development will need to focus more on the diseases of the poor and their determinants than has hitherto been the case.

The WHO and the development of HIA

On a global level, much of the impetus for WHO's work on HIA originated from the area of environmental impact assessment (EIA) and water resource development influenced partly by a joint WHO/FAO/UNEP initiative [26]. The strong body of work carried out in relation to water resource development led to the formulation and testing over a 10-year period of problem-based training courses in intersectoral decision-making skills in support of HIA of development projects, particularly in Africa but also in other regions. It was hoped that this work would lead to a comprehensive WHO programme on capacity building in HIA. By 1999, the Organization apparently had sufficient confidence in the HIA approach that it proposed to the World Commission on Dams an HIA framework [27] for use in assessing the impact of dam development.

Similar activities were developing in the Regional Offices but apparently with little coordination. Consequently in November 2000 an inter-regional meeting was organized in Arusha to harmonize and mainstream HIA in the WHO. Although all six of the Regions were invited, only three were able to attend [28]. The meeting was however able to overview HIA-related activities within the Organization and ways in which these could be more effectively coordinated. Whilst recognizing the value of the different Regions developing HIA activities suited to their particular needs it was also suggested that the WHO Collaborating Centre at the Liverpool School of Tropical Medicine could assist in preparing generic guidelines as a common WHO framework for HIA.

One of the issues highlighted during these discussions was whether HIA should be promoted as a stand-alone process or incorporated in, for example, EIA, strategic environmental assessment (SEA), or social impact assessment (SIA). During an informal consultative meeting in Columbia in 2001 [29] experience indicated that frequently EIA stakeholders were more interested in the biophysical environment and not so open to considering psychosocial impacts on health. SEA and SIA were still at a comparatively early stage of development. Depending on the country and its history of impact assessment, there appeared to be arguments both for integrating the various types of assessment and for carrying out HIA as a distinct process. Consequently, a WHO Policy Briefing note [30] for the World Summit on Sustainable Development in Johannesburg 2002 suggests that 'Local conditions will dictate a specific balance between integration and maintaining a separate profile for health in the overall impact assessment picture'.

The WHO Regional Office for Europe and the ECHP project

Similar to the experience in WHO headquarters, HIA-related activities in the WHO European Office could have profited from better coordination.

Within the Copenhagen Office and the Rome Centre, both of which have responsibility for dealing with environmental health issues, work has progressed on improving the reflection of potential health impacts in EIA. Particular attention has been given to the possibility of including consideration of health impacts of policies subject to SEA [31].

The main body of work on HIA per se was carried out by WHO's European Centre for Health Policy (ECHP) [32]. By early 1999, it was clear that a number of European countries were forging ahead with the development of approaches to HIA, and the EU was also interested in this process. Rather than reinventing the wheel, it was hoped to build on existing experience and create consensus on possible ways forward. Consequently, the ECHP HIA project was designed to:

◆ Create a common understanding of the basic concepts, and consensus on the definitions of the terms used;

◆ Build on existing knowledge by reviewing and learning from related or similar evaluation and assessment processes, and from existing models and methods of health impact assessment;

◆ Define possible principles and approaches to the implementation of health impact assessment, highlighting issues of concern and where further work is needed;

◆ Test and evaluate the results in pilot countries and regions, and revise the suggested approaches accordingly;

◆ Develop a network of decision-makers and experts, supporting each other in a continuous learning process;

◆ Raise public awareness and explore effective means of ensuring public involvement. [33]

During an EU-funded meeting organized by STAKES in Helsinki in early 1999, the ECHP agreed to set up an informal email group of people interested in HIA development. This group proved to be the backbone of the ECHP project, acting as a sounding board for new ideas, critics of ongoing work, and contributors to the further development of the project. From a small beginning of about 15 people, the network has since expanded to include over 260 people in 46 countries.

From the first rapid overview of recently published HIA models [34], it was clear that these differed considerably and that many terms were being used with a slightly different meaning in different countries. It was possible however to find a common core in these models. It was also clear that HIA was taking place on different levels from rapid HIA screening using informed opinion or available evidence to in-depth analysis involving new collection and/or analysis of primary data. At all levels, complex issues were being raised.

The first steps in the ECHP project were to clarify some of the concepts, and to search for consensus regarding the appropriate values to govern HIA. Following discussion through the email group and an international meeting organized in collaboration with the Nordic School of Public Health [35] a consensus paper [36] was prepared and widely distributed. This paper proposes the following definition for HIA:

Health Impact Assessment is a combination of procedures, methods and tools by which a policy, programme or project may be judged as to its potential effects on the health of a population, and the distribution of those effects within the population, recognizing that policies impact differently on different groups within the population.

Health impact assessment is expected to lead to policy adjustments in order to enhance potential positive and minimize possible negative health impacts. It is essential therefore that the values governing its implementation are agreed on. The consensus paper suggests that the values governing HIA should include:

- *equity*, meaning that HIA examines the impact within the population in terms of gender, age, ethnic background, and socio-economic status;
- *democracy*, emphasizing the right of people to participate in a transparent process concerning policy decisions that affect their lives;
- *sustainable development*, so that short-term and long-term, direct and indirect impacts are taken into consideration;
- *ethical use of evidence*, emphasizing a rigorous use of quantitative and qualitative evidence, based on various scientific disciplines and methodologies to ensure as comprehensive an assessment as possible.

The first step in conducting HIA is to set up a screening process, through which policies, programmes, and projects are quickly assessed in order to select those where an HIA is worth doing. Scoping is the next step, to decide into how much depth the HIA should go. Broadly speaking, from the experience so far, the consensus paper suggests that there are three levels of HIA:

- rapid HIA, relying mainly on existing evidence, knowledge, and experience
- in-depth HIA, involving the production of new information, and a wide array of disciplines and partners
- health impact review, providing a convincing summary estimation of the most significant impacts on health of policies, strategies, or clusters of policies, programmes, and projects that are so broad that an in-depth analysis seems infeasible.

Before things get started therefore, there needs to be shared understanding of the concepts, agreement on the values, and an agreed process for carrying out HIA. Who takes part, when, and how? How are suggestions for adjusting the policy to be handled? How are possible conflicts of interest dealt with?—Not an easy process to agree on, but one which can give much needed transparency to the trade-offs in policy making.

Whatever the level of HIA carried out, the evidence, including the opinions of those involved, must be presented on the basis of accepted scientific methods and the process for evaluating the adequacy of the report agreed on. This whole process raises technical issues such as risk assessment [37] and more political issues related to the communication of risks to public health [38] and the dangers of unrealistically raising public expectations regarding adjustment of the policies and programmes assessed.

Monitoring and evaluating will provide valuable insight into how the HIA process can be improved, and into the accuracy of the predictions made during the appraisal. Partly due to the newness of the HIA process, international experience as yet offers few examples of monitoring and evaluating HIA. A number of efforts are underway however to create national and international HIA databases, that should contribute to remedying this situation.

Given the complexity of the HIA process, the ECHP project took a step-wise approach, gradually building up knowledge, testing, and gaining consensus on possible pointers for good practice. To this end meetings were held on issues such as screening and scoping and taking an equity orientation. Discussion papers were developed, looking first at strategies for institutionalizing HIA [39], and a set of case studies on experience in European countries was initiated [40,41].

At the same time, the question of the possible health impact of joining the EU was felt to be of particular interest. This was due to a number of reasons: the experience of new members such as Finland and Sweden, the expansion of the Union through the expected addition of 10 new members, emerging attempts to document the health impact of the Common Agricultural Policy [18], and interest in Wales to examine the health impact of Objective 1 projects [42].

In October 2001, progress in the ECHP HIA project to date was discussed at an international meeting in Brussels in which top-level decision makers, including members of the EU High Level Committee for Health, took part. The interest in HIA and need to share information was obvious. After organizational change in the WHO European Office the HIA has been picked up again. An emphasis is being placed on finding evidence of the effectiveness of HIA and the factors enabling effective implementation.

Some developments outside Europe

Only part of the work carried out in the development of HIA is uncovered by literature searches. The WHO has therefore initiated a process to review the state-of-the-art. Here brief reference is made to some developments outside Europe.

Australia is one of the countries where emphasis has been given to making HIA part of the EIA process. A national framework [43] indicates policy areas and types of projects subject to HIA, through a participatory process and within the context of sustainable development and social justice. The United States with its long history of mandatory EIA offers considerable related experience, but more recently through the development of SIA [44], highlights the complexity of participatory processes.

In Canada the application of HIA has been highly variable across provinces. In British Columbia, for example, HIA flourished for a time in the early 1990s only to wither with political changes and the movement of key proponents to other responsibilities [39]. A report [45] submitted to the Health Promotion Development Division of Health Canada recommends that HIA should be developed in the broader framework of a systematic national goal setting process.

Developing countries are in some cases taking an innovative role in HIA development. In Zimbabwe, an HIA was carried out for a proposed small-scale irrigation and medium-size dam project in the northern part of the country. It clearly showed potential threats to health such as increased malaria and schistosomiasis and increasing sexually transmitted diseases due to an influx of migrant workers, but it also indicated improved nutrition through greater agricultural output. The cost of enhancing the positive and mitigating the negatives was estimated at about 1.8 per cent of the total project cost—not too high a price for health [30]. In Thailand, for example, a participatory learning process has been initiated by the Health System Research Institute, based on two key pillars: developing a critical mass of informed people and proper institutional design for HIA. An analytical framework for HIA has been developed based on a review of experience in other countries, and HIA case studies within five thematic networks (Industry and energy, Agriculture and rural development, Urban development and transportation, River basin management, International trade and trade agreements). A strong attempt is being made through awareness building and training, to ensure that the public and particularly civic groups throughout the country understand the concept and methodologies of HIA so that they can demand or commission HIA relating to development policies and projects in their localities.

Thoughts for the future

HIA is not a new technique. It is a new approach to a long-standing concern. What is new is the level of recognition at high levels of decision making that health can only be improved and protected by coordinated efforts in many sectors. This is evident in international organizations such as the WHO, the World Bank, and the EU. It is clear in a critical mass of countries, regions, and cities where planning for health and sustainable development is slowly being integrated. After years of discussion and lip service, through the pioneering efforts of a small number of countries and recent decisions in the EU, concrete steps are being taken to put intersectoral action for health into practice through HIA.

In the private sector in recent years, a number of multinationals, mindful of their corporate image, seem to be accepting greater responsibility for the impact of their actions on the societies in which they operate. This can be seen from discussions in the Davos Forum, for example [46], from advertisements of oil companies and others in international publications and from the dialogue set up through their websites [47]. Perhaps they could be encouraged to attempt HIA on major policies and programmes.

The combination of these developments presents a unique opportunity to assess how threats to health can be mitigated and opportunities for promoting health and reducing the health gaps can be enhanced. HIA could however be overwhelmed and discredited before it gets properly started.

First the task is potentially enormous. For example, of 359 acts and statutes approved by the Swedish Government in 1999–2000, 64 per cent were considered to have a possible impact on health, 9 per cent on the environment, and only 27 per cent were thought to have only a negligible health impact [48]. When the Netherlands started experimentally screening parliamentary documents this entailed checking 50–100 white papers, reports, and other such papers per day [40]. There will not be time or resources to carry out HIA on all policies and programmes in all sectors, neither is it necessary. Actions that potentially have a major impact, positively or negatively, particularly on the health of the disadvantaged, must however be picked up at a sufficiently early stage to allow for their adjustment. A reliable tool for screening and scoping is urgently needed to ensure such actions are not missed.

The reality of ongoing decision making will demand that HIA is carried out more quickly and on the basis of information that is less than experts would consider necessary. The experience of economic assessment where evaluation of such assessments showed large discrepancies between the predictions and what followed in reality [49], must be avoided. HIA will need to balance what one minister has called 'well-reasoned assumptions' [40], including qualitative information, with quantified data where possible and appropriate. There will need to be a rapid sharing of information and experience between HIA exponents, including experience of the process and evaluation of HIA projects.

In addition to information to enhance HIA capacity, not only basic public health skills but also skills in working with a wide range of partners, understanding the objectives of other sectors, and negotiating skills are needed. Handling the sometimes competing interests and objectives of different sectors are part of the political decision-making process. If HIA recommendations are based on the participation of a wide range of disciplines and stakeholders it can achieve both intellectual and democratic legitimacy.

References

1 Lalonde N. *A New Perspective on the Health of Canadians.* Ottawa: Health and Welfare Canada, 1974.

2 Ritsatakis A, Barnes R, Dekker E, Harrington P, Kokko S, and Makara P (eds.). *Exploring Health Policy Development in Europe* European Series No. 86. Copenhagen, WHO, Regional Office for Europe, 2000.

3 *Regional Strategy for Attaining Health for all by the Year 2000.* Copenhagen: WHO Regional Office for Europe, 1982 (document EUR/RC30/8 Rev. 2).

4 *Ottawa Charter for Health Promotion.* First International Conference on Health Promotion, Ottawa, Canada, 17–21 November 1986.

5 *A Corporate Strategy for the WHO Secretariat.* Geneva: World Health Organization, 1999 (document EB105/3).

6 *Macroeconomics and Health: Investing in Health for Economic Development.* Geneva: World Health Organization, 2001.

7 WHO. *Targets for Health for all.* Copenhagen; WHO Regional Office for Europe, 1985.

8 WHO *Health for All Targets. The health policy for Europe.* European Health for All Series No. 4. Copenhagen: WHO Regional Office for Europe, 1995.

9 *Investing for Health.* Report of the Third Annual Conference of the Regions for Health Network Bolzano, Italy, 13–14 October 1995. RHN Conference Series No. 3, Copenhagen, WHO Regional Office for Europe, (EUR/ICP/HSC 424), 1996.

10 *City Planning for Health and Sustainable Development.* European Sustainable Development and Health Series: 2, Copenhagen, WHO Regional Office for Europe, Healthy Cities Network, European Sustainable Cities & Towns Campaign, European Commission DG XI, 1997.

11 *The European Health Report 2002.* European Series, No. 97, Copenhagen: WHO Regional Office for Europe, WHO Regional Publications, 2002.

12 *Health 21—Health for All in the 21st Century.* European Health for All series No. 6, Copenhagen: WHO Regional Office for Europe, 1999.

13 *Technical Briefing. Health Impact Assessment.* Copenhagen: WHO Regional Office for Europe, (document EUR/RC52/BD/3) 6 September 2002.

14 *Consolidated version of the Treaty establishing the European Community.* Brussels: Commission of the European Communities, 1997 (Official Journal, C340), pp. 173–308.

15 *Communication from the Commission to the Council, the European Parliament, the Economic and Social Committee and the Committee of the Regions on the health Strategy of the European Community.* COM(2000) 285 final. Brussels: Commission of the European Communities, 16/5/200.

16 Mountford L. European Health Policy on the Eve of the Millennium. Working Document, Directorate General for Research, Public Health and Consumer Protection Series, SACO 102 EN. Luxembourg: European Parliament, 1998.

17 Koivusalo M and Santalahti P. Healthy public policies in Europe—Integrating health in other policies. Seminar Proceedings, Helsinki, GASPP Occasional Papers No. 6/1999, STAKES, 1999.

18 Dahlgren G, Nordgren P, and Whitehead M (eds.). *Health Impact Assessment of the EU Common Agricultural Policy.* Stockholm: The Swedish National Institute of Public Health, policy report 1997:36, November 1996.

19 Towards a healthy farming and food policy. In *European Public Health Update, No 63.* Brussels: European Public Health Alliance, pp. 8–10, July/August 2002.

20 *Communication from the Commission on impact assessment.* Brussels: Commission of the European Communities, (COM (2002) 276 final), 2002.

21 *World Development Report 1993. Investing in Health.* Washington DC: published for the World Bank by Oxford University press, 1993.

22 **Gwatkin DR and Guillot M.** *The Burden of Disease among the Global Poor—Current Situation, Future Trends and Implications for Strategy.* Washington DC: Health, Nutrition and Population Series, The World Bank, 1999.

23 *World Development Report 200/200. Attacking Poverty.* Washington, DC: published for the World Bank by Oxford University Press, 2001.

24 *The Jakarta Declaration on Leading Health promotion into the 21st Century.* Geneva, WHO (WHO/HPR/HEP/41CHP/BR/97.4), 1997.

25 http://millenniumindicators.un.org/unsd/mi/mi_goals.asp

26 *Intersectoral decision-making skills in support of health impact assessment of development projects.* Final report on the development of a course addressing health opportunities in water resources development 1988–1998. Geneva/Charlottenlund: WHO, 2000.

27 The World Health Organization's submission to the World Commission on Dams. *Human health and dams.* Geneva: World Health Organization, January 1999.

28 *Health Impact Assessment.* Report of an inter-regional meeting on harmonization and mainstreaming of HIA in the World Health Organization and of a partnership meeting on the institutionalisation of HIA capacity building in Africa Arusha, 31 October–3 November 2000. Geneva: World Health Organization (DOCUMENT WHO/SDE/WSH/01.07), 2001.

29 *Health Impact Assessment in Development Policy and Planning.* Report of an informal WHO consultative meeting. Cartagena, Columbia 28 May 2001. Geneva: World Health Organization, 2001.

30 *Health Impact Assessment WSSD Policy Brief Series.* Geneva: World Health Organization, 2002.

31 *Health Impact Assessment in Strategic Environmental Assessment.* Copenhagen: WHO Regional Office for Europe, 2001.

32 ECHP website http://www.ewo.who.int/echp

33 **Ritsatakis A.** Developing and approach to health impact assessment. In Diwan V, Douglas M, Karberg I, Lehto J, Magnusson G, and Ritsatakis A (eds.), *Health Impact Assessment: from Theory to Practice.* Report on the Leo Kaprio Workshop, Gothenburg, 28–30 October, 1999, Gothenburg: WHO and Nordic School of Public Health, NHV-Report 2000:9, September 2000.

34 **Lehto J and Ritsatakis A.** Health impact assessment as a tool for intersectoral health policy. In Diwan V, Douglas M, Karberg I, Lehto J, Magnusson G, and Ritsatakis A. (eds.), *Health Impact Assessment: from Theory to Practice.* Report on the Leo Kaprio Workshop, Gothenburg, 28–30 October, 1999, Gothenburg: WHO and Nordic School of Public Health, NHV-Report 2000:9, September 2000.

35 **Diwan V, Douglas M, Karberg I, Lehto J, Magnusson G, and Ritsatakis A** (eds.). *Health Impact Assessment: from Theory to Practice.* Report on the Leo Kaprio Workshop, Gothenburg, 28–30 October, 1990, Gothenburg: WHO and Nordic School of Public Health, NHV-Report 2000:9, September 2000.

36 European Centre for Health Policy. *Health Impact Assessment: Main Concepts and Suggested Approach. Gothenburg Consensus Paper.* Brussels: WHO Regional Office for Europe, December 1999.

37 *Methods of Assessing Risk to Health from Exposure to Hazards from Waste Landfills.* Report of a WHO meeting, Lodz, Poland 10–12 April 2000 Copenhagen, WHO Regional Office for Europe, (EUR/00/5026441), 2000.

38 *Communicating about Risks to Public Health. Pointers to Good Practice.* London: Department of Health, The Stationery Office, 1998.

39 **Banken R.** Strategies for Institutionalizing HIA. ECHP Health Impact Assessment Discussion papers, No. 1. Brussels: European Centre for Health Policy, WHO Regional Office for Europe, 2001.

40 **Put GV, den Broeder L, Penris M, and Roscam-Abbing EW.** *Experience with HIA at National Policy Level in the Netherlands. A Case Study.* Policy Learning Curve Series No. 4. Brussels: European Centre for Health Policy, WHO Regional Office for Europe, September 2001.

41 **Breeze C and Hall R.** *Health Impact Assessment in Government Policy-Making: Developments in Wales. A Case Study.* Policy Learning Curve Series No. 6. Brussels: European Centre for Health Policy, WHO Regional Office for Europe, May 2002.

42 *Building a Healthier Future by Taking Health into Account as Part of Objective 1 Projects.* Cardiff: National Assembly for Wales, 2001.

43 *A National Framework of Health Impact Assessment in Environmental Impact assessment.* Australia: University of Wollongong, 1992.

44 The Interorganisational committee on guidelines and principles for social impact assessment. In Burdge R (ed.), *A Conceptual Approach to Social Impact Assessment.* Middletown: Social Ecology Press, 1998.

45 **Frankish CJ, Green LW, Ratner PA, Chomik T, and Larsen C.** *Health Impact Assessment as a Tool for Population Health Promotion and Public Policy.* British Columbia: Institute of Health Promotion Research, University of British Columbia, May 1996. http://www.hc-sc.gc.ca/hppb/healthpomotiondevelopment/pube/impact/impact.htm

46 **Malleret, Thierry, Schwab, and Klaus.** Across the great divide—why Davos'elite must listen to the young leaders of the Birdging Europe initiative. *TIME* January 27, 2003; 56–57.

47 People, planet and profits—a summary of the Shell Report 'Tell Shell' at www.shell.com/tellshell

48 Modell och metoder för hälsokonsekvensbedomningar—återrapportering av regeringssuppdrag om Folkhälsoinstitutets arbete med hälsokonsekvensbedömningar (HKB), (A model and methods for health impact assessments (HIA)—progress report to the Swedish government on the development work on HIA performed by the National Institute of Public Health). F-serie 2001:9. Stockholm, 2001.

49 **Athanasiou L.** An outsider's cautionary observations. In Diwan V, Douglas M, Karberg I, Lehto J, Magnusson G, and Ritsatakis A (eds.), *Health Impact Assessment: from Theory to Practice. Report on the Leo Kaprio Workshop, Gothenburg, 28–30 October, 1999,* Gothenburg, WHO and Nordic School of Public Health, NHV-Report 2000:9, September 2000.

Chapter 15

HIA of policy in Canada

Reiner Banken

Introduction

Canada was one of the first countries to develop health impact assessment (HIA) during the 1980s, both as part of environmental impact assessment (EIA) and as part of strategies for healthy public policies (HPPs). Other applications of HIA in Canada are described in Chapters 13 and 27.

The present chapter will outline past and present developments of HIA at the policy level in Canada. The examples will cover the unsuccessful attempt to institutionalize HIA at cabinet level in British Columbia, the recent experience with HIA as a tool for HPPs in Québec, and the approach to international trade agreements and health at Health Canada.

The rise of HIA in British Columbia

Health impact assessment of public policies was developed as part of the health care reform in British Columbia at the beginning of the 1990s. This was probably one of the first times Policy HIA has been systematically developed anywhere in the world. This section describes the historical context, the development of specific methodological tools, the beginning of its institutionalization, and its demise with a changing health policy context. It has been described in more detail elsewhere [1].

The rise of HIA in British Columbia is closely connected to the history of health promotion. The idea of public policies influencing health determinants was present in the 1974 Lalonde Report, one of the founding documents of health promotion [2]. In 1979, McKinlay among others argued, 'prevention of disease by social and environmental management offers greater promise than any other means presently available' [3]. In the same year, Sir George Young, the British health minister made the following statement in a speech at the 4th Conference on Tobacco and Health: 'The general proposition that I wish to put to you is that the solution to many of today's medical problems will not be found in the research laboratories of our hospitals, but in our Parliaments. For the prospective patient, the answer may not be cure by incision at the operating table, but prevention by decision at the Cabinet table' [4].

The seminal work by Nancy Milio [5] provided a comprehensive overview of healthy public policies stressing the importance of public policies as an important and comprehensive tool for influencing a wide range of health determinants. Through the Ottawa Charter adopted at the First International Conference on Health Promotion [6], HPPs became one of the cornerstones of health promotion. Leonard Duhl suggested in 1986 that all proposed policies should be accompanied by HIAs, taking a lesson from the EIA process [7]. The Second International Conference on Health Promotion in 1988, known as the Adelaide Conference on Healthy Public Policy, identified the government sectors concerned with agriculture, trade, education, industry, and communications that should take into account health as an essential factor when formulating policy [8].

The thread of HIA as a tool for HPPs was picked up in British Columbia in 1989. In this year, the Ministry of Health of British Columbia created an 'Office of Health Promotion' and hosted the First National Conference on Health Promotion and Disease Prevention, which focused on healthy public policies. At this conference, Trevor Hancock reviewed the results and recommendations of the Adelaide Conference on Healthy Policies and concluded: 'The challenge, I submit, is no longer to define and debate the merits of health policy; the challenge is to identify the health impacts of current and possible alternative policies and to develop, implement and evaluate healthy public policies at all levels of government' [9].

Through the work of dedicated individuals in the Office of Health Promotion (I am indebted for this and other insights to Susan Stovel, who was one of those dedicated individuals) the idea of an institutionalized HIA as a tool for HPPs made its way into official health care policy. In its 1991 report, the British Columbia Royal Commission on Health Care and Costs [10] recommended two 'strategies for change' for achieving public policies for health:

- Evaluate the possible health effects of all proposed provincial programmes or legislation, or changes to existing programmes or legislation.

- Include studies of potential health effects in all environmental impact assessments.

The report of the Royal Commission was followed by the release of new directions for a healthy British Columbia in 1993 [11], which set a reform course for the British Columbia health care system. One of its 38 specific initiatives stated that 'a health impact assessment will be carried out for all new government policies and programs'.

Through the Deputy Minister of Health and support from the Premier's Office contacts were made with the Cabinet Planning Secretariat. The idea of integrating HIA into the formal process of policy analysis at cabinet level was accepted and in 1993 the *Guidelines for preparing cabinet submissions & documentation* were revised by adding health concerns to the list of different implications that had to be considered.

> The likely positive or negative impact of each option on the health of individuals, groups and communities, or on the health care system should be analyzed. This analysis should recognize the social, economic and physical factors affecting health, such as economic security, employment and working conditions, social support, safety, equity, education, and sense of control. The opportunity for the inclusion of individuals, communities, and other sectors in decision making on issues that affect their health should be considered. Attention should be paid to short term and long term effects. The consistency of each option with the government's objectives for improved health for British Columbians should be evaluated. It is recommended that the originating ministry contact the Ministry of Health (Office of Health Promotion), and in conjunction with this staff develop this analysis as required. [12]

The requirement of prospectively assessing the possible impacts of policies on health determinants was implemented through training sessions for the policy analysts working in the different ministries of the British Columbia government. As a result of these workshops, a toolkit was published in 1994 proposing a list of health determinants to be considered and advice on how to integrate the assessment of health implications with the other types of assessments as for example social and gender implications and sustainable development [13].

After the successful training sessions for the analysts in the different ministries, it was felt necessary to develop another tool for use in lower-level planning. The resulting *Health impact assessment guidelines* [14] were distributed at the 1994 and 1995 conferences of the Canadian Public Health Association, the 1995 National Conference on Community Health Centres, the 1994 International Public Health Association Conference, and through the British Columbia Healthy Communities Initiative. In 1995, a series of 86 workshops and 26 presentations were held across the province to increase awareness of the determinants of health and to familiarize potential users of the HIA process with the guidelines document. These sessions involved approximately 2000 service providers, educators, managers, and representatives from regional health boards and community health councils [15].

The fall of HIA in British Columbia

In 1995, the momentum of institutionalizing HIA seemed irreversible; in 1999, HIA was no longer an active issue in British Columbia's health system. Between these years the values underpinning the health care reform changed. In 1996, after the re-election of the NDP government, health care reform as laid out in

New directions for a healthy British Columbia [11] was quashed and replaced with different principles, called 'Better teamwork, better care' [16]. Davidson provides the following description of these changes.

> During the critical three years between the 1993 birth of New Directions and its funeral rites in November 1996, the British Columbia government's position on elections, taxation, local autonomy and scope of action for health authorities changed. The direction of change in each instance was consistent with the progressive abandonment of the reform principles inherent in the original policy statement. Movement was away from a perspective centred on citizen empowerment toward a policy focussing on the accountability of boards and councils to the Ministry of Health. Bound up in that change was a retreat from political accountability to the community and an advance toward managerial accountability to the ministry. [. . .] Managerial accountability refers to spending money in accordance with accepted accounting practices, providing services as efficiently as possible, and obtaining the intended results. [17]

As a result of this new policy perspective, the structure of the Ministry of Health underwent a major upheaval: the 'Office of Health Promotion' (the HIA 'think tank'), became the 'Population Health Resource Branch' and was finally abolished with a number of initiatives being moved to the new Ministry for Children and Families, and others to the Preventive Health Branch in the Ministry of Health. The HIA initiative came to rest in the 'Policy Development and Project Management Branch' in the Ministry of Health. Before and after these political and administrative upheavals, most of the dedicated individuals who had promoted and implemented HIA at cabinet level and at the community level left the Ministry of Health. Without these resource persons, adequate follow-up on HIA implementation did not occur. For example, when two key people left the Ministry in 1994, the training of policy analysts was not followed up with actual HIA practice at cabinet level.

While the *Guidelines for preparing cabinet submissions & documentation* were basically left unchanged, the wording of the section on health implications was no longer interpreted as mandatory, but rather as optional.

A complete review of the HIA guidelines was carried out in 1998. The main recommendation stated that the different guidelines should not be revised or promoted, as 'there is no reliable evidence to date that the HIA processes in place in other jurisdictions are creating policy or program changes consistent with the determinants of health perspective' [15].

The history of Policy HIA in British Columbia provides important lessons. The implementation of Policy HIA does not seem to depend on the availability of scientific knowledge or methodological tools, but rather on an appropriate policy window that can be used by social entrepreneurs [18,19]. By acknowledging the importance of health determinants and political accountability to the community, the 1991–1995 Health Reform in British Columbia opened a window of opportunity that was used by dedicated individuals in the Office of Health Promotion who acted as social entrepreneurs by joining HIA as a solution to the problem of how to implement the new orientations. The 1996 change in health policy abandoning the reform principles closed this window of opportunity.

The rapid rise and fall of Policy HIA in British Columbia may seem disheartening. However, the specific context of this development has to be kept in mind. While the development of HIA by a few social entrepreneurs at the Ministry of Health during a policy window has contributed to its rapid rise, it contributed as well to its rapid decline. When these key persons left the Ministry, HIA at cabinet level withered away. The other public health institutions in the province, such as the Public Health Association of British Columbia and the academic departments of public health, were not involved in this HIA process. A larger involvement by these other stakeholders would have probably created a greater resistance to abandoning Policy HIA. The absence of any legal basis has certainly contributed as well to the demise of Policy HIA with a changing health policy context.

Policy HIA in Québec

Québec has a long-standing history in HIA as part of EIA. After an initial use of HIA as a form of environmental health advocacy at the beginning of the 1980s, a memorandum of understanding was signed between the Ministry of Health and the Ministry of the Environment. This administrative framework has led to a very systematic and active HIA/EIA practice and a high degree of institutionalization in Québec [20]. The development of Policy HIA as an independent process from EIA has only started very recently. A new Public Health Act has created for the first time a strong legal basis for health promotion activities. The development of Policy HIA is based on Article 54, which took effect in June 2002:

> The Minister is by virtue of his or her office the advisor of the Government on any public health issue. The Minister shall give the other ministers any advice he or she considers advisable for health promotion and the adoption of policies capable of fostering the enhancement of the health and welfare of the population.
>
> In the Minister's capacity as government advisor, the Minister shall be consulted in relation to the development of the measures provided for in an Act or regulation that could have significant impact on the health of the population. [21]

Through this new disposition of the Public Health Act, the Minister of Health must be consulted on all policies that could have a significant health impact. In order to support this requirement, the Ministry of Health is currently developing a rigorous screening tool for public policies based on the potential impact on health determinants. An assessment process with the following steps is currently being discussed:

1. Screening of policies by the ministry or body that is proposing the policy.
2. Scoping and rapid HIA if necessary with support from the Ministry of Health.
3. Full HIA if necessary with support from the Ministry of Health.
4. Impact statement by the Ministry of Health.
5. Monitoring of health impacts.

The Policy HIA process is to be supported by systematic reviews of evidence on the impact of specific health determinants on the health of populations and population groups and by a specific research program to generate missing evidence. The systematic reviews are going to be commissioned with the Institut national de santé publique du Québec and may contain specific policy recommendations.

The legal obligation to submit policy proposals to an HIA process supported by the Ministry of Health and by experts from the Institut national de santé publique constitutes a very strong basis for the future development of Policy HIA in Québec. Furthermore, the new mandate of the Minister of Health of advising the other ministries on 'policies capable of fostering the enhancement of the health and welfare of the population' may have profound long-term effects on decision-making processes at cabinet level.

Assessing health effects of trade policies by Health Canada

In recent years, the issue of health impacts of trade agreements has become the object of certain publications in the scientific literature [22–24]. Opponents to such agreements often use the argument concerning potential negative effects of international trade agreements on population health.

At a meeting organized by the Pan American Health Organization in 1999 on the subject of trade and health services, Health Canada explained its strategy to integrate health concerns prospectively into trade negotiations. By establishing formal and informal communication lines and working relations with Trade Department officials, Health Canada managed to become part of all interdepartmental committees where trade issues are discussed. The overall objective of Health Canada's work is 'to insure that health concerns are considered in trade policy discussions and negotiations' [25].

Considering this overall objective, HIA seems to an ideal tool to identify health concerns and to integrate them into the decision-making process. The practice of assessing health effects of trade policies at Health Canada is, however, currently focussed on ensuring that trade policies do not, in any way, limit health policy flexibility (Personal communication with Ross Duncan, Senior Trade Policy Analyst, International Affairs Directorate, Health Canada). Canada's official negotiation position in the new round of negotiations of the General Agreement on Trade and Services ('Doha round') aims to 'preserve the ability of Canada and Canadians to maintain or establish regulations, subsidies, administrative practices or other measures in sectors such as health, public education, and social services' [26].

Limiting the assessment of health effects of trade policies at Health Canada to health services rather than the complete range of health determinants is contrasting with the official Canadian policy on population health [27]. The gap between the Canadian population health policy based on a model of health determinants and the very limited assessment of health implications of trade polices may be explained by the lack of integration of health concerns into the strategic environmental assessment process.

The current framework guiding the strategic environmental assessments of trade negotiations [28] is based on the guidelines on implementing the '1999 Cabinet Directive on the Environmental Assessment of Policy, Plan and Program Proposals', which established that all initiatives forwarded for consideration to the Cabinet of the Canadian Federal Government that are likely to have import-ant environmental effects, either positive or negative, must be subject to a strategic environmental assessment [29]. In principle, the guidelines consider human beings as part of the concept of environment [28 annex 8 page 1]; in practice, health and health determinants are only rarely taken into account in policy assessments, contrasting with the assessments of projects, where health effects as a result of environmental contamination are systematically studied (see Chapter 27).

In the present revision of the Canadian Environmental Assessment Act, the current scope of environmental assessments becomes very explicit as the govern-ment refuses to widen its breadth: in the words of the president of the Canadian Environmental Assessment Agency

> Sustainability impact assessment, as a means to understand the interrelationship among economic, social, and environmental factors, is a complex issue. Changing the focus of EIA or strategic environmental assessment (SEA) to address sustainability more broadly could profoundly impact their current function. In the process of attempting to address not only environmental but also social, economic, health, gender, or cultural issues dur-ing assessments, environmental considerations could lose the important emphasis that they currently receive.

> The idea that environmental assessment tools should be broadened to assess sustain-ability is an issue that has some currency. Recently during hearings on proposed

amendments to the Canadian Environmental Assessment Act, it was suggested that the definition of environment in the legislation be amended to include social, economic, and cultural considerations. As the Minister of the Environment noted in his response to the Parliamentary Standing Committee on Environment and Sustainable Development, modifying EIA to deal with economic and social considerations could pose some risks, 'you could tend to get flooded out by these other issues to the detriment of environmental values'. [30]

The current perceptions of environmental assessments as they are described in this citation hinder comprehensive efforts in Canada for assessing the health effects of trade agreements. Indeed, as a result of this concept, the Department of Foreign Affairs and International Trade views health effects as results of environmental effects, which are the result of economic impacts [28,31]. Health Canada is presently conducting methodological research on HIA of trade policies based on a framework of health determinants in order to see what that may encompass, and to assess the feasibility of its use in the future.

Conclusions

The Canadian experience with Policy HIA provides a number of lessons for the international community. The experience of British Columbia stresses the importance of institutionalizing HIA during a window of opportunity. The current evolution in Québec can become a success story if the actors in public health will make an effective use of the lever of Article 54 of its new Public Health Act. The practice of assessing the health consequences of trade policy at Health Canada illustrates the lack of integration of health dimensions into the strategic assessment process of policies.

Policy HIA in Canada departs from the traditional public health approach for fostering HPPs, which has been based on a process of advocacy. Advocacy is one of the three major strategies of health promotion, defined as 'a combination of individuals and social actions designed to gain political commitment, policy support, social acceptance and systems support for a particular health goal' [32]. Public health advocacy is not a form of advocacy with a defined constituency and accountability towards this constituency. It is rather a form of professional advocacy based on the value of health and the value of public service [33]. In an advocacy model, public health is one of many interest groups trying to influence the decision-making process.

Kemm [34] provides a good description of the tensions between the use of HIA as a means of advocacy and its use for informing decision making:

Health advocacy needs to be informed by Health Impact Assessment, which supplies the evidence that advocates can use to argue that the measures they favour will produce beneficial consequences. The practice of Health Impact Assessment creates a favourable climate for health advocacy by putting health high on the agenda and encouraging an open and participative process. However there is a tension between Health Impact Assessment, which seeks to make an impartial assessment of the health consequences of different policy options, and health advocacy, which is usually committed to one option.

Both in the past HIA practice in British Columbia and in the present model in Québec, Policy HIA is being based on a model of empowering non-health actors to produce public health knowledge and the Ministry of Health assuring the quality of routine HIA by the decision-making body. In this model, the public health experts and institutions that are involved in HIA cannot be involved in advocacy at the same time. Other public health actors may however want to use a strategy of advocacy to increase the decision makers' receptiveness to the knowledge on possible health impacts.

The link to the public health institutions outside the Ministry of Health appears however to be crucial for the long-term success of Policy HIA. Without an independent control as to its quality, including its scientific validity, HIA, like all impact assessment, may become a symbolic function without real-world effectiveness. 'The politics of bureaucracy provides an environment in which the effectiveness of impact assessment can be tempered, subverted, and broken in the absence of adequate provisions for external accountability' [35].

The current assessment of the health consequences of trade policy at Health Canada illustrates the lack of integration of health dimensions into the strategic assessment process. If the scope of strategic assessments in Canada will not be widened to include not only environmental aspects but also social aspects, Health Canada will have to work towards a parallel HIA of policies, including trade policies, in order to stay coherent with its policy perspective on health determinants and population health.

References

1 **Banken R.** Strategies for institutionalizing HIA. Brussels: European centre for Health Policy, 2001. Access Date: 2003. S.I. Available at http://www.who.dk/document/E75552.pdf.

2 Canada, Dept. of National Health and Welfare, Lalonde M. A new perspective on the health of Canadians a working document. Nouvelle perspective de la santé des canadiens; un document de travail. Ottawa: Government of Canada, 1974. Access Date: 2003.3.18. Available at http://www.hc-sc.gc.ca/hppb/phdd/pube/perintrod.htm.

3 **McKinlay JB.** Epidemiological and political determinants of social policies regarding the public health. *Social Science and Medicine* 1979; **13A**:541–558.

4 **Daube M.** The politics of smoking: thoughts on the Labour record. *Community Medicine* 1979; **1**:306–314.

5 **Milio N.** *Promoting Health Through Public Policy.* Philadelphia: Davis, 1981.

6 World Health Organization. Ottawa charter for health promotion; an international conference on health promotion—the move towards a new public health. Health and Welfare Canada, 1986. Access Date: 2003.3.17. Available at http://www.who.int/hpr/archive/docs/ottawa.html.

7 **Duhl LJ and Tamer J.** *Health Planning and Social Change.* New York, NY: Human Sciences Press, 1986.

8 World Health Organization. The Adelaide Recommendations: Conference Statement of the 2nd International Conference on Health Promotion, Adelaide South Australia, 5–9 April 1988. Access Date: 2003.3.17. WHO/HPR/HEP/95.2. Available at http://www.who.int/hpr/archives/docs/adelaide.html.

9 **Hancock T.** Where the rubber meets the road—from health promotion rhetoric to health promotion action. In National Symposium on Health Promotion and Disease Prevention, 1989. Presentation available from the author at thancock@on.aibn.com.

10 **Seaton PD.** *Closer to Home: The Report of the British Columbia Royal Commission on Health Care and Costs.* Victoria: The Commission. Crown Publications Distributor, 1991.

11 Ministry of Health and Ministry Responsible for Seniors. *New Directions for a Healthy British Columbia: Meeting the Challenge, Action for a Healthy Society.* Victoria: Province of British Columbia, Ministry of Health and Ministry Responsible for Seniors, 1993.

12 Government of British Columbia Guide to the Cabinet Committee System. Part 4— Guidelines for preparing cabinet submissions and documentation. In *The Cabinet Document System.* Victoria Province of British Columbia, Cabinet Office 1993.

13 Population Health Resource Branch. *Health Impact Assessment Toolkit—A Resource for Government Analysts.* Victoria, BC: Province of British Columbia, Ministry of Health and Ministry Responsible for Seniors, 1994.

14 Population Health Resource Branch. *Health Impact Assessment Guidelines—A Resource for Program Planning and Development.* Victoria, BC: Province of British Columbia, Ministry of Health and Ministry Responsible for Seniors, 1994.

15 **Lewis C.** *Health Impact Assessment Guidelines.* Document: Review and Recommendations. Victoria, BC: Ministry of Health, 1998.

16 British Columbia, Office of the Auditor General. *A Review of Governance and Accountability in the Regionalization of Health Services.* Victoria: The Office, 1998. (Report; vol 1997/98: 3). Access Date: 2003.3.17. Available at http://bcauditor.com/PUBS/1997–98/report-3/HealthServices.pdf.

17 **Davidson AR.** British Columbia's health reform: 'new directions' and accountability. *Canadian Journal of Public Health* 1999; **90**(Suppl 1):S35–S38.

18. **de Leeuw E.** Healthy cities: urban social entrepreneurship for health. *Health Promotion International* 1999; **14**:261–269.

19 **Kingdon JW.** *Agendas, Alternatives, and Public Policies.* 2nd edn. New York, NY: HarperCollins College Publishers, 1995.

20 **Banken R.** Information management in HIA: knowledge, methodological and organizational challenges. In *Health Impact Assessment: From Theory to Practice.* pp. 231–252. Gothenburg: Nordic School of Public Health, 2000.

21 Gouvernement du Québec. Public Health Act. Chapter S-2.2: 2001:Article 54.

22 **Bloche MG.** Introduction: Health and the WTO. *Journal of International Economic Law* 2002; **5**:821–823. Access Date: 2003.3.25. Available at http://www3.oup.co.uk/jielaw/hdb/Volume_05/Issue_04/pdf/050821.pdf.

23 **Bloche MG.** WTO deference to national health policy: toward an interpretive principle. *Journal of International Economic Law* 2002; **5**(4):825–848. Access Date: 2003.3.24. Available at http://www3.oup.co.uk/jielaw/hdb/Volume_05/Issue_04/pdf/050825.pdf.

24 **Lang T.** The new GATT round: whose development? Whose health? *Journal of Epidemiology and Community Health* 1999; **53**:681–682.

25 **Vellinga J.** An approach to trade and health at Health Canada. Vieira C (ed.), *Trade in Health Services: Global, Regional, and Country Perspectives.* Washington, DC: Pan American Health Organization, 2002. Access Date: 2003.3.25. Available at http://www.paho.org/English/HDP/HDD/trade.htm.

26 Canada, Dept. of Foreign Affairs and International Trade. *Canada's Initial Negotiating Position for the General Agreement on Trade in Services* (GATS). Ottawa: Dept. of Foreign Affairs and International Trade, 2001. Access Date: 2003.3.25. Available at http://strategis.ic.gc.ca/epic/internet/instp-pcs.nsf/vwGeneratedInterE/sk00096e.html.

27 Federal, provincial and territorial advisory committee on population health. *Strategies for Population Health: Investing in the Health of Canadians* (adopted at the meeting of the ministers of health, sept 14–15, 1994, Halifax, Nova Scotia). Ottawa: Minister of Supply and Services, 1994. Access Date: 2003.3.25. Available at http://www.hc-sc.gc.ca/hppb/phdd/pdf/e_strateg.pdf.

28 Canada, Dept. of Foreign Affairs and International Trade. *Handbook for Conducting Environmental Assessments of Trade Negotiations.* Ottawa: Dept. of Foreign Affairs and International Trade, 2002. Access Date: 2003.3.25. Available at http://www.dfait-maeci.gc.ca/tna-nac/documents/handbook-e.pdf.

29 Government of Canada, Privy Council Office. Guide to the Regulatory Process—Developing a Regulatory Proposal and Seeking its Approval. Ottawa, 2002. Access Date: 2003.3.25. Available at http://www.pco-bcp.gc.ca/raoics-srdc/default.asp?Language=E&Page=Publications&Sub=RegulatoryProcessGuide.

30 **Gershberg S.** Sustainable Development, Sustainable Governance and Environmental Impact Assessments in Canada. Intergovernmental Policy Forum 17 June 2002 held in conjunction with 22nd Annual Conference of the International Association for Impact Assessment, The Hague, Netherlands. Presentation available from vanessa.cook@ceaa-acee.gc.ca.

31 **Gillmore T.** Strategic Environmental Assessment of Trade Negociations. 22nd Annual Conference of the International Association for Impact Assessment. Presentation available from the author at thomas.gillmore@dfait-maeci.gc.ca.

32 World Health Organization. Health Promotion Glossary. Geneva: WHO, 1998. Access Date: 2003.3.17. Available at http://www.who.int/hpr/archive/docs/glossary.html.

33 **Wynia MK, Latham SR, Kao AC, Berg JW, and Emanuel LL.** Medical professionalism in society. *New England Journal of Medicine* 1999; **341**:1612–1616.

34 **Kemm J.** Health impact assessment: a tool for healthy public policy. *Health Promotion International* 2001; **16**:79–85.

35 **Bartlett RV.** Policy through impact assessment: institutionalized analysis as a policy strategy. (Contributions in political science; no. 235). New York, NY: Greenwood Press, 1989.

Chapter 16

HIA and national policy in the Netherlands

Ernst W Roscam Abbing

The development of HIA in the Netherlands

The development of health impact assessment (HIA) of national policy in the
Netherlands has to be seen in the context of a pluralistic health system with an
emphasis on health service policy. Health policy analysts and health researchers
were becoming increasingly interested in factors outside the health sector that
had an influence on health. This interest was evidenced by an active research
programme on socio-economic health differences and publication of 'Public
health projections reports' which explored a set of development scenarios. In
1986, intersectoral policy was mentioned in a comprehensive memorandum
[1] presented by the Secretary of State for Health. The document 'Prevention
policy for public health [2]' was sent to Parliament in 1992. This included a
chapter on priorities for intersectoral policy and suggested that HIA be used as
a tool for intersectoral policy, especially for addressing socio-economic
inequalities in health.

An expert report [3] commissioned in 1993 by the Ministry of Health recommended that screening of national policy proposals for health impacts should be
started on an experimental basis. First of all a particular effort should be made
to answer the following questions:

- What were the difficulties in assessing the risk of unwanted health effects?
- How much time was needed to assess proposals?
- What documentation was needed?
- How many proposals had a possible impact on health?
- How often did the official policy process recognize the possible health
  consequences of such proposals?
- What types of policy had possible health impacts?
- Which health determinants were most commonly influenced by policies?

At a further workshop in 1994, people experienced in environmental and
other impact assessments met with scientists and policy makers to discuss how

the report could be taken forward. They tried to clarify what HIA was and how it could be practically applied to policy making. The workshop concluded that, despite a number of uncertainties, HIA was potentially a useful tool that should be developed. It also concluded that at that time a formal legal procedure was neither feasible nor desirable. Following the expert report and workshop, the Minister of Health issued a policy document 'Safe and sound [4]'. This suggested that HIA was a methodology that could result in the 'estimation of impacts of policy measures on health status' of the Netherlands population. Following this an action plan to develop HIA [5] was presented to Parliament in 1995.

The action plan for HIA

The action plan for HIA proposed five work programmes:

+ Make an inventory of methods and tools that had been used for impact assessment both in the Netherlands and in other countries;

+ Develop methods for estimating the size and significance of impacts on health of policy proposals;

+ Develop procedures for HIA;

+ Undertake on an experimental basis some HIAs of national policy;

+ Investigate whether HIA could be institutionalized in the Netherlands.

An annual sum of €230,000 was made available for carrying out the plan of action. The Dean of the Netherlands School of Public health (NSPH) was asked to give technical assistance to this plan of action by setting up a help desk to support the ministry. In 1996, the Intersectoral Policy Office (IPO), led by the author together with a small steering committee, was established within the NSPH.

A report 'Health impact screening [6]' published in 1996 compared several impact assessments and tests being developed in the Netherlands. It concluded that HIA could be carried out using either a 'rational' or an 'incremental' process. The 'rational' process followed fixed protocols and procedures, and produced quantitative measures of health effects. The 'incremental' process used procedures and methods chosen in interaction and negotiation with different actors and produced a consensus about health interests and likely health effects. The choice between 'rational' and 'incremental' process depends on the specific situation in which the HIA is to be performed.

This report was sent to Parliament as an appendix to the second report to Parliament on HIA [7] together with a letter from the Minister of Health. She argued that, although many cause–effect relations could not be quantified,

HIA was still possible and worthwhile. For each policy subjected to HIA, the different options should be discussed with a view to identifying the option that would be expected to have the best health impact.

Since 1996 the sole purpose of IPO has been to help the Ministry of Health to develop HIA. The office staff has grown from a project manager, a staff assistant, and a secretary in 1996 to four staff members and two secretaries in 2001. The total annual budget has increased from €230,000 to 340,000. The IPO is financed by the Ministry, but has independent control over its budget.

It is important that the IPO is located in the NSPH rather than in the Ministry of Health because:

◆ Health impact assessment is still an experimental procedure;

◆ Health impact assessments commissioned by an independent external organization might be more acceptable to other ministries;

◆ The NSPH can be relatively independent in commissioning HIAs because it has close links to numerous academic institutes, non-governmental organizations, and research institutes;

◆ The NSPH is not a research institute and so has no conflict of interest when commissioning HIAs.

The IPO performs two major functions. First it is responsible for commissioning experimental HIAs on national policy proposals. Second it develops HIA methodology and builds capacity to undertake HIA. Those involved in doing HIAs include organizations in the (public) health sector as well as organizations in education, finance, defence, environment, and social affairs sectors.

The Ministry of Health and the IPO try to screen the policies of all ministries to identify those that might have an impact on health. It was hoped that this screening would mostly be done by a particular department of the Ministry, with the IPO only supplying technical and methodological support. In reality it is the IPO that plays the major role in screening.

After the Ministry of Health and the IPO have agreed that a topic should be subject to HIA, an official request is sent from the Ministry to the NSPH in which the NSPH is asked to set up, coordinate, execute, or commission an assessment. The assessment may be a health impact screening (HIS) or an HIA. An HIS is a more superficial assessment, more or less comparable to the HIA rapid appraisal as discussed in the Gothenburg Consensus Paper [8] while an HIA is a more extensive, focussed, and in-depth assessment. The IPO would then find appropriate research institutes to conduct the HIS or HIA.

Table 16.1 shows the HIAs and HISs that have been performed since 1996. Nearly all of them were ex ante (prospective) evaluations of national policy proposals and the majority were HISs.

Table 16.1 HIAs and HISs produced or coordinated by the NSPH/IPO

Year	Subject	Type
1996	Energy tax regulation (Ecotax)	HIS
1996	High-speed railway	HIS
1997	Tobacco policy (2 reports)	HIS
1997	Alcohol and Catering Act	HIS
1997	Reduction of the dental care package	HIA
1998	National Budget 1997/Annual survey of care	HIS
1998	Tobacco policy	HIA
1998	Election programmes of political parties	HIS
1999	Housing forecast 2030	HIS
1998	ICES ('Operation Interdepartmental Commission for Economic Structural Reinforcement')	HIS (2 reports)
1998	Identification of policy areas influencing determinants of five major health problems	HIA
1999	Occupational Health and Safety Act and Monitoring 24-h economy	HIS
1999	Coalition agreement 1998	HIS
1999	Employment policy proposals and health effect screening	HIS
1999	National Budget 1999	HIS
1999	Regional development policy	HIS
2000	National Budget 2000	HIS
2001	National Budget 2001	HIS
2002	Housing policy	HIA
2002	National Budget 2002—Ministry of Education	HIS
2002	Coalition agreement	HIS
2003	Long-range development programme of 4 Dutch cities 1999–2003	HIS
2003	Coalition agreement	HIS

Over the years, there has been a tendency to focus more on policies sponsored by other ministries and less on Ministry of Health policies. There has also been a trend away from short assessments towards more comprehensive HIAs.

In 2003 the IPO was moved from the NSPH to the National Institute of Public Health and Environment (RIVM).

Building HIA into the political process

The aims of the activity were to

- influence policy deliberations in favour of health,
- increase long-term awareness of the impacts of policies on health,
- set the agenda (get and keep health on the agenda),
- increase the probability of health interests receiving structural attention.

A three-phase model was used:

- case-finding—identifying policies that could impact on health,
- assessing impacts on health,
- applying the results of the first two steps to influence intersectoral political–administrative decision making in favour of health.

Figure 16.1 shows how HIA can be integrated with the policy development process. A legal framework for this model was neither present nor desired. This model can be used as a framework for discussions within the Ministry of Health, for discussions between the Ministry of Health and other ministries (the proponents), and even as part of the parliamentary process.

The first step is a preliminary screening (quick scan or rapid appraisal) using a checklist to find policies that might be health-relevant (case-finding). When a

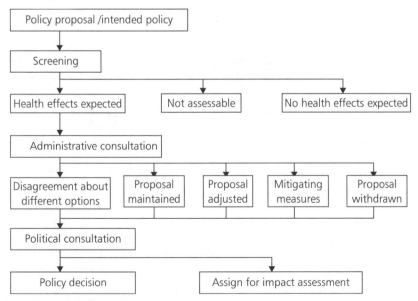

Fig. 16.1 Integration of HIA into the policy-making process.

policy proposal that could have health impacts is detected the HIA process provides a framework in which the Ministry of Health can consider the next steps. It may decide a full HIA is needed before further discussions with the proponent ministry informed by the outcome of the HIA (reactive intersectoral policy). It may think it more appropriate to tell the proponent about the results of the HIS and invite them to work jointly on the proposal involved, to find ways of preventing negative health effects or developing mitigating measures (proactive intersectoral policy).

The course chosen depends on the administrative and political context. Relations in the Ministry of Health and between them and the other ministries (and the position of the Ministry in the Cabinet) are important. Timing is crucial to the success of any kind of intervention and therefore determines how impact assessment is used. A study by the University of Nijmegen of the effectiveness of the gender impact assessment [9] concluded that as the proposal moves from draft version to formal proposal (green paper) windows of opportunity close.

When health impacts are discussed the proposal may be accepted unchanged, accepted with amendments, or withdrawn. There will probably be some measure of negotiation between the different and possibly incompatible interests involved. Sometimes the parties will persist in their different points of view and the civil servants involved will be unable to agree. The issue is then raised with the political representatives for decision. The politicians may require an health impact analysis or health impact review before reaching a decision.

In her letter to Parliament in 1995 (5) the Minister of Health stated that HIA should not be limited to easily assessable and quantifiable issues (causal relationships). She contended that it should clearly be possible to describe health effects in a qualitative and/or tentative way using case studies or even well-reasoned assumptions. Although in theory this point of view provides the opportunity for a wide range of assessments, in practice this is much more difficult. There is often a need for quantitative data to convince other parties of potential negative health effects or health benefits.

Timing of the HIA

Conflicting realities make it difficult to determine the best time to do an HIA in the context of national policy-making. On the one hand, the willingness of the proponent to co-operate with an HIA process and to act on the HIA report decreases with time. Most policy-making processes include some intersectoral aspects. In the idea-forming stage, a logical suggestion in a positive atmosphere can easily result in the proponent recognizing the health impact.

However, if the same concerns are raised when the proposal is finally before Parliament after two years of hard work, the proponent may be very resistant to change. A Minister of Health will need a very solid argument backed up by a detailed analysis of expected impacts to obtain changes at this stage. On the other hand, many proposals remain vague and abstract with rapidly changing ideas until a very late stage of development. Assessing the proposal for health issues is hardly possible in such situations.

A good illustration of this dilemma was the screening and scoping phase of the HIA on housing policy. When this HIA was begun in April 1999, two green papers identifying relevant trends and problems were available but neither contained any form of proposal. In December 1999 a confidential conceptual white paper was circulated but this was still subject to major change. In March 2000 the first official draft was sent to the Ministry of Health for consultation, asking for comments within a few weeks. In May the draft white paper including 69 proposals was published and sent to a number of advisory boards, which did not include those of the Ministry of Health. A month later the IPO, the Ministry of Health, and the Ministry of Housing met to discuss scoping the HIA, but decided that an HIA could no longer have any effect on the draft proposal. It was therefore decided to do the HIA on the white paper, which had already been accepted.

Thus, when there were opportunities to bring forward health aspects in the general public debate preceding the green paper on Housing Policy, there were no proposals to be assessed. By the time the proposals were becoming clear there was no time or willingness to cooperate with an HIA that could still have an impact on the draft policy. This example illustrates the importance of the timing of HIA in the policy-making process, especially when the aim is to assess specific proposals and not major trends. If an HIA has to be conducted at this late stage, it must be supported at the highest ministerial level if it is to have any hope of influencing the decision.

Fortunately, the draft proposal appears to have a quite positive approach to health (especially concerning well-being and health care in relation to aging) but there were still different opinions on the concept of promoting health. For example, during the intersectoral consultation it suddenly became clear to the other party (i.e., civil servants of the Ministry of Housing) that their overall policy would have negative effects for physical exercise. Elevators, for example, are a prominent feature of buildings, while staircases are perceived as a secondary facility and are sometimes even placed out of sight. Another example is that the housing policy aims to get people as quickly and easily as possible from one place to another and consequently neglects opportunities for physical exercise.

After this conceptual breakthrough, HIA was judged in a more positive way and the health targets of the Ministry also became better understood. During 2000, representatives of both ministries and the NSPH finally agreed on an in-depth HI review on the Housing Policy.

The case finding and scoping step

In the Netherlands the terminology of HIA is slightly different from elsewhere. The first inspection of policies to identify those with possible health impacts, which might be called screening elsewhere, is called case finding. The term HIS is used to describe a fairly quick and superficial assessment, which elsewhere might be called a rapid appraisal. Scoping is defined as the process of determining which potential direct and indirect health effects of the proposed policy, programme, or project need to be further considered, with regard to which population, and by which methods with which resources, with whose participation, and by which time [8]. A checklist was developed as a practical instrument for case finding and scoping. As well as the paper version of the checklist, customized software was written in order to have an electronic version that could be used on personal computers. It was also intended that this checklist could be used as a rapid instrument for HIA by officials at both national and local levels.

The checklist contains a structured list of questions and check boxes in order to describe and evaluate a proposal. One set of questions covers the potential influence of the proposed policy on the determinants of health (see Table 16.2). Any proposal is judged as being relevant if one or more of the health determinants is ticked. Although the checklist is intended to be used to screen policy proposals, it has not yet been proved that it is suitable for evaluating policies, programmes, and/or projects. Tests on specific proposals show that it is difficult or even impossible to get clear-cut answers.

Table 16.2 Determinants of health as used by NSPH/IPO

Lifestyle	Physical environment	Socio-economic environment	Health care system
Diet	Environment	Income	
Alcohol, tobacco, drugs, and gambling	Housing conditions	Education	
	Safety	Employment	
Exercise	Other	Social contacts and welfare	
Safe sex		Recreation	
Other		Other	

Case finding was done by scrutinizing a number of sources for health relevant proposals:

♦ Parliamentary documents (white papers, committee reports),

♦ National budget,

♦ Report of advisory bodies to ministries.

Parliamentary documents (white papers, reports of committee meetings, etc.) available through Internet were passed through case finding daily. This is done in two steps: first an inspection of the title only (50–100 per day) and a subsequent inspection of possible health-relevant documents. A useful 'side effect' of this activity is that it provides an opportunity to see how (old and new) policies develop. The minutes of parliamentary meetings are also helpful in evaluating the impact of HIAs in political debates, for example, if HIA reports are referred to or if the need for HIA on a specific topic is expressed.

The National Budget is systematically screened. This provides a good idea of upcoming major governmental policies that might be relevant for health. Close scrutiny of the annual parliamentary debate on the National Budget is also useful. Up to now four screenings have been conducted (see Table 16.1). A subsequent analysis revealed that a two-year frequency was sufficient, and was reliable enough. It was recently decided to screen about half of the ministries each year.

In the Netherlands each ministry has one or more advisory bodies. The reports of these bodies are often used as input for new policy (or at least one can assume that they will be followed up). Thus screening the programmes of these bodies and screening specific reports can help to track potential new policies at a very early stage, perhaps even before they are drafted.

HIA and inequalities

Until recently, health inequalities were not systematically included in HIA of national policies in the Netherlands. Only one HIA (tobacco) explicitly refers to the impact of the policy proposal on socio-economic differences in health. In a more implicit way, however, health inequalities have been addressed in the work carried out. The checklist, discussed earlier, includes key determinants of socio-economic inequalities in health: employment, education, and income. It also contains questions focused on vulnerable groups such as migrants, the unemployed, and the poor, although explicit questions related to the distribution of health effects between different groups are lacking.

The IPO is starting to work out ways of implementing an equity approach in HIA. In the Health Impact Review on National Housing Policy, for instance, specific attention is being paid to equity matters (including gender-based inequalities). A supportive factor is that currently, in national as well as local health policies, reducing socio-economic health differences is considered an important issue.

Who has responsibility for HIA

The Minister of Health should be responsible for assessing the health effects of national policies outside the direct health domain. The most appropriate way of exercising this responsibility would be to implement HIA throughout the Ministry and make it part of the daily work of all departments. However in reality civil servants are occupied with urgent matters and may give a low priority to HIA, which is perceived to be new and difficult. A strong influence outside the Ministry of Health could help promote the development of HIA, especially at a stage when it is still of unproven value. However creating an independent screening and assessment facility outside the Ministry of Health might suggest that the Ministry is released from this responsibility.

Dealing with health issues within intersectoral policy and using HIA requires the commitment and efforts of the Ministry of Health. Up to now there has been very little follow-up by the Ministry of issues identified by HIA. The early HIAs coordinated by the IPO (see Table 16.1) were mainly of Ministry of Health topics or of proposals that were already at a final stage. The Ministry made an effort to influence the policy proposals of proponents. The Minister of Health in assessing the effectiveness of the earliest HIAs found it to be luke-warm. This was not surprising given the experimental nature and the early stage of development.

Other changes within the Ministry of Health are also needed to achieve the necessary awareness of health in intersectoral policy. The Ministry has been responsible only for the health care system with issues such as financing, waiting lists, and adequate response to emergencies. This distracted attention from new concepts about the broader determinants of health and the importance of intersectoral policies. This is reflected in the small proportion of the national budget for health that is spent on public health in general and on intersectoral policy and HIA as a part of public health.

Increasing awareness within the Ministry of HIA and of intersectoral policy implies a new way of thinking and a readiness to look at health effects outside the Ministry's domain. The Council for Health and Care has recommended that the Ministry of Health should not have the monopoly on commissioning HIA but that other national, regional, and local authorities should also be able to commission HIAs [10]. The Ministry of Health has also urged that HIAs should be done at a local level [11].

The initial screening of policies (case finding) could be done by civil servants in the Ministry of Health or even in the initiating ministry, who know about new policies at an early stage. Ministries already evaluate their own draft proposals for gender impact assessments. However in the Netherlands HIA screening is usually done outside the ministries by the NSPH, which screens most policy proposals at the national level.

The Ministry of Health is responsible for acting when HIAs suggest a policy could have adverse health impacts. This could mean making efforts to ensure adjustments or modifications, or even cancellation. The IPO has no significant role to play at this stage, but it remains to be seen if this will be the case in the future.

Did the HIAs make a difference?

More than 20 reports have now been produced (see Table 16.1). These HIAs are extremely varied, in subject matter, in the way they were initiated, and their timing in relation to the policy-making process. It is clear from the Netherlands experience that the reality of HIA differs from the theory. It is difficult to assess how HIA has influenced policy in the Netherlands since the programme has attempted to apply and to develop better methods of HIA at the same time. The HIS/HIA reports produced so far are concerned both with indicating and assessing health-related policy proposals (first phase of intersectoral policy) and with further description and estimation of potential health effects (second phase). The reports do not all provide an equally clear account of influence on policy. In reporting on the effectiveness of HIA to the Lower House the Minister of Health described several examples [12].

The HIS on the regulatory levy on energy (Ecotax) was one of three research reports sent to the State Secretary for Finance to the Upper House, in 1996. One effect it had was to lead the cabinet to reconsider the income situation of the chronically ill and the disabled, in terms of income. In response to this the Cabinet modified the tax allowances for these groups and increased the benefits available to them.

In considering amendments to the Tobacco Act in 1999 the Lower House explicitly referred to the HIA on policies to discourage smoking and discussed its findings and conclusions. The results of the HIA played a part in the interministerial consultations on striking a balance between economic and health-related interests, part of the policy preparations for this legislative proposal. The ministers took particular note of the finding that negative business impact on the tobacco sector due to a reduction in consumption were compensated for at the macroeconomic level and of the favourable effects shown for sectors such as the police and fire services.

The findings of the HIS on the Licensing Act were incorporated into the Explanatory Memorandum accompanying the Bill for Amendments to this bill. The HIS served to counterbalance the Effect Assessment of Alcohol Legislation Proposals carried out by the Committee on Market Deregulation and Legislative Quality, which was primarily concerned with the consequences for business. The legislative proposal is currently awaiting passage through the Upper House.

Within the framework of the housing policy of the Ministry of Housing, Spatial Planning and the Environment, and the previously published *Projections for residential requirements 2030*, an HIS was carried out to gain an insight into the potential effects of the envisaged scenarios on health and policy. This HIS has recently been discussed by a group of 30 experts from various disciplines with the aim of selecting specific issues that might lend themselves to further study, in the form of either an HIS or an HIA. One result of the discussion was that it gave the experts the opportunity to view their own policy areas from a different perspective.

HISs were also performed on national budgets (1997, 1999), a number of 1998 party electoral manifestos, and the 1998 Coalition Agreement. These screenings belong to the initial indication phase and resulted in an analysis of policy proposals relevant to health.

At the moment the Ministry of Health, Welfare and Sport is not really convinced that HIA is an applicable and usable tool with regard to the protection and improvement of health. The Ministry is concerned that doing HIA on the policies of other ministries might be seen as checking up on them and so intends to put responsibility for HIA with these other ministries but support them in ensuring that their policies are in keeping with public health policy. The ministry will intensify co-operation with other ministries in the fields of labour and health, environment and health, lifestyle, food, youth, and reducing health inequalities. The Ministry has also set out new policy for HIA to be developed at the local level and tends to put the responsibility for HIA and its implementation with the other ministries.

The way forward for HIA

In order to determine future actions the following questions and issues need to be resolved.

- Should HIA be applied to a broad range of policy areas or should it be limited to the priorities of the Ministry of Health (such as lifestyles and health inequalities)?

- Should HIA be a key instrument for the Ministry of Health's intersectoral policy role or should other possible strategies be preferred?

- Are methodological tools available (or can they be developed) for appropriate selection of policy proposals, which could have health impacts?
- Should HIA at national policy level be institutionalized more formally (e.g., by a Cabinet white paper)?
- Should responsibility for HIA be located at a higher (more strategic) administrative level?
- Should HIA be the sole responsibility of the Ministry of Health or should other ministries share responsibility?
- Should HIA be implemented at the national or local level or both?

The way in which HIA develops will depend on the answers that politicians give to these questions.

References

1 Nota 2000 [Memorandum 2000]. The Hague, Ministry of Welfare, Health and Culture, 1986.

2 State Secretary of Welfare, Health and Culture. Preventiebeleid voor de volksgezondheid [Prevention policy for public health]. The Hague, SDU, 1992 (Parliamentary document 22 894, No. 1).

3 Roscam Abbing E, *et al*. Gezondheidseffectscreening. Verkennend rapport en verslag van een workshop [Health impact assessment. Report and proceedings of an expert meeting]. The Hague, Ministry of Health, Welfare and Sports, 1995.

4 Gezond en wel. Kader voor het volksgezondheidsbeleid 1995–1998 [Safe and sound. Framework for the national health policy 1995–1998]. The Hague, Ministry of Health, Welfare and Sports, 1995 (Parliamentary document 24 126, No. 2).

5 Gezondheidseffectscreening [Health impact assessment]. The Hague, Ministry of Health, Welfare and Sports, 1995 (Parliamentary document 24 126, No. 3).

6 Putters K. *Health Impact Screening. Rational models in their Administrative Contexts*. The Hague, Ministry of Health, Welfare and Sports, 1996.

7 Gezondheidseffectscreening [Health impact assessment]. The Hague, Ministry of Health, Welfare and Sports, 1995 (Parliamentary document 24 126, No. 14).

8 *Health Impact Assessment: Main Concepts and suggested Approach*. Gothenburg: consensus paper. Gothenburg: Nordic School of Public Health, 2000.

9 *Van de EER geleerd* [lessons learnt from the emancipation impact assessment]. Nijmegen: Catholic University, 1998.

10 Gezond zonder zorg [Healthy without care]. Zoetermeer, Council for Public Health and Health Care, 2000 (available in English).

11 Nationaal contract openbare gezondheidszorg [National contract for public health]. The Hague, Ministry of Health, Welfare and Sports, 2001.

12 *Zorgnota 2000* [Health care memorandum 2000]. The Hague, Ministry of Health, Welfare and Sports, 1999 (Parliamentary document 26 801, No. 5).

Chapter 17

HIA in Scotland

Margaret Douglas and Jill Muirie

The Scottish context

Scotland is part of the United Kingdom but also has its own distinct national identity and, since 1999, its own Parliament. Scotland has five million people who live in a diverse geographical area that covers 78,133 km^2[1]. This includes a relatively densely populated area in the 'central belt' between the cities of Edinburgh and Glasgow, rural areas that are more sparsely populated, and populated islands.

The health status of the Scottish population is the worst in the United Kingdom, and is worse than most other Western European countries [2]. It is known that deprivation is one of the key reasons for Scotland's poor health. However, recent analyses suggest that deprivation cannot explain all of the difference in health between Scotland and England. Current research is investigating this apparent 'Scottish Effect' further [2]. There are also health inequalities within Scotland, indeed the mortality gradient between affluent and deprived groups is sharper than the difference seen in the rest of the United Kingdom [3].

The Scottish people voted in favour of establishing a Scottish Parliament with devolved powers in 1997, and the Parliament sat for the first time in 1999 [4]. The new existence of the Parliament with the Scottish Executive, its government, brings both challenges and opportunities for public health. Having a parliament physically located in Scotland has raised the profile of public policy-making and should enable Scottish policies to reflect the needs and priorities of Scottish people. However there is also potential for fragmentation of decision making in relation to public health. The Scottish Executive has been granted devolved powers and responsibility for policy areas including health, but the Westminster government retains responsibility for many other areas that impact on health, for example, fiscal policy, and the benefits system. This could lead to more emphasis on delivery of health services than the wider public health. Problems within health care services are often given a high profile by the media. Despite this, the Scottish Executive has implemented a range of policy initiatives that explicitly recognize multiple and interlinked influences on health and well-being. These include the *Social justice strategy* [5], *Towards a healthier Scotland: a white paper on public health* [6], the *Review of the public health function* [7], *Nursing for health*, a review of public health nursing [8], and most recently *Improving health in Scotland: the challenge* [9].

Scotland has 15 National Health Service (NHS) boards responsible for commissioning health services for their resident populations, each of which has a public health department led by a Director of Public Health. It has 32 local authorities with a wide range of duties and powers. Both local authorities and NHS boards are increasingly recognized as public health organizations and local authorities have been given new powers to take action that improve the well-being of their population [10].

One way that NHS boards, local authorities, and other agencies are engaging in joint work is through the community planning process. Local authorities lead this process, which brings together local organizations to plan, provide, and promote the well-being of their communities. This results in a shared vision and priorities in the form of a community plan, and an ongoing process of partnership working [11]. This is a key arena for partnership working to address health determinants.

Public health capacity and networks

Scotland has several national public health bodies, including the Scottish Centre for Infection and Environmental Health (SCIEH), the Information and Statistics Division (ISD), and the newly created NHS Health Scotland, which was formed by merging the Health Education Board for Scotland (HEBS) and the Public Health Institute of Scotland (PHIS).

PHIS had HIA as an explicit part of its remit and coordinated the Scottish HIA Network (see later). This network is now supported by the new organization. There is a strong history of public health networks in Scotland. For example, the Scottish Needs Assessment Programme (SNAP) was a network of public health specialists and others who drafted needs assessment reports on topics of common interest, and led early work on HIA. The HIA network has adopted a similar approach.

The public health workforce is difficult to define as so many different disciplines both within and outside the NHS make a contribution. The 15 NHS boards, and the national public health organizations just listed, employ public health physicians and specialists from a range of disciplines. The workforce has also recently been augmented by new groups of staff. Following the *Nursing for health* review of public health nursing, Public Health Practitioner (PHP) posts were created to lead public health work in primary care. There is now a PHP working in each Local Healthcare Cooperative (LHCC). The PHPs come from health visiting and a range of other backgrounds. In local authorities, Health Improvement Officers (HIOs) have been appointed to take on a similar role. These new practitioners come from a range of backgrounds and training and development programmes are being arranged for them. Ultimately, they should be able to offer the skills needed to contribute to public health work including support for HIA.

Policy statements on HIA

The 1998 green paper *Working together for a healthier Scotland* recognized the need to address life circumstances as well as lifestyles and disease topics. It proposed that HIAs should be carried out on evolving national policies; and that NHS boards' Directors of Public Health should assist local authorities in preparing HIAs in relation to key local policy proposals and initiatives [12].

Following this, the 1999 white paper *Towards a healthier Scotland* described HIA as an 'essential step' when formulating policy at national and local levels in order to 'place health at the centre of planning and decision making' [6]. Among the actions, it asked SNAP to develop guidance on HIA and said that the Minister for Health would 'promote widespread use of HIA when formulating government policies'.

The Scottish *Review of the public health function* (1999) envisaged a national HIA network that would have a role in development, training, and education and also carry out national HIAs [7]. The review recommended the formation of PHIS and that HIA should be one of its areas of work.

Most recently, *Improving health in Scotland: the challenge* notes work to develop and use an integrated policy impact assessment tool [9].

HIA in practice

Scottish Needs Assessment Programme HIA pilots

In the wake of *Towards a healthier Scotland*, the Scottish Executive commissioned SNAP to pilot approaches to HIA in a Scottish context. Two groups were established and two case studies were conducted. One of these was an HIA of the City of Edinburgh Council's Local Transport Strategy [13,14]. The other was an HIA of North Edinburgh Area Renewal Housing Strategy [15]. The two groups took different approaches but both were carried out in partnership with the relevant planners. A discussion paper, *HIA: piloting the process in Scotland*, reported on the lessons learnt from the two case studies and made recommendations for using and developing HIA in Scotland [16]. The report found that no single blueprint for HIA could be appropriate for all situations. Instead it proposed a set of 'key principles' for HIA [17]. The pilot studies and key principles were presented at a national seminar early in 2000.

The Scottish HIA Network

PHIS set up the Scottish HIA Network in June 2001. The network consists of approximately 50 public health professionals from around Scotland. It aims to encourage and support the use of HIA by bringing together people with similar interests to share experiences and disseminate learning. As part of this, there are plans to link together people working on HIAs of similar topics to create informal learning sets. The network also has current task groups working on:

Developing and piloting screening tools for HIA. This will draw on work commissioned by the Scottish Executive to develop an integrated impact assessment tool for use on government policies. It is planned to adapt this to create similar tool(s) for use by local authorities.

Drafting a guide to HIA of housing proposals. This brings together a systematic review of the effects of housing interventions on health [18] with other evidence and outlines how this evidence can be used in HIAs of housing proposals [19,20].

A survey of and report on the use of HIA in Scotland to date. This is summarized in the following section.

The whole network currently meets twice a year and shares information by email between these meetings. It is chaired by a Consultant in Public Health Medicine (CPHM) from Lothian NHS Board and supported by a PHIS Public Health Project Manager (the authors). Neither of these has dedicated time assigned to HIA or to the network. As discussed above, Scotland has a diverse and growing public health workforce but unlike the other UK countries we do not have anyone who is employed solely to carry out or support HIA.

Current position across Scotland: 'still on the runway'

There is an overwhelming impression that HIA has not 'taken off' in Scotland as it has in other parts of the United Kingdom. The SNAP pilots were widely discussed and well received after their publication in 2000. There has been considerable interest in HIA and the network from a wide range of people, but most members of the network are not actively involved in HIA. In June 2002 the Scottish HIA Network decided to carry out a survey of HIA activity in Scotland to date and to identify reasons why so few assessments were being done.

The group wrote to all NHS Boards, local authorities, and Local Health Care Co-operatives as well as relevant networks and national organizations with a very simple questionnaire asking them if they had done any HIAs or if they were actively planning to do any. There were 87 responses, 53 from NHS organizations, 20 from local authorities, 7 from Social Inclusion Partnerships, and 7 others. Findings confirmed the impression of very little HIA activity. Only three completed HIAs were identified, including the two SNAP pilots mentioned above. Two additional responses reported on HIA screening exercises. Three respondents believed they had undertaken an HIA, which on further investigation turned out to be health needs assessment or evaluation rather than HIA. In addition, of the five responses that stated that their organizations were actively considering HIA, three were found to refer to the same piece of work, which upon further discussion was found to be a long-term piece of evaluative research into the impact of a regeneration programme. These findings confirm that there is still confusion about HIA.

The group followed up this brief questionnaire by conducting telephone interviews with a sample of respondents. The sample was chosen to reflect the range of organizations, contexts, and staff groups who had responded to the initial questionnaire. A semi-structured interview schedule was used to highlight barriers and opportunities for HIA.

From the interviews conducted, a range of barriers was identified. First, knowledge of HIA is generally poor and there are very different impressions of what it is, what it involves, and how it should be done. Most respondents from organizations that had considered HIA had a broadly accurate understanding of HIA but reported poor awareness amongst their colleagues and particularly senior staff.

A wider understanding of the broad definition of health and health inequalities is needed to underpin development of HIA, especially amongst more senior members of staff and elected members in local authorities, and senior medical staff in NHS organizations.

Some areas reported that there was commitment from senior levels but that the work needs to be done by officer-level staff who already feel overburdened. In our survey, heavy workloads and numerous new initiatives were frequently cited as reasons why HIA was not done, despite a generally favourable political climate. One local authority respondent stated that there was a 'sense that it is yet another thing we have to do'. Another said 'There are not lots of staff mulling around to do it'. Two respondents stated that 'unless it is mandated there will always be other priorities'.

Health impact assessment is often regarded as a complex technique, requiring highly skilled staff. There may be dependence on one or two key staff to carry out HIA. One senior NHS professional reported that HIA ' . . . was under active consideration then the member of staff with expertise obtained a substantive post. No free time available to undertake a substantial piece of work like this'.

Although many respondents wanted practical tools that were not time consuming, a few respondents expressed a preference for comprehensive, large-scale HIAs. They questioned whether 'quick and dirty' approaches were robust enough for decision making.

Though most respondents recognized the importance of linking HIA with the planning process, many were unclear as to the best way of doing this. Some reported concerns that negative results from an HIA might prevent a project from proceeding or cause considerable delay. There was also a fear that the HIA may raise expectations that cannot be realized due to resource implications or competing political agendas.

Those who had been involved in HIAs, or had plans to incorporate HIA into their planning process, identified facilitating factors or support that they felt was helpful in promoting the approach in their organization. These included:

Willing partners. People from a range of agencies who already have a well-developed working relationship and are able to work together on the HIA. However, it is important that there is an identified leader for the process.

A local champion. People who maintain the profile of HIA at the local level by providing support, training, and help to identify resources. One of the authors has established a local HIA steering group, which has developed an HIA screening tool and is training a wide range of people to use the tool.

Incorporating HIA into other impact assessment frameworks. Two local authorities reported that they are moving towards the use of HIA as part of the broader policy development process, for example, by incorporating health impact questions into their Sustainability and Social Inclusion Impact Assessment checklist and guidance.

Tools. Checklists and integrated tools are needed that are not time consuming, can be easily incorporated into the policy-making process and can be adapted according to local circumstances. It is important that these are practical to use but also meaningful: not a 'tick-box exercise'. Some respondents to our survey had tried to identify tools but had not found existing tools that they considered useful. One suggestion was for a generic checklist supported by more detailed guidance that covered issues and good practice specific to different services (e.g., housing).

Training. Overall there appeared to be considerable interest but a relatively poor understanding of HIA. This demonstrates the need for training and awareness raising. In addition, local authority respondents wanted training and guidance on how to support others in their organization to use HIA as part of the policy-making process. Where work on HIA was progressing, respondents cited training as one of the reasons for their understanding of its importance and their efforts to initiate its use within their organization.

Good outcomes were reported for most of the HIAs and screening exercises that respondents had been involved in. Though most found it difficult to specify these exactly, direct and indirect outcomes that were reported included revisions and additions to the proposal or strategy being considered, improved understanding of the relevant issues by partners, and improved working relationships between partners and with the local community.

Where do we want to go?

Members of the Scottish HIA Network believe that to make the maximum difference to the health of Scottish people, HIA should be a routine part of planning and policy making in all sectors [21]. This belief is supported by experience in other settings [22]. Community planning and other partnerships provide a way to identify people with appropriate knowledge and skills, carry out HIA, and then engage with decision makers to implement recommendations. The survey showed that time constraints were an important barrier to using HIA. We want to enable HIA to be conducted routinely without creating a large and unnecessary burden of work. We therefore want to support two complementary approaches to HIA:

1. 'Health impact screening' of proposals, led by the planners or policy makers developing them. The aims of screening are to:

   ◆ encourage planners to think more broadly about the effects of proposals, including unintended effects;

   ◆ raise awareness of potential impacts on health;

   ◆ identify changes to maximize health improvement;

   ◆ identify those proposals that need more detailed assessment to inform recommendations to maximize health improvement.

2. More detailed HIA of selected proposals. The SNAP case studies suggested that HIA does not require *new* methods. Rather, it is a way to use available research evidence and a variety of existing methods to inform HPP. The skills required are similar to those used in other public health approaches: to interpret and collate different forms of evidence and make recommendations based on them.

Achieving 'take off'

The results of the survey suggest that in Scotland work is needed to raise awareness of HIA, and how it differs from other forms of impact assessment, research, or evaluation of initiatives intended to improve the health of a defined population. There is also a need to continue to promote an awareness, especially at senior levels, of the broad concept of health, the public health role of statutory organizations, and the part HIA can play in planning and policy making. Since our survey, several local authorities and NHS Boards have invited the network to give presentations, to hold workshops about HIA, and to help them use HIA in their own planning processes. This interest is very welcome, but there is a need for more people with the appropriate expertise and time to support this work.

A key challenge is to develop screening tools that can quickly and easily be incorporated into existing planning processes, and are used in a meaningful way. The network has developed screening tools that identify impacts on health and health inequalities by considering impacts on different population groups and on a range of health determinants. These tools are now being piloted in several local authorities and NHS Boards, and will be evaluated. The evaluations aim to inform development of the tools; establish whether they need to be adapted for different sectors; identify how and when they should be used in planning processes; and determine what training and support are needed for planners and policy makers to use the tools. Eventually the network hopes to build on the Scottish Executive's integrated impact assessment tool to develop a version that is suitable for use in local authorities and other organizations.

Finally, we need to develop the capacity of the public health workforce to do more detailed HIAs when needed. The network is exploring ways to support members and develop the skills and confidence of the Scottish public health workforce to carry out HIA. Ultimately, we hope HIAs will be seen as part of the 'core business' of public health.

References

1 Randall J. Scotland's population 2001—the Registrar General's Annual Report of demographic trends. Edinburgh: Registrar General for Scotland, 2002.

2 Hanlon P, Walsh D, Buchanan D, et al. Chasing the Scottish Effect. Glasgow: Public Health Institute of Scotland, 2001.

3 Blamey A, Hanlon P, Judge K, et al. Health Inequalities in the New Scotland. Glasgow: Public Health Institute of Scotland, 2002.

4 Devolution Factsheet Scottish Executive. 2003.

5 Scottish Executive. Social Justice—A Scotland Where Everyone Matters. Edinburgh: The Stationary Office, 1999.

6 Secretary of State for Scotland. *Towards a Healthier Scotland: A White Paper on Health.* Edinburgh: The Stationary Office, 1999. Cm 4269.

7 Scottish Executive. *Review of the Public Health Function in Scotland.* Edinburgh: The Stationary Office, 1999.

8 Scottish Executive. *Nursing for Health: A Review of the Contribution of Nurses, Midwives and Health Visitors to Improving Public Health in Scotland.* Edinburgh: The Stationary Office, 2001.

9 Scottish Executive. *Improving Health in Scotland: The Challenge.* Edinburgh: The Stationary Office, 2003.

10 Scottish Parliament. *Local Government in Scotland Bill 2003.* Edinburgh: The Stationery Office 2003.

11 **Roger S, Smith M, Sullivan H, *et al.* Community Planning in Scotland: An Evaluation of** *the Pathfinder Projects Commissioned by CoSLA.* Edinburgh: Convention of Scottish Local Authorities and Scottish Executive, 2000.

12 Secretary of State for Scotland. Edinburgh, (ed). *Working Together for a Healthier Scotland.* The Stationary Office. 1998.

13 **Gorman D, Douglas MJ, Conway L, Noble P, and Hanlon P.** Transport policy and health inequalities: a health impact assessment of Edinburgh's transport policy. *Public Health* 2003; **117**:15–24.

14 Scottish Needs Assessment Programme. *Health Impact Assessment of the City of Edinburgh Council's Urban Transport Strategy.* Glasgow: Scottish Needs Assessment Programme, 2000.

15 Scottish Needs Assessment Programme. *Health Impact Assessment of the North Edinburgh Area Renewal (NEAR) Housing Strategy.* Glasgow: Scottish Needs Assessment Programme, 2000.

16 Scottish Needs Assessment Programme. *Health Impact Assessment: Piloting the Process.* Scottish Needs Assessment Programme, 2000.

17 **Douglas MJ, Conway L, Gorman D, Gavin S, and Hanlon P.** Developing principles for health impact assessment. *Journal of Public Health Medicine* 2001; **23**(2):148–154.

18 **Thomson H, Petticrew M, and Morrison D.** Health effects of housing improvement: systematic review of intervention studies. *British Medical Journal* 2001; **323**:187–190.

19 **Thomson H, Petticrew M, and Douglas M.** Health impact assessment of housing improvements: incorporating research evidence. *Journal of Epidemiology and Community Health* 2003; **57**:11–16.

20 **Douglas M, Thomson H, and Gaughan M.** *Health Impact of Housing Proposals: A guide.* Glasgow: Public Health Institute of Scotland, 2003.

21 **Douglas MJ, Conway L, Gorman D, Gavin S, and Hanlon P.** Achieving better health through health impact assessment. *Health Bulletin* 2001; **59**(5):300–305.

22 **Banken R.** Strategies for institutionalising HIA. ECHP Health Impact Assessment Discussion papers; 1. Brussels: WHO European Centre for Health Policy, 2001.

Chapter 18

The experience of HIA in Wales

Ceri Breeze

Introduction

The development of HIA in Wales has taken place within a context of significant constitutional change and a new approach to policy making. This has provided opportunities to explore the usefulness of HIA and to identify the practical issues around its application.

This chapter describes the Welsh Assembly Government's work to develop the use of HIA. It highlights issues that are relevant to its use in policy making by government and by other organizations both nationally and locally.

Policy context

The National Assembly for Wales was established on 1 July, 1999 following a public referendum as part of the UK government's pledge to modernize British politics and to decentralize power [1]. Prior to this, the Secretary of State for Wales, a member of the Cabinet of the UK government, held responsibility for policies and public services for Wales. Services were delivered directly by the Welsh Office and indirectly through local authorities, health service providers, non-Departmental Public Bodies, and agencies. The National Assembly has given Wales more control over its own affairs and enabled it to adjust its policies to meet its specific needs.

The Assembly has 60 elected Members. The First Minister heads a Cabinet of eight ministers. Seven subject committees cover the policy areas listed in Table 18.1 and are supplemented by four regional committees and several standing committees.

The creation of the Assembly required the transition of the Welsh Office—a territorial government department—into the effective permanent secretariat of a new elected body [2]. It has altered fundamentally the way business is transacted with the committees as vehicles for policy development and formal partnership arrangements replacing *ad hoc* arrangements with external bodies [3]. A broader approach has emerged to address issues such as

Table 18.1 Broad policy responsibilities of the National Assembly for Wales

Agriculture	Social services
Industrial and economic development	Arts and cultural heritage
Education and training	Sport
Health and health services	Tourism
Housing	Transport, planning, and environment
Local government	

social inclusion, equal opportunities, sustainable development, and health that cut across functional policy areas.

Inherent in the Assembly's first strategic plan [4] was the commitment for the Government to tackle the relatively poor levels of health in Wales and the inequalities in health that exist between some communities [5]. Better health and well-being was one of the Assembly Government's five strategic priorities along with a better, stronger economy, better opportunities for learning, and better quality of life, which are health determinants in their own right. A fifth priority—better, simpler government—included a commitment to better policy-making. This provided a strong hook for the development of HIA and was highlighted in the Cabinet's vision for the longer term [6].

Developing HIA in Wales

The foundations for HIA were set by the green paper 'Better Health Better Wales' [7] and the 'Better Health Better Wales Strategic Framework' [8]. Both documents reflected recognition at the political level of the social, economic, and environmental determinants of health and the need to address them as a fundamental part of health strategy [9] and HIA was seen as something that could help. The Assembly Government made a public commitment to develop its use and set about encouraging other organizations in Wales to do the same.

The Assembly Government set out its plans to develop the use of HIA in a guidance document [10]. The purpose of the document, which was endorsed by all members of the Assembly Government's Cabinet, was twofold. First, it sought to raise awareness of the HIA concept. Second, it provided a base for a development programme to explore and test its usefulness.

The guidance document did much to raise awareness of HIA in Wales and provided a solid foundation for the development programme. The programme included a series of awareness raising events, pilot projects, and other activity. It had two broad strands. First, action to develop the use of HIA within the Assembly Government and second, action designed to encourage other organizations in Wales to use it. A pragmatic approach was adopted and some of the projects undertaken are described here. Learning from the experience of applying HIA has been one of the programme's main aims.

The Objective 1 Programme

The use of HIA for the Objective 1 Programme in Wales illustrates well a number of issues. The aim of the Objective 1 Programme, which is part of the European Community's Structural Funds, is to stimulate and support economic regeneration. As a major strategic programme, it was complex with relatively long timescales for both development and implementation.

The Programme could have been an important opportunity to test the use of HIA but its development commenced well before the Assembly Government's plans for HIA were in place. A prospective assessment was therefore not possible. Although an assessment did not take place during its development, action on health still became part of the Programme, which illustrates the fact that HIA is by no means the only way of taking health into account. That said, experience suggests that the use of HIA would have made the negotiations on the Programme's health dimension easier.

A health group was one of several working groups established by stakeholders to develop the programme. Unfortunately, its suggestions were not included in the initial draft of the Programme Document. The precise reasons for this are not clear but it appears that at that time health was not considered sufficiently relevant by other sectors. Scope for action to improve people's health as part of community and economic regeneration was built into the programme at a relatively late stage. The absence of health in the early draft of the programme triggered discussion between officials internally within the Assembly Government and externally in the health sector, which lobbied for the inclusion of health. But the most influential factor was the political will of ministers to achieve a more integrated approach for the Programme.

The decision to apply HIA after the Programme had been finalized was made for two reasons. First, it was clear that the majority of stakeholders in other sectors still did not understand or perhaps accept the Programme's relevance to health. Second, programme implementation could be influenced so that action to improve people's health could be an integrated part of wider social and economic regeneration activity. Health needed to be placed on the agenda and in the minds of any organization or group that could develop projects.

The approach chosen for the assessment reflects realities that policy makers face regularly. There was a very short timescale or 'window of opportunity' if deadlines were to be met. Only in-house resources could be used as the timescale ruled out the option of commissioning the work from an external agency.

The work had to be balanced against other priorities. However, undertaking the assessment in this way had some advantages as the assessors had detailed knowledge of the Programme obtained through negotiating its health components. This significantly reduced the time needed for the assessment.

The assessment [11] used available evidence to highlight the Programme's potential impacts on health. It mapped the Programme's priorities and measures, and their relationship to people's health and well-being. While the preferred option would have been to involve stakeholders directly, this was not possible as many stakeholders were under pressure themselves to finalize the Programme document. Instead, stakeholders were involved at the draft report stage and subsequently in discussion following the publication of the assessment. Feedback suggests that nothing was lost through this approach.

The assessment report was instrumental in stimulating discussion and consideration of health by non-health sector organizations and groups. It was followed by publication of a much shorter guide for potential bidders. This included a simple assessment tool to help organizations and groups consider the relevance of health to their ideas [12]. Some project developments featured health as an integrated part of wider economic regeneration activity although the picture varies across Wales.

National Skills and Employment Action Plan

The National Skills and Employment Action Plan was circulated around the Assembly Government's policy divisions in early draft form. Health officials were included in the loop reflecting progress towards a more joined-up approach on policies and programmes. Encouragingly, officials considered that health might be relevant to the Plan and circulation of the draft was accompanied by a request for advice.

For a number of reasons—including time and resources—a rapid HIA was undertaken and a short report produced. The report used available evidence to explain in broad terms the links between the Plan and people's health. Action on some health issues, such as drugs and alcohol, already featured in the plan and the assessment helped to identify how health could be further built into the proposals. The main benefit of the assessment has been stronger links between officials from both policy areas and greater co-operation, as evidenced by joint action to address issues of health and economic inactivity.

Other pilot projects

Recognizing the potential of HIA, one of Wales' local authorities wished to explore the use of HIA as part of its development of a housing regeneration strategy for one of its smaller communities [13]. The involvement of stakeholders—including the public—is a desirable feature of HIA and this project was undertaken as a way of exploring how it could be achieved. The Assembly Government commissioned support for the assessment process as part of its development programme. This work is described in Chapter 8.

As part of the development programme, several other pilot projects have been undertaken to test the HIA process. An assessment of the New Home Energy Efficiency Scheme in Wales prior to its implementation [14] drew out the relevance of the programme to people's health and raised awareness among stakeholders. Some unexpected issues were uncovered during the process including, for example, the programme's relevance to crime prevention and issues related to unsafe domestic appliances.

An HIA of the National Botanic Garden of Wales took place long after major decisions had been taken and was undertaken as a systematic examination of the relevance of health to its future development. Over and above practical issues around its application, the project highlighted the importance of small-area statistics to HIA [15].

Resources for HIA

Lack of time and resources is frequently cited as a barrier to HIA and unless spare capacity exists, it has to compete with other priorities. Perhaps this is no bad thing as it forces consideration of the reasons for the assessment, its scope, the potential benefits, and justification for using the resources required. Difficulties may be caused through not understanding the component parts of the HIA process, or not knowing how much time may be required. Information on the approaches used for, and resources required by, other HIA is therefore helpful.

Some of the examples of assessments undertaken have used the time and effort of existing staff while others have required dedicated funding to enable work to be commissioned. Table 18.2 indicates the resources used in some of the assessments undertaken to date. The estimates of person days include the preparation of reports to final draft stage. Because the projects were seen as part of a learning curve, every project included a reflective element, which among other things reviewed the resources used. This is a discipline that should be continued as part of the continuous improvement of the HIA approach.

Developments

Within the Assembly Government, action has focused on building health and HIA into the mechanisms and strategies of other policy areas. At the corporate

Table 18.2 Resources required for selected HIAs

HIA	Person days	Comments
Objective 1 Programme	12.5	Split between two people, but assessment aided by existing knowledge of complex programme
National Skills and Employment Action Plan	2	Rapid assessment
Housing regeneration strategy	67	Included 49 days of commissioned researcher time for literature review, and for interviews with local people and stakeholder organizations
National Botanical Garden of Wales	20	Included approximately 4 days total time for steering/stakeholder group
New Home Energy Efficiency Scheme	20	Split between 5 people

level, work on HIA has informed the development of an integrated high-level screening tool by the Assembly's Policy Unit. When finalized, the tool will facilitate the initial screening of new policies against all the Assembly's priorities. The tool is being trialled as this book goes to print.

As there is scope for co-operation between policy areas, so too is there scope for co-operation on other issues such as research programmes. The proposals for the Assembly Government's Economic Research Programme were screened within the very short timescale available and components relevant to health were identified and highlighted. A short briefing paper was produced to enable the Minister for Health and Social Services to contribute to consultation between the Assembly's Health and Social Services Committee and its Economic Development Committee. The briefing paper also provided the base for subsequent discussion between officials on common ground and opportunities for joint working.

Modern policy-making requires a holistic view, which looks beyond institutional, policy, and professional boundaries [16] and these factors constitute barriers to, or opportunities for, that development. Many organizational factors relevant to the development of HIA have been identified by work inside the Assembly Government. Such factors—which include process factors, culture and working practices, and the knowledge and perspectives of individuals—will also be found in other organizations [17]. The key to joined-up government is most certainly about learning about shared purpose, teamwork, partnerships, and building relationships and is built around the knowledge and know-how of people, which differs from the organization model of the past, which was built around tasks, units, and titles [18].

Structural change within the health system in Wales has also provided opportunities to encourage the use of HIA. In April 2003, the Welsh Assembly Government implemented a new structure for the National Health Service (NHS) in Wales. A new National Public Health Service was established to reinforce efforts to protect and to improve people's health. In addition, the arrangements included a joint statutory duty on Local Health Boards and local authorities to produce local Health, Social Care, and Well Being Strategies in conjunction with others and in consultation with local people. Written into the guidance for the development of the strategies is an expectation that HIA will be used to help develop a more integrated approach to policies and programmes.

Over and above projects designed to test out HIA, the development programme has delivered presentations and workshops to increase awareness and understanding of HIA. Such action, which has covered officers and elected members of local authorities, health service staff, and voluntary sector organizations, is ongoing. Action is supported by the enthusiastic efforts of a number of professionals and practitioners across Wales. This helps to develop a critical mass of expertise and facilitates the multidisciplinary inputs required for successful assessments.

One of the most significant developments to support the development of HIA in Wales has been the establishment of the Welsh Health Impact Assessment Support Unit. Funded directly by the Assembly Government, its role is to develop and support the use of HIA by organizations in Wales. A core component of its programme is work with local authorities and community organizations to increase their awareness and understanding of the approach and to help them test its usefulness as an integral part of their work. Although limited in size, the Unit provides dedicated resources for advice, guidance, and assistance. Use of policy levers to encourage the use of HIA, such as the guidance for local Health, Social Care and Well Being Strategies, have increased the demand for its services and its capacity is being considered.

Perspectives on HIA

The HIA development programme has uncovered considerable support for the concept and genuine enthusiasm to try it out. It has also identified potential barriers to its use with some seeing it as complex, technical, the domain of experts, rigid, always requiring detailed assessments, taking a long time (months if not years), and expensive. Some had unrealistic expectations thinking HIA always produced quantified data on impacts and highly accurate forecasts. The term 'HIA' itself has posed some difficulties with 'health' sometimes read as 'hospitals' and 'treatment' or interpreted very narrowly in terms of specific environmental risk factors to health. The effect of these misconceptions has sometimes been an early, and instant, dismissal of health as something that is not relevant or worth considering.

Sometimes, not only the actual 'impacts' of policies, programmes, or other developments on people's health, but also the wider 'implications' for existing services and facilities are important. One of the strengths of HIA is being a systematic process. The first stage of the HIA process, usually known as 'screening', is perhaps the most important. This stage is often the trigger for subsequent stages and it needs to be embedded in organizations' policy making and programme development processes. It requires relatively little time and just ensures that someone asks of every policy the simple question 'Is this something that is, or could be, relevant to people's health?' If the answer is 'No, definitely not', then the HIA process goes no further. If the answer is 'Yes', 'Possibly', or 'Don't know', then further assessment is needed. This may involve a couple of hours work talking to people who can advise on how health might be relevant, a rapid assessment approach typically requiring a few days, or perhaps an in-depth assessment requiring much longer.

There is a need to continue increasing awareness and to generating a common understanding of HIA. This means demystifying the concept and overcoming the jargon and terminology that tend to characterize different policy areas and can present potential barriers to effective intersectoral working.

Benefits

The development of HIA has provided a rich source of experience and learning. Other benefits include highlighting the relationships between health and other policy areas, ensuring that the potential health consequences of decisions—positive or negative—are not overlooked, and identifying new opportunities to protect and improve health and to inform decisions on appropriate action. Its use in Wales has helped to increase awareness and understanding and has stimulated discussion between organizations and individuals working in different sectors.

The increased awareness of health across policy areas and sectors is welcome given that effective health strategy needs to include action to address the social, economic, and environmental determinants of health. However, perhaps one of the main benefits of action has been the opportunity to apply HIA. The value of this should not be underestimated in terms of the learning it delivers on the use of such a tool within the realities, and often the constraints, of policy and programme development and more general decision-making processes.

One of the strengths of HIA is its philosophy; that is, as a general way of thinking about policy making that facilitates consideration of the impacts or implications for other policy areas. It has been described as a classic means of helping policy makers to 'think outside the box'. It has also been described as 'common sense', and there is much to support that view. At the same time however, a process that may be 'common sense' to one person may be 'innovation' to another. HIA can be thought of as a systematic means of applying 'common sense'. In Wales, it has helped to change the way that 'health' is viewed by other sectors and policy areas and to develop a health strategy that is much wider than health services alone.

Informing future developments

Experience in Wales to date has taught many lessons, but three main themes are worth emphasizing.

First, the key to progress is to apply it. Developing a sound understanding of the approach and learning from others are essential, but there is no substitute to trying it out yourself and learning from the experience. The temptation to focus on achieving the perfect approach should be avoided. Health impact assessment is a flexible approach and should be thought of as a framework of components and principles. However, there is a need to avoid overselling the concept thereby creating unrealistic expectations of HIA.

Second, the need to recognize that HIA is not the only way in which health can be taken into account in decision-making processes [19]. Experience points to the need for HIA to complement and add value to, other policy-making mechanisms and processes. The development of HIA in policy-making cannot be seen in isolation and should not be viewed as something 'special' or separate. In developing HIA one has to recognize that several other mechanisms exist in government to foster joined-up working—Ministerial committees, interdepartmental working groups, and task forces for example. Similar mechanisms exist in other organizations at national, regional, and local levels. Some of these mechanisms are not necessarily systematic but nevertheless are examples of joint work between departments and ways in which health can be taken into account across policy areas. HIA needs to dovetail with these mechanisms. Indeed, the establishment of, for example, a joint ministerial committee or task group should be seen as an opportunity to test out HIA as a tool and systematic approach that can assist its work.

Third, the development and use of HIA needs to recognize the realities of policy making. New policies emerge for a variety of reasons, are developed in different ways, and to different timescales. Organizations vary in structure and size and while there are common elements, planning and decision-making processes also vary. These may constitute both barriers to, and opportunities for, the use HIA. Either way, its deployment needs to take account of such factors for if it cannot deliver within such circumstances, then it won't be used. Embedding the philosophy of HIA in government policy-making, or 'institutionalizing' HIA as it has also been called [20], is both the main aim and the main challenge.

Experience has shown that the use of HIA by government at the national level is important but that alone is not sufficient. Efforts to utilize HIA at the national level need to be reinforced by efforts to promote the take-up of HIA at a local level. Some policies are implemented directly by government while others—local government for example—have a major role to play given the breadth and depth of their responsibilities and services.

There are benefits to be gained from developing the use of HIA and, providing its principles can be followed, it may be used on a stand-alone basis or as part of other forms of impact assessment. Experience in Wales suggests that HIA can help to achieve a more integrated approach to policies and programmes, but it is a means to an end not an end in itself. The real goal is to improve policy making and policy implementation with the ultimate goal in mind; that is, to help people to improve their health and to reduce inequalities in health.

Reasonable progress has been made in Wales but the journey is by no means over. There is still much to do to exploit its full potential.

References

1 Welsh Office. *A Voice for Wales: The Government's Proposals for a Welsh Assembly* (Cm3718) London: Stationery Office, 1997.

2 National Assembly for Wales. *Preparing for the Assembly: A Management Strategy.* Cardiff: National Assembly for Wales, 1999.

3 National Assembly for Wales. *Policy Background: Notes for Assembly Members.* Cardiff: National Assembly for Wales, 1999.

4 National Assembly for Wales. *Strategic Plan of the National Assembly for Wales.* Cardiff: National Assembly for Wales, 2000.

5 National Assembly for Wales. *Welsh Health: Annual Report of the Chief Medical Officer.* Cardiff: National Assembly for Wales, 1999.

6 The Government of the National Assembly for Wales. *Plan for Wales 2001.* Cardiff: National Assembly for Wales, 2001.

7 Welsh Office. *Better Health Better Wales: A Consultation Paper.* Cardiff: Welsh Office, 1998.

8 Welsh Office. *Better Health Better Wales Strategic Framework.* Cardiff: Welsh Office, 1998.

9 The Official Record, Plenary Business, 30th November 2000 (www.wales.gov.uk/assemblydata) Cardiff: National Assembly for Wales, 2000.

10 National Assembly for Wales. *Developing Health Impact Assessment in Wales.* Cardiff: National Assembly for Wales, 1999.

11 **Breeze CH and Kemm JR.** *The Health Potential of the Objective 1 Programme for West Wales and the Valleys: A Preliminary Health Impact Assessment.* Cardiff: National Assembly for Wales, 2001.

12 National Assembly for Wales. *Building a Healthier Future by Taking Health into Account as Part of Objective 1 Projects.* Cardiff: National Assembly for Wales, 2001.

13 **Elliott E and Williams G.** *Housing, Health and Well Being in Llangeinor, Garw Valley: A Health Impact Assessment.* Cardiff: Welsh Assembly Government, 2002.

14 **Kemm J, Ballard S, and Harmer M.** *Health Impact Assessment of the New Home Energy Efficiency Scheme.* Cardiff: National Assembly for Wales, 2001.

15 **Kemm J.** *National Botanic Garden of Wales: Health Impact Assessment Report.* Carmarthen: National Botanic Garden of Wales, 2001.

16 Cabinet Office. *Professional Policy Making for the Twenty First Century.* London: Cabinet Office, 1999.

17 **Breeze CH and Hall R.** *Health Impact Assessment in Government Policymaking: Developments in Wales.* Policy Learning Curve Series Number 6. Brussels: World Health Organization European Centre for Health Policy, 2002.

18 Cabinet Office. *Peer review of Cabinet Office* (www.cabinet-office.gov.uk/moderngovpeer_review.htm). London: Cabinet Office, 2000.

19 Welsh Assembly Government. *Health Impact Assessment and Government Policymaking in European Countries: A Position Report.* Cardiff: Welsh Assembly Government, 2003.

20 **Banken R.** *Strategies for institutionalizing health impact assessment. European Centre for Health Policy Health Impact Assessment Discussion Paper Number 1.* Copenhagen: World Health Organization, 2001.

Chapter 19

HIA at the local level in Sweden

Karin Berensson

Background

The importance of assessing the impact of political measures on health has been stipulated internationally by the European Union (EU) in Article 152 of the Treaty of Amsterdam [1] and in its public health programme (2003–2008) [2], and by the WHO in 'Health 21'. In Sweden too, the field of public health has been broadened so that, in a more deliberate way, it covers the impact of political decisions on public health and seeks to integrate public health aspects into political and other activities at both the national [3] and local levels.

In Sweden, the county councils/regions and local authorities have responsibilities and powers that cover a number of areas that are important to public health. The 290 local authorities are responsible for tasks such as schools, social care, care of the elderly and the environment. The county councils/regions are responsible for health care and public health, some public transport, and regional development. The local authorities, like the county councils/regions, have directly elected politicians and have independent taxation rights. In the mid-1990s, leading politicians in the local government sector initiated the development of health impact assessment (HIA) in relation to political decisions at the local and regional levels. The development work was carried out within the framework of a political programme that aimed to increase the focus on public health and 'to contribute to favourable and equitable health development'.

It was felt that HIA would make it possible to systemize knowledge about health and ill-health so that the impact of political decisions on health could be clarified. This would in turn make it possible to weigh health impacts against other interests before making political decisions. This does not mean that health impacts would always be allowed to tip the balance, but that HIA would guide the way to more 'healthy decisions'.

Initially, and as part of the political programme, the literature on HIA was reviewed in the spring of 1996, and telephone interviews and a seminar [4] were conducted to compile experience and proposals regarding the development programme. It emerged that the experience available was difficult to apply to the Swedish programme, but that there was a great deal of interest in HIA. A development project was started that aimed to encourage work on HIA and to develop a tool for HIA at the local level. The hopes or objectives for HIA were that it would place health issues on the political agenda, contribute to a reduction in health inequalities, and promote the revitalization of policy making at the local level.

Development of the HIA tool

Two processes were central in the development of the HIA tool. One was to identify the determinants of health in Sweden in the late 1990s, the other to find out the characteristics the tool required if it was to be accepted by players in the political process (Fig. 19.1). The process used to identify health determinants came from a project on indicators for public health [5]. Briefly, interviews with a range of population groups, for example, politicians, the general public (including immigrants, unemployed people, women, and young people throughout Sweden), experts and scientists were conducted to explore the question 'what is health and what are the factors that determine health?' The results from these groups were processed to arrive at a list of health determinants (the determinants that the population groups identified were very similar to those identified by the politicians and the experts). These and other factors (see later), for example, the impact of proposals on various different population groups, were then operationalized to form the HIA tool.

During the development phase study visits were made to the Netherlands, the European Commission in Luxembourg, and Canada [6], and the politicians, who were highly involved in the process of development, discussed HIA and the relationship between health and policies with leading scientists and researchers in the field [7]. The politicians also presented specific demands

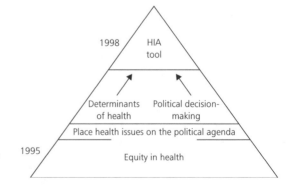

Fig. 19.1 Development of the Swedish HIA tool.

regarding the nature of the tool almost from the start. The tool was to reflect the overall objective of the public health programme—'favourable and equitable health development'—and it needed to cover social impacts as well as impacts on the environment and equality. Specifically, it was felt that important elements to be included in the tool were (a) the factors determining health, (b) the impact of proposals on various population groups, (c) what a decision would mean in the short and long term, and (d) any alternatives to the proposal concerned. The politicians also emphasized that the tool must be easy to use and that it should consider whether proposals are in accord with the overall objectives of the municipality or county council.

With regard to the decision-making process, the politicians believed that the earlier the health impacts of a policy proposal were discussed, the greater the opportunities would be to affect the final decision. That is, if consideration of health aspects is introduced at a late stage, so much has often already been 'invested' in the proposal that it is hard to effect a change in direction.

The HIA tool was published [8] in the spring of 1998 and is available on the Internet. The tool that was developed poses a basic question: 'how is the health of different groups affected by the proposed policy decision in question?' but consists of three parts depending on the 'ambition' of different policy users and areas:

- the health question, a simple itemized list of health impacts (Fig. 19.2),
- the health matrix, a matrix showing how a decision may impact on various conditions, groups, and so on (Fig. 19.3),
- the health analysis, a foundation for HIA prior to decision making (Fig. 19.4).

Encouraging the use of HIA

The HIA development project covered both the development of a tool and efforts to encourage the use of HIA. These two tasks were integrated, although the development focus switched from one to the other as the project progressed. The following were initiated to encourage the use of HIA.

Network HIA. A network designed to stimulate the use of HIA was formed for public officials and experts in the municipalities and county councils. This network also contributed to the development of the tool, and has arranged annual, national conferences that are open to anyone with an interest in HIA.

Printed material/home page. One important ambition has been to make HIA as accessible and simple as possible. The main brochure consists of only 27 pages and includes the tool. Background material has been compiled in a file, 'The HIA File', that can be ordered. A dedicated Internet home page has been created

This simple option can be adopted prior to consideration of an individual policy proposal. It can also be used before collective decisions are made at meetings of local boards/committees.

A. Will the proposal promote health development for various groups/the population in relation to the social environment (e.g. opportunity to exert influence, mutual work and support)?

Yes ☐   No ☐

B. Will the proposal promote health development for various groups/the population with regard to certain risk factors (e.g. the physical environment or living habits)?

Yes ☐   No ☐

C. Is the proposal consistent with overall municipality/county health targets and objectives?

Yes ☐   No ☐

Comments/justification:

_____

_____

Alternative proposal:

_____

_____

Our assessment is that:

_____

_____

Fig. 19.2 The health question

containing the HIA tool, information on the network, examples of practical application, and international links. The tool and its component parts can be downloaded and adapted in line with the needs of the user.

Kick-off conference.   A national conference was attended by nearly 400 leading politicians, officials, experts, and scientists. The tool was presented and the HIA brochure was distributed.

International co-operation.   The development of HIA has been conducted in parallel in many countries, and international co-operation has been important. A seminar in Brussels was attended by Swedish and international experts [9], and material has been submitted to the journal *Eurohealth* [10]. There has

	Prioritized group		Entire population	
	Long term	**Short term**	**Long term**	**Short term**
Democracy/opportunity to exert influence/equality				
Financial security				
Employment/meaningful pursuits/education				
Social network				
Access to healthcare and welfare services				
Belief in the future/life goals and meaning				
Physical environment				
Living habits				

Is the proposal in accordance with the overall targets of the municipality/county council?

Yes ☐   No ☐

Comments/justification: _____

Alternative proposal: _____

Our assessment is that: _____

Fig. 19.3 The health matrix.

also been close co-operation with the European Centre for Health Policy, WHO [11], and other players in the field.

Support for continued implementation. An Internet-based training package entitled 'Democracy and health' has been produced for new politicians. This material highlights the link between health and the policies adopted. Development work on a Local Welfare Management System (LWMS) for the municipalities has been carried out in parallel with the HIA development project. In both projects, an attempt has been made to systematically integrate health and welfare aspects into the political/administrative planning process. Both HIA and LWMS are based on an understanding of what factors are important to the health of the population and how public health is affected by the policies pursued. Other action includes the appointment of an official to support HIA work in the county councils and municipalities, and conferences, seminars, and training courses have been held at the national and local levels.

Health Impact Analysis is guided by a number of key questions. They may, for example, be appropriate to raise prior to analyses of strategic policy decisions.

General questions

1a: What does the local Public Health Report show regarding the health conditions of different groups within the municipality/county? Are there groups which are particularly vulnerable or already exposed to numerous health risks, or are there groups with evident health-trend problems?

1b: Are there defined health-policy targets?

Questions linked to the matter at hand

2. Are there particular health risks which can be expected to decrease or increase as a result of the proposal? Will impacts become apparent in the short term (within 5 years) or in the long term.

3. For the distribution of ill-health within a population, it is of decisive importance which groups are subjected to decreased/increased health risks, and whether any decision will affect these groups' capacity either to deal with difficulties or, by contrast, increase their vulnerability.

4. In what way will the social environment in the local community be affected by the proposal?

5. Is there a risk that a proposal may have a 'double' impact on certain groups, i.e. that both their health risks increase and their social environment deteriorates?

6. Are there alternative policies which might result in better health for exposed groups and the population as a whole?

7. Summary

Fig. 19.4 The health impact analysis

Application of HIA

In 2001, the Federation of Swedish County Councils and the Swedish Association of Local Authorities carried out a survey to investigate to what extent HIA was being used in the municipalities and county councils. This survey revealed that 10 per cent of the municipalities had decided to use HIA, 22 per cent had begun to introduce it, and 4 per cent were already using it. The corresponding figures for the county councils were 60, 70, and 48 per cent. In these statistics the county councils are reported as a whole. However, this can have the effect, for example, that the reported figures can refer to merely one of six districts. The survey also showed that 12 per cent of the municipalities had decided to use a LWMS, 20 per cent had begun to introduce such a system, and 4 per cent had actually compiled 'welfare accounts'. An important element of the introduction process was the training of employees and elected representatives. In addition, integrating HIA into the everyday work of the municipalities and county councils may require adapting the tool to local needs. The needs of an urban health district may be very different to those of a municipality for the latter is responsible for a greater range of areas that relate to health determinants than the former.

The use of HIA in a health district authority

One of the health district authorities in the County of Stockholm, the South West Stockholm Health District Authority (SWS HD) began using HIA at a very early stage. One of the main tasks of the SWS HD board, which includes directly elected politicians, was to ensure the provision of 'good and proper health and medical care that contributes to the security of individuals, reduces health inequalities and provides the best possible results'.

The SWS HD began its involvement in HIA in the autumn of 1998. The politicians on the board and its officials were given training on health determinants, how policies affect public health, and how important the role of a politician is for public health. In addition the tool has been adapted to local condition. The latest version is a checklist based on the determinants of health and is structured under four headings: health promotion, disease prevention, health care, and rehabilitation. A start-up conference was held, and since September 1999 the SWS HD has conducted HIAs on all political proposals. The role of the officials has been to carry out the HIAs, while the role of the politicians has been to ask for HIAs and to use them in the decision-making process. Initially, the HIAs were documents attached to the draft proposals to the board. Later they became part of the proposals to the board based on an assessment using the checklist.

The Swedish Federation of County Councils has commissioned the Karolinska Institute to conduct an evaluation on the practical application of the HIA tool within the SWS HD [12]. Some of the conclusions were as follows.

The single most important factor for success in implementing HIA in SWS HD has been the unity that has existed between management at the political and administrative levels. This unity, combined with a clear decision, was arrived at after a thorough process of participation, with recurrent opportunities for training and opportunities to influence the process.

A positive factor is that the process has put public health issues on the agenda and vitalized internal discussions.

One of the lessons learned from this study is that the HIA tool must be developed and adjusted to the direction and needs of the local activities, that the application and practical application of HIA is a process of collaboration between politicians and civil servants, which initially requires certain training, and that subsequently a recurrent dialogue takes place in order to safeguard the quality of the tool and develop it. Furthermore, civil servants feel that the application of HIA in the preparation of cases has not taken very much time.

HIA has also developed as a support for the board to provide a better basis for decisions, to systematically analyze the health consequences of all proposals for decision, and to systematically take the health perspective into account throughout the lifecycle of the case concerned.

HIA at the county council level

In one medium-sized county council, the County Council of Gävleborg, managers and politicians have been given basic training in public health in order to prepare the ground for further training and the introduction of HIA. It has taken a relatively long time to gain acceptance for HIA in the departments concerned, but a training programme has now been agreed on. Following the provision of general information to all the officials at the County Council, those officials who work with the drafting of policy proposals will attend two days of training with an intervening period. In this intervening period, the participants will practice applying what they have learned to the cases they are working on and report their experience on the second training day.

The use of HIA combined with the UN Convention on the Rights of the Child

The City of Malmö is testing a combination of HIA with the UN Convention on the Rights of the Child. The City has merged parts of the HIA tool with questions from the Convention to form a Children's Checklist. Experience to date varies. Some officials have, with good results, tried asking various groups of children about their views on the political proposals in the pipeline. The critics, on the other hand, feel that political decisions often have to be processed quickly, which means that there is too little time to ask children for their views. They see the Children's Checklist as adding to the stress involved in dealing with cases. A follow up is in progress.

The LWMS

The LWMS for local authorities is still another way of adapting the HIA concept, and this has proved popular with many municipalities. Like HIA, this system is based on health determinants. The main difference lies in the fact that HIA is carried out at the start of a process, while the LWMS entails compiling 'welfare accounts' for all of a municipality's activities over the course of a year.

Where are we now, what have we learned?

The development of a tool for HIA has taken place in parallel with efforts to encourage the use of HIA. This has probably helped to generate interest in the concept and to ensure the practical applicability of the final product. It may also have helped to keep interest in HIA alive following the launch of the tool, as so many people have been involved in the work and see the project as 'theirs'.

Two processes, identifying the health determinants and understanding the political process, were central to the development of the tool. Members of the public, as well as experts and scientists, were involved in identifying factors that affect health and thus the knowledge and experience of the population at large was considered in conjunction with that of the academic world. With respect to the decision-making process, politicians initiated the HIA development projects, applied their knowledge and experience to the work, and were its 'owners'. This increased the chances of the tool being in line with current political and administrative structures, and the likelihood that the tool would be able to contribute relevant information [13].

The implementation of HIA in the everyday work of the county councils/regions and municipalities requires training programmes for both elected representatives and public officials. A knowledge and understanding of the factors that determine health are central in this context. Participating actively in the process of adapting the tool to local conditions gives rise to new questions and new learning.

It is likely that there will be further development of HIA for specific operational areas. The use of HIA combined with the UN Convention on the Rights of the Child indicates one possible direction for development. The LWMS is another. It is hoped, however, that the broad view of health that characterizes the HIA concept will provide a common platform for public health work in a variety of operational areas.

References

1 European union. Treaty of Amsterdam amending the Treaty on European Union, the Treaties establishing the European Communities and certain related acts. Luxemburg: Official Journal no C 340, 10.11.1997.

2 Decision no 1786/2002/EC of the European Parliament and of the Council of 23 September 2002 adopting a programme of Community action in the field of public health (2003–2008). Official Journal no L 271, 9.10.2002. pp. 1–12.

3 Health on equal terms: national goals for public health. English version of SOU 2000:91. Final report by the Swedish National Committee for Public Health. Stockholm: Fritzes Offentliga Publikationer; 2000. Governmental Official Reports 2000:91 Ministry of Health and Social affairs.

4 **Landstingsförbundet**. Hälsokonsekvensbeskrivningar: Dokumentation av ett arbetsseminarium den 17 juni 1996. [Health Impact Assessments: Documentation from

a seminar 17 June 1996]. Stockholm: Swedish Federation of County Councils, 1996. [In Swedish.]

5 **Ader M, Berensson K, Carlsson P, Granath M, and Urwitz V.** Quality indicators for health promotion programmes. *Health Promotion International* 2001; **16**:187–195.

6 **Frankish CJ,** *et al.* Health impact assessment as a tool for population health promotion and public policy. Victoria: University of British Columbia, 1996.

7 **Diderichsen F.** Public health and policy decisions. In Federation of Swedish County Councils, Swedish Association of Local Authorities, *Focusing on Health—HIA—How can the Health Impact of Policy Decisions be Assessed?* p. 10–14. Stockholm: Federation of Swedish County Councils, 1998.

8 Federation of Swedish County Councils, Swedish Association of Local Authorities. Focusing on Health—HIA—How can the health impact of policy decisions be assessed? Federation of Swedish County Councils. Stockholm, 1998. http://www.lf.se/hkb/engelskversion/enghkb.htm

9 Federation of Swedish County Councils, Swedish Association of Local Authorities. Idea-Exchange Day on Health Impact Assessments. Documentation from Brussels meeting, 26 November 1998. Stockholm: Federation of Swedish County Councils, 1998.

10 **Berensson K.** Focusing on health on the political arena. *Eurohealth* 1998; **4**:34–36.

11 WHO European Centre for Health Policy. Health Impact Assessment. Main concepts and suggested approach. Gothenburg Consensus Paper, December1999. Copenhagen: WHO Regional Office for Europe, 1999.

12 **Finer D, Tillgren P, and Haglund B.** *Hälsokonsekvensbeskrivningar i ett sjukvårdsområde. En utvärdering av praktisk tillämpning och genomförande* [Health Impact Assessments in a health district. An evaluation of practical application and implementation]. Stockholm: Karolinska Institute, Swedish Federation of County Councils; 2001. [In Swedish.]

13 **Kemm J.** Health impact assessment: a tool for healthy public policy. *Health Promotion International* 2001; **16**:79–85.

Chapter 20

HIA in Australia

John SF Wright

The Australian model of health impact assessment (HIA) is different from most European models. By definition, Australian HIA is not a 'combination of procedures, methods and tools' that applies to any 'policy, program or project'. It judges neither the potential effects of impacts on 'the health of a population', nor the 'distribution of those effects within the population' [1]. In Australia, HIA is 'the process of estimating the potential impact of a chemical, biological, physical, or social agent on a specified human population under a specified set of conditions and for a certain time frame' [2]. Both the state and Commonwealth governments conduct HIA as part of environmental impact assessment (EIA). While the administrative ease with which one assessment process can be included within the other has certainly contributed to Australia's preference for HIA/EIA, its liking for the practice can also be attributed to the nation's long-established culture of environmental protection, which dates from the Whitlam government's 1974 *Environment Protection Impact of Proposals Act* to its repeal and replacement with the Howard government's 1999 *Environment Protection and Biodiversity Conservation Act*.

Australian HIA is distinctive from European models in four key respects. In the first place, the installation of HIA within the edifice of EIA suggests that all issues of public health have an environmental component. Although Australian governments distinguish between public and environmental health, they legislate HIA from the paradigm of environmental health: a subset of public health, within which health and environment are inextricably related. Second, the practice of EIA/HIA colours the latter with the legislative characteristics of the former. For example, in European circles HIA is often touted as a decision-making device; in Australia, however, HIA shares the role of EIA, and exists as a decision-support mechanism rather than as a decision-making tool in its own right. Most significantly, the conduct of EIA/HIA in the states has forced the federal government to publish national regulations for HIA in order to standardize its practice throughout the Commonwealth. Although the story of HIA begins in Tasmania, Australia operates under a federal system of government, with the federal constitution awarding specific powers to the Commonwealth government and leaving the remaining responsibilities to the states. If state and Commonwealth law should conflict on an issue, the state legislation is invalidated. Lastly, the practice of EIA/HIA casts a quantitative, or scientific, shadow over Australian HIA. For both the state and Commonwealth governments, HIA is about hard facts and rigorous analysis. And while the primacy of the quantitative approach has circumvented the methodological disputes currently plaguing HIA in Europe, it typically precludes consideration of economic, educational, and other like determinants that the European community often associates with a wider notion of public health.

Environmental impact assessment

As state and Commonwealth governments apply HIA within the existing structure of EIA, the major pieces of legislation governing the conduct of EIA in Australia are of some importance. Until 1999, EIA was administered under the Whitlam government's 1974 *Environment Protection Impact of Proposals Act* (EPIP). Managed by the Department of the Environment, EPIP applied to Commonwealth actions with environmental significance, or with the potential to threaten the extinction of a native species. The EPIP ensured that these would receive adequate attention through federal government processes. The EPIP also applied to proposals in which the federal government had a role. For example, developments with Commonwealth funding—railways, roads, airports, use of crown land and water—or state projects with specific purpose federal grants, or projects requiring approval under the government's foreign investment review process—mining, manufacturing, real estate, and tourist development—were subject to assessment under EPIP. Private developments not in receipt of Commonwealth funds or without the requirement for Commonwealth approval escaped the attention of EPIP, though remaining subject to relevant state or territory legislation. Proposals first became subject to EPIP in 1975. Some early examples were the construction of rail terminal facilities at Alice Springs through the Australian National Railways Commission, the development of recreation reserves in Queensland at Wellington Point and Victoria Point through the Redland Shire Council, the construction of a television relay station on Canberra's Mount Taylor through the Australian Telecommunications Commission, and a Morphettville bus depot through the Adelaide State Transport Authority.

Under EPIP, responsibility for initiating EIA lay with the minister of the relevant Commonwealth department. If the proposal was deemed 'environmentally significant', or thought capable of being improved through EIA, the minister would refer the proposal to the Department of the Environment. Thereupon, the developer would supply the Commonwealth with preliminary information on the proposal, more commonly known as the Notice of Intention (NOI), which provided a brief summary of the proposal, a description of the project, the alternatives, the current stage of development, a description of the environment, an indication of the potential impacts on the environment, and details of any safeguards and standards required to protect the environment. Ultimately, the NOI enabled the Commonwealth to make a decision about the level of assessment appropriate to the proposal.

In 1999 the Howard government replaced the EPIP with the *Environment Protection And Biodiversity Conservation Act* (EPBC). The purpose of the Act was to define more clearly the role of the Commonwealth in the EIA process. The EPBC established a criterion of direct triggers based on the environmental significance of the project that allowed the Commonwealth to determine whether or not it would become involved, and if so, at what stage. Under the old EPIP, Commonwealth involvement in EIA had depended upon indirect triggers such as federal funding or foreign investment, which often engaged the Commonwealth in inappropriate developments, or even prevented it from becoming involved at all. In other words, EPIP forced the Commonwealth's hand on EIA. And indeed, while the construction of National Rail facilities in Alice Springs would have certainly interested the Commonwealth, the new Morphettville bus depot might not have piqued its curiosity. The EPBC also improved transparency, provided opportunities for public participation, and enhanced enforcement mechanisms. Significantly, it awarded responsibility for declaring the necessity of EIA to the Commonwealth Environment Minister rather than the relevant resources minister. Like EPIP, the EPBC required that actions likely to impact on matters of environmental significance be assessed for the purposes of decision making, and allowed the Commonwealth to select from a range of assessment options: preliminary documentation, public environment report (PER), environmental impact statement (EIS), or public inquiry. However, unlike EPIP, the EPBC permitted the states and territories to conduct their own EIA and receive Federal Government accreditation. Thus, under EPBC a single assessment process could satisfy both State and Commonwealth requirements.

Public and environmental health

Although there are managerial and institutional advantages to carrying out HIA within EIA, Australian governments also favour the practice because they associate public and environmental health. For example, section 528 of the EPBC defines human health in terms of the environment. According to the Act, an ecosystem includes 'the social, economic and cultural aspects' of 'people and communities; natural and physical resources' together with the 'qualities and characteristics of locations, places and areas' [3]. Environmental health is a subset of public health. According to the *National Environmental Health Strategy* (NEHS), managing environmental health is about 'creating and maintaining environments which promote good public health' [4].

Australia has a long tradition of associating public and environmental health. The Australian public health movement has its origins in environmental issues, beginning in the colonial era with the sanitation movement, whose platform was adequate sewerage, waste removal, and water supply. Over the twentieth century the movement enjoyed several victories against disease and mortality [4]. Today, the focus on the environment remains, but emphasis is placed on delivering a more integrated and sophisticated approach to the management of the environment. For example, according to the National Environmental Health Council (Enhealth), the authority responsible for drafting and implementing the national guidelines for HIA, the challenge for the twenty-first century is to safeguard public health through the responsible and national management of the environment consistent with Australia's international obligations. For Enhealth, efficient environmental management provides 'the basic infrastructure on which public health is built', and maintaining a high level of public health means 'addressing emerging health risks that arise from pressures of development on the human environment' [4].

The Enhealth Council is a component of the 1999 NEHS. Published by the Commonwealth Department of Health and Family Services, the NEHS describes environmental health as 'the corner stone of public health', and associates key improvements in environmental health—'sanitation, drinking water quality, food safety, disease control, and housing conditions'—with 'the massive improvement of quality of life and longevity experienced in this century' [4]. The strategy maintains that protecting public health is a matter of providing for the assessment, correction, control, and prevention of environmental hazards that might damage human health. Hazards can originate from either a lack of development, or unsustainable development; and threats to environmental health increase as society becomes urbanized, 'more populous, more complex'. Thus, consumption patterns, urban settlements, and the interaction of human lifestyles, together with the state of the environment, must be closely monitored. Managing waste, emissions, and modern lifestyles is essential to the minimization of air, chemical, water, and soil contamination, and, as a consequence, reductions in the rates of respiratory and cardiovascular diseases, physiological and neurological disorders, and the range of cancers [4].

The state and territory governments also link public and environmental health. New South Wales public health authorities interpret environmental health as 'the interaction between the environment and the health of populations of people', and outlines health determinants as the 'physical, biological, and social factors in the environment'. Distinct from environmental protection, environmental health regards aspects of human health influenced by 'drinking water supplies, recreational use of water, sewage management, public swimming pools, toxicology, microbial control, skin penetration industries, funeral industries, arbovirus control, air quality, waste management, and basic hygiene'[6].

The Tasmania government also closely associates public health and environmental health. In 1996, the state government introduced legislative requirements for the conduct of HIA as part of the EIA process through the *Environmental Management and Pollution Control Act*. With local governments holding responsibility for public health, the state government released the *Public and Environmental Health Services' Local Government Manual* to aid local officers in carrying out their duties [7]. Published in 1998, the manual formally tied public health to environmental factors, arguing that maintaining public health necessitated public health surveillance, the ongoing systematic collection, analysis, interpretation, and dissemination of information, and in particular human pathogens and contaminants in food, drinking water, and recreational water.

Decision-support tool

Comfortably installed within the EIA process, HIA shares many of its procedural characteristics. In particular, HIA is deployed as a decision-support tool rather than as a decision-making tool in its own right. By contrast, European models of HIA are often decision-making aides that require an intimate connection to the policy process. Australian EIA/HIA has no connection to the policy process. Rather, it provides a means of communication between the developer and the decision maker in which health impacts receive attention 'as part of the overall impact assessment process' [2].

The story of HIA as a decision-support device begins in Tasmania. In 1992 the National Health and Medical Research Council (NHMRC), a national research body tasked with raising the standard of public health in Australia, advocated the inclusion of HIA within existing EIA processes. In 1994 the Council published the *National Framework for Health and Environmental Impact Assessment*, which outlined a formal model for the conduct of EIA/HIA.

In 1996, following a review of its public health legislation, the Tasmanian government, consistent with the NHMRC guidelines, introduced the *Environmental Management and Pollution Control Act* (EMPCA), which subjected all activities currently requiring an EIA to the additional test of HIA. By incorporating HIA within the existing EIA process, the Tasmanian government had intended that health impacts should receive adequate attention without duplicating procedures. However, a related consequence was that HIA adopted the procedural characteristics of EIA. Like EIA, HIA did not penetrate Tasmanian policy-making circles, but became an additional test to which developments were subjected. For example, the Act outlined three categories for developments that might require EIA/HIA:

+ level one, developments likely to cause minor environmental harm;

+ level two, developments likely to cause significant damage;

+ level three, developments likely to cause harm of state-wide significance.

Level one developments were subject to EIA/HIA at the discretion of the Director of Public Health; however, a mandatory EIA/HIA was carried out on all Level two and three developments. In addition, the Tasmanian Environmental Health Service published guidelines to assist practitioners in carrying out their responsibilities under the EMPCA. The guidelines structured the HIA process in terms of the perceived threats to environmental health—environmental degradation and industrial and chemical pollutants—and provided methodological guidance on risk assessment, ongoing monitoring and post-project evaluation.

Commonwealth interest in HIA followed closely on the heels of the Tasmanian legislation. Through the NEHS, the Commonwealth government declared its intention to uphold human and environmental health through national safeguards. However, in some ways, the strategy gave national impetus to both the NHMRC guidelines and the Tasmanian approach. Following the guidelines, Enhealth established HIA as part of the EIA process, and explicitly described the process as a 'decision support tool and not a decision making tool'. HIA should 'not have the power of veto over a development, but will provide advice and recommendations to whatever statutory body is ultimately responsible' [2]. Similarly, as Tasmanian HIA had engaged professionals across a variety of disciplines and involved a range of public health and related skills, Enhealth concluded that HIA was better construed as an exercise in lateral thinking than as a new discipline in itself. Accordingly, Enhealth maintained that individuals with broad experience in health, environment, regulatory, and land-use planning policy were best equipped to conduct HIA. Similar to the Tasmanian Handbook, the NEHS also associated public and environmental health through ecological indicators such as drinking water, sanitation, waste disposal, occupational hazards, disease prevention, defor-estation, climate change, and air pollution.

According to the NEHS, bureaucracy, wasted resources, and 'the fragmentation of management across government and key organisations' are equally significant threats to public health. Australia operates under a federal structure of govern-ment, and individual jurisdictions have different operational approaches to envi-ronmental health that, according to the NEHS, result in 'reduced awareness of existing activities, lack of coordinated actions and duplication of effort' [4]. Arguing that the enhancement of environmental health demanded a coordinated approach from all sectors of society, the NEHS established the Enhealth Council to provide national leadership and a coordinated focus 'on all environmental health issues' [4].

National Guidelines

The *Health Impact Assessment Guidelines* (HIAG) incorporate HIA into the existing process of EIA rather than launching it as an evaluation process in its own right [2]. Anticipating that HIA might develop in practice, the guidelines do not explicitly detail how HIA should be carried out, and only allocate cer-tain parties with general responsibilities. The guidelines make the developer responsible for preparing a HIS, the health authority responsible for advising the proponent on the conduct of the assessment, and the decision-making body responsible for advising the health authority on the findings of the assess-ment. The guidelines describe the HIA/EIA process as a largely risk-based exer-cise, beginning with an 'objective risk assessment', that establishes the degrees of scientific uncertainty associated with impacts. Following the results of the risk assessment, the guidelines request stakeholder input regarding options for managing the risk, but suggest that any regulatory measures adopted must be proportionate to the amount of risk to be reduced or eliminated.

The guidelines do not specify the types of issues that the developer must include in the assessment. It is for the proponent to judge what data should be included. The HIAG offer assistance on issues that might be relevant, but insist neither that particular issues be considered, nor that the omission of issues outlined in the guidelines be explained in the assessment. The level of detail and the range of issues canvassed depend entirely upon the health impacts identified during the scoping stage. Herein, the jurisdictional health authority is responsible for assisting in the progress of the HIA and advising on the required level of detail. In particular, the health authority must discuss the HIA, the methodology, and the available sources of evidence with the proponent. It must set out the relevant data, the health dangers, and ultimately ensure that the developer invests the appropriate amount of resources in preparing the statement. Following the completion of the assessment, the decision-making body provides the relevant health authority with the results of the evaluation.

The HIAG outline the HIA process in terms of what the impact statement might include. They suggest that the statement should cover details of the development, the site history, the climate, and the proximate community. For example, the assessment might profile the local community through demographic data, and consider smaller components of the community in terms of indigenous population, the young, the elderly, and the poor [2]. The statement might also include standardized mortality and morbidity data for diseases related to potential health impacts, and may even need to cross reference these statistics with elements of the population at greater risk of adverse health effects. Environmental factors, notably, food, water, air quality, and waste disposal, should also be considered. Social impacts are relevant to HIA, but only insofar as they intertwine with health impacts. Critically, the HIA process should not become an economic assessment. Economic impacts are only relevant where they have significant health impacts; and generally, their analysis should be independent from the HIA.

Quantitative discipline

Published in 1994, the NHMRC's *National Framework for Health and Environmental Impact Assessment* recommended the inclusion of HIA within EIA in order that it more fully consider impacts on human health. Under EIA, health impacts were often measured in an unstructured way, and usually from a negative aspect. The NHMRC argued that including HIA within EIA would enhance the depth and effectiveness of the impact assessment process. The HIAG developed and expanded the NHMRC's approach. It is perhaps fair to say that EIA/HIA may advantage the developer over the community and the environment. The inclusion of HIA within EIA, the requirement for evidence-based conclusions and risk-weighted options necessitate the use of scientific methods of analysis, which weights the process in favour of the quantitative sciences. Accordingly, those with scientific training and sophistication, and, those most able to collect quantitative evidence and apply rigorous methods of analysis, are most likely to influence the EIA/HIA process. The important characteristics of the Australian EIA/HIA process are that primacy is placed on evidence-based conclusions, evidence is selected on the basis of its amenability to quantification, and conclusions are made upon the basis of risk assessment.

Nonetheless, it would be a mistake to allege that the federal guidelines are responsible for the quantitative character of Australian EIA/HIA. Quantitative methods were well established in the states long before the publication of the Enhealth guidelines. Indeed, the first formal HIA conducted in Australia provides a useful example of both the quantitative character of Australian EIA/HIA, and how the process developed within the states from the NHMRC's original guidelines.

In 1995, researchers from the University of Melbourne, Monash University, and the Victorian Road Research Board conducted an HIA on a proposed highway development in inner suburban Melbourne [8]. The point is that the team consciously approached the assessment in the light of the NHMRC guidelines, and thus anticipated, albeit unknowingly, the methods that would be outlined in the EnHealth Guidelines some years later. Through the guidelines and the practice with the states, HIA became a quantitative and broadly scientific discipline. From the outset of the assessment, the team maintained that studies of health impacts were largely 'empirical studies that seek to inform the policy process during the planning phase of construction' [8]. In Melbourne, the assessment commenced with the scoping of health impacts namely stress and anxiety, traffic injury, and respiratory diseases. Amenability to quantification determined the sources of evidence that the team associated with each impact.

The assessors defined stress and anxiety as 'community disruption, visual intrusion and the symbolic threat of a freeway'[8]. As stress represented the psychological impact of the freeway on an individual, precise knowledge of its impact required that both the freeway exist, and that it should hold significance for the individual. However, the freeway did not exist, and its significance for individuals could not therefore be known. Thus, the team argued that they could only account for stress based on 'the magnitude and characteristics' of the freeway [8]. Noise became the principal measure for stress. After all, noise was 'a well understood phenomenon amenable to precise measurement'. In addition, having associated increased risk of traffic injury with increased traffic volume, the team consulted routine traffic accident statistics, collected by state police, to which they could apply 'sophisticated computer-based models' that would relate 'traffic injury and accident levels to traffic movements by flow and type' [8]. Using these techniques, the team was able to establish health levels in the affected area in both the presence and absence of the freeway. However, the assessors had difficulty finding data sets with which to measure the impact of the freeway on respiratory health. Routine statistics on relationships between environmental pollutants and respiratory diseases were not available, and models for relating disease to traffic flows did not exist. In this case, the lack of availability of quantitative measures influenced the choice of evidence, with the assessors being forced to employ 'proxies'. The Victorian Environmental Protection Agency (EPA) routinely measured levels of pollutants associated with traffic volumes, and models for projecting increases in these pollutants also existed. By associating these volumes with standard threshold exposure data for vehicular emissions the team was able to estimate the impacts of pollutants on health.

The point is that throughout the assessment quantitative evidence was deemed superior to qualitative evidence. For example, the team consulted the community through a series of public meetings intended to identify popular concerns. Groups opposed to the construction of the freeway argued that 'the large proportion of the population in the affected area suffering social disadvantage and consequently health disadvantage . . . would have a lower average health status and be more susceptible than other populations to the health related effects of the freeway' [8]. However, the characteristics of the data sets routed popular concerns: 'this argument ignores the fact that community standards for environmental pollutants are traditionally defined in terms of those groups most susceptible to exposure as a result of prior disease or other condition . . . provided standards are so defined, the effects of the freeway will not be underestimated' [8]. Similarly, when elements opposed to the freeway questioned the accuracy of the traffic flow estimations, a sensitivity analysis quelled any doubts. Altering the estimations in terms of best- and worst-case scenarios, the team found that, regardless of the accuracy of the projections, 'the freeway would reduce vehicle travel on major roads in the area and thus reduce the overall number of accidents. In any event, risk assessment confirmed that: "The reduction in traffic injury and noise-related problems (including stress) outweigh the small risk for an increase of slight respiratory problems associated with atmospheric pollution, generated by the freeway" ' [8].

Although the HIAG are sufficiently broad to allow HIA to develop further, the Melbourne study provides a useful example of the major features of Australian EIA/HIA. Australian HIA is typically quantitative and scientific, it associates public and environmental health, it functions as a decision-support tool rather than as a policy-making device, and it is subject to national regulations. For the future, Australian governments appear comfortable with the practice of EIA/HIA, and environmental authorities are currently encouraging public health professionals to develop and expand the evidence base for linking environmental and human health. While the practice of EIA/HIA offers Australian governments procedural and institutional advantages, the process is vulnerable to criticism as being quantitative, elitist, and environmentally dependent. Some critics may even argue that Australian HIA is 'not really HIA'. Nonetheless,

Australian governments conduct HIA, Australian HIAs are published and disseminated, they can be generalized from one to the other, and they can withstand legal challenge; the HIA process is institutionalized within established legislative and administrative procedures, and it influences the design of projects. While Australian HIA may not be 'what-is-really-HIA', certainly it is real in ways that 'what-is-really-HIA' still remains very much unreal.

References

1 European Centre for Health Policy. *Gothernburg Consensus Paper*. Copenhagen: World Health Organization: 4, 1999.

2 EnHealth. *Health Impact Assessment Guidelines*. Canberra: Commonwealth Department of Health and Aged Care: v, 2001.

3 Environment Australia. 2003: http://www.ea.gov.au/epbc/index.html. Access Date 24.04.03.

4 EnHealth. *National Environmental Health Strategy*. Canberra: Commonwealth Department of Health and Aged Care, 1999.

5 EnHealth. (2003). http://enhealth.nphp.gov.au/what_enh.htm. Access Date: 24.04.03.

6 Health NSW. (2003). http://www.health.nsw.gov.au/public-health/ehb/about/about.html. Access Date: 24.04.03.

7 Public and Environmental Health Service. *Local Government Manual*. Tasmania: Department of Health and Human Services: 1, 1998.

8 **Dunt David R, Abramson Michael J, and Andreassen David C.** Assessment of the future impact on health of a proposed freeway development. *Australian Journal of Public Health* 1995; **19**(4): 347.

Chapter 21

HIA and policy development in London: using HIA as a tool to integrate health considerations into strategy

Caron Bowen

Background

In 1998 the newly elected Labour government began to discuss the creation of an elected regional assembly for London, as part of the delivery of its manifesto. Across London a partnership of organizations interested in health and the wider determinants of health came together to lobby the government to include health in the Greater London Authority (GLA) Act. As a result the GLA Act 1999 section 30 states that the Authority must exercise its power in a manner calculated 'to promote improvements in the health of persons in Greater London'. The GLA, which came into being in 2000, has an elected assembly and a separately elected mayor. The mayor has a statutory responsibility to develop and implement eight high-level strategies for Greater London on a range of issues, including economic development, transport, and spatial planning.

During the period of campaigning before the election the partnership of organizations continued its lobbying role, meeting with all the mayoral candidates to develop their understanding of health and ensure their commitment to addressing the wider determinants of health and addressing health inequalities. Alongside this work, the partnership drew in a wider range of people and organizations and developed the London Health Strategy (LHS) [1]. The LHS was published in March 2000 following wide-ranging consultation with over 1,500 people from more than 300 organizations. Its development was initially overseen by a multi-sectoral steering group, which eventually formed the Coalition for Health and Regeneration.

The LHS has four headline priorities: transport, black and ethnic minority health, health inequalities, and regeneration. The LHS also has four underpinning themes designed to help deliver on the headline priorities. These are health impact assessment (HIA), the development of the London Health Observatory (LHO), community involvement and engagement, and communicating London's health work. All the mayoral candidates agreed to support the delivery of the LHS and when Ken Livingstone was elected mayor he used the already formed partnership of London organizations to form the London Health Commission (LHC), an independent organization with a chair appointed by the mayor. Alongside this, the GLA created a strategic health policy post and seconded an employee from National Health service (NHS) London regional office to work within the GLA to incorporate health into their programme.

Getting HIA into strategy development

Ensuring that health was considered in the development of the statutory strategies has been viewed as a real opportunity to embed health into the work of the GLA and its functional bodies. The group of organizations for which the GLA is responsible (called the GLA group) include Transport *for* London (T*f*L), London Development Agency, the Metropolitan Police Authority, and the London Fire and Emergency Planning Authority. It was decided to use HIA as a tool at the point where the draft of the document was complete and had been delivered to the Assembly and functional bodies responsible for scrutiny. Thus, HIA has become part of the scrutiny process. The GLA recognized that it needed to address health and its wider determinants, and through discussions with the member organizations of the LHC it was decided that employing HIA could be the way forward.

The first strategy to be developed was the draft Transport Strategy [2]. Although in principle there was agreement for an HIA there was concern about the possible recommendations and what it could mean for the planned congestion charging. Initially a public health specialist did a desktop appraisal of congestion charging [3] and then after a number of meetings between a range of public health specialists and T*f*L it was agreed that the HIA would include the findings of public health evidence and a stakeholder workshop. This would ensure that the HIA addressed the whole strategy rather than just congestion charging.

This first HIA made a series of positive recommendations for incorporating health into the strategy and other strategy development teams were made aware that an HIA would be part of the process for scrutinizing their strategies before they went out for public consultation. The completed HIA report of the draft Transport Strategy was very supportive about most of the proposals within the document, which ensured that the recommendations made were viewed positively. This was helpful in ensuring that the mayor and GLA officers were accepting of HIA and viewed it as a useful tool.

The mayoral strategies have been developed by teams working either within the GLA, or by the organizations within the GLA group where the remit for the work falls (e.g., the London Development Agency was responsible for the writing of the Economic Development Strategy). These teams are specialists in the area they are involved but do not always have an understanding of how their work impacts on the health of Londoners. While the report is being developed they have input from a public health specialist working at the GLA.

The HIA process in relation to the mayoral strategies

As already mentioned, the HIAs have taken place at the stage where the draft strategy goes to the Assembly and the functional bodies for scrutiny. This is prior to public consultation. Usually there has been no more than eight weeks in which to complete the whole HIA process, from scoping right through to delivering the final report to the mayor and the Assembly. Due to the high-level nature of the strategies there was no specific screening process for it was agreed that all the strategies would meet most criteria that could be set and that all tackled at least one (and in most cases several) important determinant of health.

A core team of people came together as a steering group to develop the process and organise the HIAs. This consisted of representatives from the GLA, the LHC, the London NHS Executive (which then became the Department of Health and Social Care—DHSC), where possible, the public health specialist who was responsible for reviewing the evidence, and (from the second HIA onwards) an external consultant employed to write the report. Once the LHO was set up, the LHO's HIA Facilitation Manager joined the core team.

For the majority of the HIAs the core team scoped the main topic areas that the HIA would focus on. This scoping exercise was undertaken via a series of meetings where the strategy and the related evidence base were discussed, and topic areas where there was the greatest opportunity to increase health gain and address inequalities, identified. These topic areas were then developed further to form the basis of the small-group discussions held as part of the rapid appraisal workshop (see below). With some of the strategies, such as the draft Waste Management Strategy [4] the identification of topic areas was relatively easy as groups could be organized on the basis of each of the waste management types that were discussed in the document. With others it was more difficult as the strategy covered a much wider remit. When this was the case the core team decided the most important topic areas to concentrate on. An exception to this was when the HIA for the draft London Plan [5] (the Spatial Development Strategy for London 2002–2015) was being organized. There was agreement within the team that a larger scoping meeting would be beneficial, as this strategy is overarching and creates the planning framework for the delivery of all the other strategies. A wide range of people, including LHC members, public health specialists, and representatives of specific interest groups were invited to an initial meeting to discuss the possible areas on which the HIA would focus. The core team then developed the ideas from this meeting to ensure the range of topics identified was addressed.

During the scoping phase the planning for the workshop and report writing also took place. The list of invitees for each workshop was compiled by the core team with input from members of the LHC and other individuals, including in some cases members of the strategy teams. For each of the appraisal workshops as wide a range of people as possible were invited. This included local authority employees (both from the topic area the strategy covered and from health policy), public health specialists, voluntary organizations representing specific groups (e.g., Age Concern), and environmental groups, who would have an interest in the strategy (e.g., Friends of the Earth). People from the private sector were also invited since the strategies are likely to have implications for this sector. The invitee list has usually been quite large, as the importance of having as wide a cross-section of participants/stakeholders as possible at each workshop has been central to the process.

The core team organized the commissioning of a rapid review of the evidence relating to the topic areas and the links to health for each of the strategies. For this work, public health specialists with an interest in the strategy area were identified and given a specification of what topic areas were likely to appear in the strategy. They usually began work before the strategy was published and a summary of the evidence was prepared once the strategy was published and it was clear what parts of the strategy the HIA would concentrate on. This work has been refined over time and in an effort to make the evidence as accessible as possible for lay people who have attended the appraisal workshop, the summaries of evidence have been made clearer and easier to understand.

Alongside the summarizing of the evidence it has been recognized that it is important to make the key policy proposals within the strategy as accessible and as clear as possible. As the HIA process developed it became apparent that it was not feasible to assume that all the participants attending the workshops had a clear understanding of the strategy. Therefore, it has been essential to summarize the policy proposals to ensure that participants are aware of the key issues.

The majority of the rapid appraisal workshops have been half-day events; however, the HIAs on the Air Quality and Bio-diversity Strategies were run on the same day and the HIA on the London Plan was a full-day event. The events have begun with a presentation about the strategy and the major policy objectives within the strategy, a short presentation about the public health evidence related to the strategy and an explanation of what HIA is. There have been criticisms of this format as some participants have felt that there was too much input at the beginning of the event and this meant there was less time for the small-group discussion. For the most recent HIA on the London Plan more time within the day was given for the small groups to look at the evidence and there was only a very short presentation at the beginning of the day giving the major headlines in terms of evidence.

Once the presentations have been given, the participants are then asked to elect to be part of a small, facilitated group looking at a specific area of the strategy. The small groups have about 90 min to work through the questions and to decide the main points they wish to feed back (a rapporteur is present to record the discussion). The facilitator is responsible for feeding back to the plenary session at the end of the workshop, which is chaired by an executive member of the LHC.

During the early HIAs the facilitator was asked to lead the group through a series of questions. These were:

- Which determinants of health are likely to be affected by the strategy?
- How may health determinants change as a result of the strategy?
- How might the expected changes affect the health of people?
- What might be the outcomes for health?
- What do you think should be recommended in this area?

These questions create discussion and many worthwhile inputs. However, as the HIA process has developed, it has become clear that when those attending the workshop are not particularly knowledgeable about the strategy, discussion is better supported by using more specific questions, tailored for each group.

During the workshop, the facilitators have themselves wished to participate in the discussion rather than just playing the role of group facilitator. The core team has recognized that the implications of this, especially in terms of power dynamics and the need to ensure that all the participants in the group have the opportunity to voice their view. For later workshops, facilitators have been given a written brief outlining their role and responsibilities. However, discussions with the external evaluation team who have been commissioned to complete a qualitative evaluation of four of the HIAs suggest this may not be enough and there may be a need to work more closely with those facilitating the small groups before the workshop.

After the workshop an external consultant, who is part of the core team and has been involved in the organization and attends the workshop, writes the report using the notes taken by the rapporteurs and the summary of the evidence base. Recommendations are then formulated on the basis of the workshop discussions and where there is evidence to support them this is reflected in the report. There is a continuing need to review how evidence is collated and used. This is especially true in the light of possible legal challenges, which has already been an issue for the GLA in relation to the Transport Strategy, where there was a legal challenge from a local authority on the issue of congestion charging. There are ongoing debates within the core team who organize the rapid appraisals about how to meet this challenge, although there has not been any resolution to date. This is of course a wider issue and is being debated elsewhere as well [6,7].

Once the report has been drafted it is then sent to all the workshop participants for comments. If there are any responses these are incorporated into the report before it goes to the LHC where it is debated and any additional comments or rewording of recommendations are made. There is often lively debating at these meetings and much constructive feedback and criticism.

The report is then sent to the mayor, the Assembly, and the strategy development team. The recommendations are then, where feasible, incorporated into the draft strategy before it goes out for public consultation. The LHC asks the strategy development team to report back on which of the recommendations have been incorporated into the strategy and in the case of recommendations that have not been incorporated to explain the reasons why. This has been successful, in that it ensures dialogue continues between the strategy development team and the LHC, and that health remains on their agenda.

The outcomes of incorporating HIA into strategy development in London

The early draft strategies had not made consideration of health a priority to any great degree. The conduct of HIAs was an important exercise for ensuring that the strategies reflected health concerns and for raising awareness about health and its determinants within the GLA more generally. The first two strategies to be developed were the draft Transport Strategy and the draft Economic Strategy [8]. The draft Economic Development Strategy did not overtly reflect health concerns. The draft Transport strategy's main health emphasis was on air pollution. Modes of transport such as walking and cycling were mentioned, however, the HIA ensured that these received more emphasis in later versions. The HIAs ensured that health was more integral to both strategies. Some of the recommendations reported as being included in the strategies after the HIAs were conducted are as follows.

In the Transport Strategy:

- The objective that the transport strategy should help to address social exclusion, including addressing the travel requirements of groups with specific needs.
- Encouraging more sustainable modes of transport including workplace travel and school travel plans.
- Emphasis on involving boroughs in the development and implementation of plans to improve transport including ensuring that greater use is made of borough powers to introduce 20 mph zones and speed limits.
- Developing a walking plan for London

In the Economic Development Strategy:

- Promoting Londoners' health was incorporated into a new Charter objective.
- The links between economic development and health was acknowledged.
- A broad definition of health was adopted in the strategy.
- Promoting social inclusion and renewal amongst all London's communities was included as a revised charter objective.
- Addition of a commitment for the London Development Agency to undertake further work to fund breakfast clubs in schools to promote healthy eating.

In the Bio-diversity Strategy:

- An additional proposal has been added regarding working with custodians of green spaces (both public and private) to ensure that Londoners are aware of the capital's green spaces and waterways.
- The need to address perceived safety risks of accessing green space has been added to a proposal.
- A proposal has been added on promoting environmental education

With later strategies it became increasingly apparent that the incorporation of HIA into the scrutiny process has led the strategy development teams to consider health and its wider determinants at an earlier stage in the strategy development process. For example, when the draft London Plan was being developed members of the team worked with a group of public health specialists, explicitly wishing to ensure that health was central to the strategy.

The GLA is paying for an external evaluation of the process and findings will be disseminated when the report is finished The early findings of the evaluation are largely positive and show that the process of engaging stakeholders through the HIA process has been successful. Furthermore, it is possible to say that incorporating HIA into the strategy development has had two main benefits: it has helped the mayor to fulfil his duties to consider health as a central theme and it has ensured that many officers working at the GLA are more aware of health and its determinants.

References

1 www.londonshealth.gov.uk. Access Date: 22.12.2002.
2 Final strategy available at http://www.london.gov.uk/approot/mayor/strategies/index. jsp Access Date: 18.12.2002.
3 **Mindell J, Sheridan L, Joffe M, Samson-Barry H, and Atkinson S.** Health Impact Assessment as an agent of policy change: Improving the health impacts of the Mayor of London's draft transport strategy. *Journal of Epidemiology and Community Health* (accepted for publication).
4 Final strategy available at http://www.london.gov.uk/approot/mayor/strategies/index. jsp Access Date: 18.12.2002.
5 Final strategy available at http://www.london.gov.uk/approot/mayor/strategies/ index.jsp Access Date: 18.12.2002.
6 **Joffe M and Mindell, J.** A framework for the evidence base to support health impact assessment. *Journal of Epidemiology and Community Health* 2002; **56**:132–138.
7 **Mindell J, Hansell A, Morrison D, Douglas M, and Joffe M.** Participants in the quantifiable HIA discussion group, What do we need for robust, quantitative health impact assessment? *Journal of Public Health Medicine* 2001; **23**:173–178.
8 Final strategy available at http://www.london.gov.uk/approot/mayor/strategies/ index.jsp Access Date: 18.12.2002.

Chapter 22

Using HIA in local government

Susan J Milner

This chapter will draw on work carried out with local authorities in the North East of England, to demonstrate how health impact assessment (HIA) can be brought into the mainstream of local policy-making and planning activities. It will consider the relationship between HIA and other appraisal and assessment requirements, such as sustainability impact assessment and rural proofing. It will explore how local government can use HIA to help them achieve a broad range of corporate objectives. It will consider some of the practical issues that need to be addressed if HIA is to become embedded in local authority business. Consideration will also be given to problems associated with trying to involve the public in the assessment process.

The focus of HIA

An HIA is designed to identify aspects of a proposal or activity that could affect the health and well-being of defined populations. These health impacts are most likely to occur because the proposal or activity affects the key determinants of health, rather than because the proposal impacts directly on human health (though this may happen occasionally, e.g., exposure to physical or chemical hazards). An HIA is, therefore, focused on the changes to the key determinants of health that are predicted to occur as a result of the proposed activity.

Health impact assessment is predicated on a social model of health. Within that model the key determinants of health are factors such as employment, education, housing, and transport and these are largely under the control of government organizations. National and international policies in these areas are important but, at the local level, it is local government organizations that are instrumental in controlling the determinants of health. These local organizations include the regional development agencies, regional assemblies, and government offices, but it is the local authorities that have most influence over the factors that affect the health and well-being of their local communities. All local authority decisions impact directly or indirectly on the health of the populations they serve in ways that might not be immediately obvious to the decision makers. Because of this influence, the local authority setting is extremely important for the development of HIA. Within local authorities, some policy makers and planners are unaware of the important relationship between the work they do and the health of the population. They often see 'health' in very narrow biomedical terms and consider it to be the sole concern of the NHS.

Local authorities as key in a setting for HIA

Local government has a tradition of involvement in the provision of health and social services and retains statutory responsibility for key areas of public health, for example, maintaining specific environmental standards. Writers such as Ashton and Seymour [1], Betts [2], and Webster [3] give insights into this valuable role of local government and the advances made since the nineteenth-century public health movement. Whilst changes in the organization of the NHS in the 1970s reduced some of the disease prevention role for local authorities, the 1980s and 1990s have seen this local authority involvement in health-related matters rekindled. Initiatives such as the WHO's Health for All (HFA) [4], the Healthy Cities movement [5], and 'Agenda 21' [6] together with national initiatives, such as the 'Health of the Nation' [7] and 'Our Healthier Nation' [8], have all helped to fuel this momentum for the development of local government strategies for the improvement of health.

> Local governments have a prominent and vital role to play in promoting the health and well-being of the local population and that all councillors and every officer, in every department, at every level, have an important contribution to make. [9]

Many local authorities are now actively involved in developing policies, programmes, or projects that specifically seek to improve the health of the local communities and reduce health inequalities. They are invariably in partnership with the National Health Service (NHS) and other relevant agencies in this work. Local authorities have a duty 'to promote the economic, social and environmental wellbeing of their communities' [10].

Local authorities are very well placed to act as key agents in improving health within their local communities because they:

- are democratically accountable,
- have a consultative relationship with local people,
- are key players in multi-, single-focus, and cross-cutting strategic partnerships,
- have community planning responsibilities,

- have a new legal power of 'well-being',
- are legally empowered, in many instances, to take systematic action to promote or protect the public's health,
- have scrutiny functions, including that of health scrutiny,
- have responsibility for the governance of social care,
- are well placed to institute the monitoring and evaluation of interventions to improve people's health.

Within local authorities the health impacts of key decisions need to be assessed. This could ensure a more coordinated approach to tackling threats to health within the local authority and promote more effective inter-agency work. HIA may also facilitate the development of effective communication channels between local residents and those departments [within the local authority] having responsibility for specific activities regarding health.

In order to achieve this, all policy makers and planners should consider the potential positive and negative impacts (including health impacts) of new and substantially amended policies and proposals across a wide range of parameters. Health impact assessment is one way of ensuring that during planning and decision making the overall, long-term well-being of the population is one of the criteria routinely taken into account. In this way the HIA process puts health on the non-health policy agenda in local authorities.

The local government association stated

> Health Impact Assessment may develop into an important tool for both central and local government in the fight against health inequalities. However it is essential that as this tool is developed that it ensures that the voices of the local communities are heard and given weight in making decisions about the health impacts of particular policies or projects. [11]

In addition there are bureaucratic requirements within the local authorities that lend themselves to the inclusion of HIA, that is, 'best value' processes and community planning.

Other forms of impact assessment undertaken by local authorities

Policy makers and planners working in local authorities are often required or requested to undertake other forms of impact assessment or policy appraisal such as environmental, sustainability, economic, equal opportunities, human rights appraisals, and rural proofing (considering impacts on rural areas). When these forms of assessment are compared, large areas of overlap are found in the questions posed to assessors by the respective assessment tools.

In practice, this could lead to policy makers and planners being expected to undertake several types of, potentially overlapping, impact assessments on the same proposal. This means duplication of effort, wastage of scarce resources (time and money), and considerable irritation on the part of those being asked to undertake multiple assessments.

Multiple assessment of the same proposal is inefficient and unpopular and has led to calls for the integration of various forms of impact assessment to create 'Integrated Impact Assessment' tools. The integration of HIA into other forms of assessment is becoming more commonplace [12]. The integration of HIA and environmental impact assessment (EIA) has been discussed in several chapters of this book. Integrated assessment tools usually incorporate the basic aspects of health, health inequality, environmental, sustainability, equal opportunities and economic impact assessment. Such integrated tools may not feature the term 'health' in their title but be described by terms such as 'well-being impact assessment' or 'policy-makers checklist'. Some HIA practitioners regard such integration with suspicion. They feel that it represents a dilution or weakening of the assessment of health impacts in the process. Other HIA practitioners believe that such integration is inevitable and represents a pragmatic solution to the problem of persuading decision makers to use any form of impact assessment in their planning and policy making.

The process of making it happen—theoretical and practical considerations

From the work undertaken with local government organizations in the North East of England by Northumbria University, it is possible to identify a number of theoretical and practical considerations, which impinge upon the adoption of HIA (or integrated impact assessment) within these organizations.

Defining health within the organization

The conceptualization of health is a complex and often contested process. This has important implications for any attempt to assess the 'health impact' of a given set of activities. In order for such an assessment to be accepted as useful to those concerned with it, a shared (and negotiated) concept of health is needed. The production of population health is influenced by a wide variety of factors, most of which lie outside the formal health care setting. The perceived importance of each of these health influences varies across professional groups and individuals. This variation needs to be acknowledged in the HIA process.

Although much is known, in broad terms, about the production of health (as incorporated in a social model of health), it is often difficult to specify the causal relationships in a way that satisfies local policy needs. For many of the interventions of greatest local relevance, such as those to create new employment opportunities, the identification and valuation of forecast outcomes is difficult. This is because it is possible to identify both health 'gainers' and 'losers' in the population, and because account must be taken of desired outcomes other than health.

What should be assessed and when?

To mainstream HIA in local authority decision-making it is necessary to understand how each part of the organization operates. Local authorities are very complex organizations and each department within the authority will have a different way of working. It is necessary to establish what the policy and decision-making cycles are and to help the decision makers determine how best to insert the assessment process into these cycles.

Health impact assessments can be conducted with different degrees of thoroughness varying from rapid appraisal to in-depth HIA. Policies and decisions can be subject to prospective HIA, concurrent monitoring, or retrospective evaluation. Commonsense dictates that not all policy and planning decisions can be subjected to in-depth HIA or evaluation. There needs to be some way of prioritizing proposals for assessment. HIA screening can do this.

HIA screening is the systematic selection of proposed activities for more in-depth analysis. Once a proposed activity has been selected for further HIA, through the screening process, the assessors have to decide how much further analysis is required. Because of resource and time constraints, it is likely that most HIA-type activity will be at the rapid appraisal end of the HIA spectrum. Screening tools can be developed for local authorities that could eventually be applied to all policies, programmes, and projects. If, in the longer term, screening for health impacts is to become a normal part of policy development, it will need to be grounded in the everyday reality of local authority business.

HIA can be used at any point in the policy-making and planning cycle. Some would argue that it is best to use it at the beginning of the cycle in order to ensure that the general direction of the proposal is sound with regard to its health impacts. Further, HIA could be used as part of an initial option appraisal to determine the strategic direction or overall aim of the proposal. But HIA can then be used at any further point in the cycle to ensure that the proposal is as health enhancing as possible. Impacts can be monitored during the implementation of the proposal and impacts can be retrospectively evaluated. In an ideal situation, retrospective impact evaluation could be applied to a proposal that has undergone a prospective HIA in order to determine the accuracy of predictions made in the impact assessment and assess the 'impact of the HIA' in reducing negative and enhancing positive health impacts.

Who should undertake the HIA?

If one accepts that most HIA-type activity in local authorities will be at the screening/rapid appraisal end of the HIA spectrum in order to achieve maximum coverage, then time and resource constraints dictate that such assessments are undertaken by the decision makers and planners themselves. Then HIA becomes an 'in-line' planning and policy-making tool. In this sense, HIA should be seen as a broad mapping exercise rather than a 'scientifically robust' predictive tool. Thus, introducing HIA into local authorities is more about engaging people with the broad concepts of health, facilitating the process of it becoming a 'mind set', an automatic way of thinking within the authority. It is about raising the health consciousness of the organization.

Critics of such an approach would argue that HIA, undertaken this way, would be biased and tokenistic. They believe that assessing one's own proposals would not be 'objective'. They would prefer external assessors to be used to provide a more 'objective' assessment. But, if HIA is too technically difficult (or resource intensive) the opportunity to put health on the non-health agenda may be lost. It may be impossible to fully engage the policy makers and planners who influence the key determinants of health locally. It is simply not feasible for external assessors to be used to screen decisions in this way. External assessors may be used on more comprehensive, in-depth HIAs, where a wide range of expertise may be required and in-house staff could not be expected to have such in-depth knowledge.

Infrastructure and capacity building

Within each authority the HIA development process requires a clearly focused structure to manage it and drive it forward. A core group of committed individuals needs to be established. The right infrastructure needs to be put in place to make it happen. This infrastructure includes:

- Development of HIA tools for use by local authority staff (especially at the screening/rapid appraisal end of the HIA spectrum). These may include IT-based HIA tools.
- Basic training of the staff who are likely to be asked to use these tools.
- Identification and further training of departmental 'champions' of the HIA process, who can offer advice and guidance to colleagues.

- A defined policy for the use of HIA, that is, the circumstances under which HIA should be used and how the conclusions of the assessment are reported and used in the decision-making process.
- Identification of resource requirements. Funding mechanisms need to be established if HIAs are to be commissioned from outside the organization.
- Commissioning guidelines for use when external assessors need to be brought in to undertake an in-depth HIA.
- An audit of existing data sources.

A large amount of data is already routinely collected by a number of different agencies for a wide range of purposes, for example, mortality statistics, census data, EIA data, and the like. In addition, a body of evidence exists concerning causal relationships, effective (and ineffective) health and social interventions, and past case studies. Some of these data will be useful for HIA, but it cannot be assumed that such data can be readily absorbed into the HIA process. Having said this it makes sense to use already available data whenever possible.

Timescales

In-depth HIAs may take months (or even years) to complete. Such lengthy timescales may not always be feasible within the normal planning and policy cycles and a balance between pragmatism and rigour has to be found.

What is the impact of the impact assessment?

To fully engage local authority staff in the HIA process, they need to be convinced that HIA is a worthwhile process. Ways are needed of evaluating the implementation of HIA in order to determine if the process adds value to existing health-gain activities within the authority.

Involving the public in the HIA process

It is not easy to involve 'the public' in the HIA process. To date such involvement has been largely confined to the better-resourced HIAs, which have been able to identify a broad range of stakeholders in the assessment process. Members of the population groups affected by the proposal being assessed are stakeholders and should therefore, be consulted. However there are real practical difficulties in accessing such groups. Ison (Chapter 11, in this volume) suggests some ways in which this might be done.

In reality, the quicker the HIA the more difficult it is to include 'the public' in the process. Community 'representatives' may be invited to HIA workshops and related activities, but it is difficult for the assessors to access a broad range of lay stakeholders in a short space of time and with limited resources. With in-house HIA screening, the chance of involving the public in this stage of the HIA process would be very slim indeed. Local authorities do have a duty to consult their resident populations over a variety of issues, so it would be possible to think more creatively about how the information obtained from such consultations could be used in the HIA process.

Summary

At this early stage in the methodological development and practical application of HIA we make the assumption that HIA can deliver benefits in terms of health gain and, therefore, is worth doing. If it is worth doing it is worth doing across a broad range of activities in order to get adequate coverage of key decisions. To achieve a broad range of HIA 'coverage' the process has to be embedded in the ordinary planning and decision-making processes of organizations (not an add on or external activity). Given time and resource constraints within local authorities it is reasonable to assume that most HIA activity will be at the screening/rapid appraisal end of the HIA spectrum.

In the absence of regulatory or nationally defined HIA frameworks it is sensible to assume that a range of approaches to this task will be developed. Over time 'best practice' will emerge—what works and what does not in a given set of circumstances. A knowledge base will develop from an increasing number of prospective and retrospective HIAs undertaken. To ensure broad participation in HIA it needs to be as pain-free as possible for those people who will have to incorporate this into an already heavy workload.

Introducing HIA into local authorities is about changing attitudes of staff, winning hearts and minds, and raising health consciousness, rather than developing precise predictions of the size of health impacts. As such, its introduction and implementation is governed as much by organizational development theories as public health and epidemiological theories.

References

1 Ashton J and Seymour C (eds.). *The New Public Health*. Buckingham: Open University Press, 1993.

2 Betts G. *Local Government and Inequalities in Health*. Avebury, 1993. Aldershot, U.K.

3 Webster C. Local government and Health Care; the historical perspective. *British Medical Journal* 1995; **310**:1584–1586.

4 World Health Organisation. *Health for All by the Year 2000*. Geneva: WHO, 1977.

5 World Health Organisation. *Targets for Health for All*. Copenhagen: WHO, Regional Office for Europe, 1985.

6 United Nations Conference on Environment and Development. The Earth Summit (Agenda 21). Rio de Janeiro, Brazil. United Nations Organisation, 1993.

7 Department of Health. *The Health of the Nation.* London: HMSO, 1992.

8 Department of Health. *Saving Lives: Our Healthier Nation.* London: The Stationery Office, 1999.

9 Local Government Management Board Promoting health and local government: A report prepared for the Health Education Authority and the Local Government Management Board by Ged Moran, Leeds Metropolitan University. Health Education Authority, 1996.

10 Local Government Act 2000. www.legislation.hmso.gov.uk/acta/acts2000/20000022.htm

11 Local Government Association. *Briefing—The White Paper on Public Health: Saving Lives—Our Healthier Nation.* p. 5. Local Government Association, 1999. London.

12 **Bailey C, Deans J, and Pettigrew D**. Integrated Impact Assessment. UK Mapping Project Report. Newcastle. Northumbria University, 2003.

Chapter 23

HIA: the German perspective

Rainer Fehr, Odile Mekel, and
Rudolf Welteke

Introduction

The basic idea of health impact assessment (HIA) as a task for public health professionals and for the public health service in Germany was introduced in the late 1980s. Since then, the idea was pursued both within as well as independent from existing impact assessment procedures in the environment sector. By now, it is widely held that impact assessment is a key instrument to link science and decision making, offering unique opportunities for the protection and promotion of human health.

In starting to present the German situation it seems appropriate to acknowledge some major international strands of development that were explored in the past and stimulated the HIA development in Germany. Such approaches included WHO's 'Health & safety component of environmental impact assessment', 'Baseline risk assessment' of the US-Environmental Protection Agency, the Dutch 'Effectvoorspelling' and 'Gezondheidseffectrapportage', 'Health aspects of environmental impact assessment' in Canada, 'Public Health assessment' of the US-Agency for Toxic Substances and Disease Registry, the Australian 'Environmental & health impact assessment', New Zealand's 'Health impact assessment', and the 'Health risk assessment' of the California Air Pollution Control Officers Association [1]. The latter approach, in our perspective, was among the first to apply the concepts and methodology of quantitative risk assessment prospectively to development projects.

It was noticed in Germany that prominent policy documents including Agenda 21, the WHO Health for All (HFA), and Health 21 programmes as well as many of the National Environmental Health Action Plans (NEHAPs) in Europe call for improved impact assessments. Today, HIA is seen as a 'smart' combination of consultation and modelling, and is gaining recognition as an attempt to intelligently apply existing knowledge for practical decision-making. This chapter summarizes key lines of HIA development in Germany, highlighting some topics of the current debate.

Project HIA in Germany

Environmental impact assessments (EIAs) were performed in Germany on a voluntary basis since the 1980s. At an early stage, Wodarg [2] published a paper on HIA ('Gesundheitsverträglichkeitsprüfung') and thus brought the issue to the attention of the public health community in Germany. In 1990, the EU directive on EIA was transformed into German legislation (Umweltverträglichkeitsprüfungsgesetz, UVPG). As prescribed in the EU directive, the act includes 'humans' in the list of items to be protected. It soon became apparent, however, that the coverage of human health issues was limited and often did not satisfy the expectations. In 1992, the Conference of the German Ministers of Health passed a resolution on HIA in the context of EIA.

A first HIA project (research and development project) was conducted during 1992–1996, as a cooperative project of the University of Bielefeld and the Institute of Public Health North Rhine-Westphalia. The project was part of a North Rhine-Westphalian Research Consortium, and was funded by the Federal Ministry of Research and Technology. It aimed to improve the coverage of human health in the process of EIA and included the following components: analysis of status quo concerning HIA, including legal basis and existing approaches; survey of current practice and involvement of public health departments concerning HIA; analysis of HIA documents with respect to coverage of health aspects; comparison and evaluation of existing HIA approaches; development of a 'generic' HIA concept, which would be broadly acceptable from scientific as well as from practical perspectives; deployment of quantitative risk assessment as a key methodology for HIA; and evaluation of this concept in model applications. This approach was applied to various topics, including the transport sector.

Within this project, the current HIA situation in Germany was analyzed by means of document analysis and postal survey. In an existing collection of EIA documents, all documents dealing either with transportation or waste disposal projects were analyzed. This set contained 51 EIA documents concerning transportation including 46 highway projects and 5 rail projects, as well as 20 documents concerning waste disposal, including 8 dump site projects, 10 incinerator projects, and 2 recycling plants. The document analysis was performed as a screening version for all documents and then as an in-depth version for a subset of documents. The screening analysis found limited or missing coverage of human health aspects in the majority of documents. The in-depth analysis confirmed a lack of systematic approaches. In summary, the coverage of human health aspects in the documents tended to be highly incomplete.

In order to investigate the involvement of the Public Health Service in prospective impact assessment, a postal survey was performed covering the local health departments in the state of North Rhine-Westphalia. In summary, the survey showed the Public Health Service to be highly motivated to engage in HIA. At the same time, due to the inherent complexity of HIA, it demonstrated the need to provide guidance. The demand of a feasible procedure for inclusion into the 'tool-box' of local health departments became obvious and triggered the refinement of the integrated HIA approach.

The project identified several essential HIA components, especially the following:

◆ Prognosis of a future situation or course of events.

◆ Assessment: attaching value statements to the predicted future(s).

◆ Communication and participation: communicate the prediction(s) and assessment(s)—HIA, by its very nature, requires public participation which is not possible without adequate communication.

◆ Evaluation: evaluate the predictions, the assessments, and the communication.

Originating from the context of EIA, a ten-step model was designed (Fig. 23.1) [3]; it covers the broad categories of steps necessary in most HIAs:

◆ Step 1: Project analysis characterizes expected hazards, including acute toxicity and carcinogenicity during normal operation as well as accidental releases.

◆ Step 2: Regional analysis describes physio-geography, meteorology, natural features, and land use, and includes a definition of the study area for further investigation.

◆ Step 3: The population is described by size, age, gender, health status, and behavioural patterns, for example, food consumption patterns and leisure activities.

◆ Step 4: The background situation is characterized based on the preceding three steps and on environmental monitoring of existing pollution.

◆ Utilizing analogies and dispersion modelling, step 5 involves the prognosis of future pollution, including air, surface and ground water, soil, flora, and fauna.

◆ The key step, step 6, is the prediction of health impacts. It consists of both a qualitative assessment of changes concerning neighbourhood features, quality of life, and citizen concerns and a quantitative risk assessment.

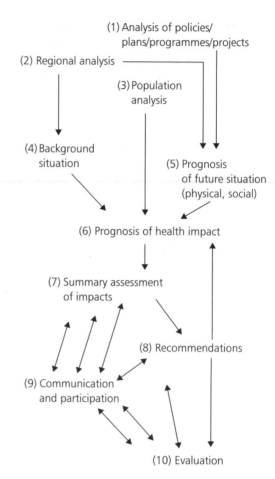

Fig. 23.1 Ten-step model of HIA.

- Step 7: A summary assessment of the predicted health impacts is given.

- Step 8: Based on all the information of the preceding steps, recommendations are made concerning planning alternatives, emission control, monitoring, public information, post-project analysis, and so on.

- Step 9: Communication of the results and underlying assumptions to all parties involved including planners, decision makers, and the public at large. This is no easy task given the complexity of the overall procedure, the numerous details of the methods, and the range of assumptions involved.

Whenever one of the proposed project alternatives scrutinized by HIA is actually implemented, the opportunity arises to evaluate HIA methods and assumptions (step 10), by comparing the predicted impact to the actual situation in terms of the state of the environment as well as human exposures and health outcomes.

An example where the ten-step procedure was applied is the enlargement of a waste disposal facility. Regarding the planned extension of a non-toxic waste disposal site, a task force on HIA was formed, and the ten-step HIA model approach was applied. Even in the 'common', non-toxic waste disposal site, complex physico-chemical processes take place, depending on the waste composition including solubility and volatility of components, on humidity, acidity, and temperature. These processes last for long time spans (decades) beyond the filling phase of the disposal site. They strongly change over time and involve discharges of gases, dust, microbial contamination, and fluids (leachate). In the typical case, gases are collected and incinerated, resulting in stack emissions composed of a variety of anorganic and (chlorinated) organic compounds. In addition, trucks delivering waste will travel to and from the waste disposal site, so traffic emissions (chemicals, noise) and traffic-related injuries also need to be considered.

The second HIA application reported here refers to a planned major by-pass road in the City of Krefeld [4]. Due to a long-standing problem of traffic congestion within inner-city areas of Krefeld, plans were made to build a by-pass road, relieving inner-city areas partially from traffic flows. The EIA procedure for the by-pass road took six different routes into consideration. Changes of traffic flow had been computed using two different planning scenarios, the first of which implied constant numbers of employees in the area, whereas the second scenario implied slightly increased numbers in future years. The EIA had led to the recommendation of one specific route alternative.

The Chief Officers of the German State Health Departments installed a 'Committee on HIA in Environmental Impact Assessment'; the committee developed a concept concerning procedure and contents of HIA [5]. Several states, especially Hamburg [6], have since been active in performing HIAs. A first German monograph on HIA was published in 1997 [7]. A survey on HIA training programmes in Germany was conducted in 1997 [8], and a survey on the status quo of HIA in North Rhine-Westphalia in 1998 [9].

In the process of preparations for the 2nd European Ministerial Conference on Environment and Health, held in London 1999, the crucial role of the local and regional level for implementing HIAs became obvious, and it was decided to give special attention to this topic. A scientific meeting was held in Bielefeld (Germany) in May 1999, which examined the HIA of transportation with a focus on the regional and local perspective [10]. The meeting identified knowledge gaps and research priorities, including insufficient knowledge on exposure–effect relationships, and on interaction of various factors with each other; lack of precise knowledge on the vulnerability of groups such as the young, the elderly, and the sick; difficulty to include socio-economic factors formally in HIAs; and need of a better basis for 'scoping' decisions.

The meeting also identified a range of policy priorities. Priorities concerning procedural and methodological aspects of HIAs included the following: (1) Efforts should be taken to complete the range of health outcomes covered in HIAs, and equity should be taken into account. Mechanisms for timely public participation and stakeholders involvement as well as communication should be improved. (2) The knowledge base supporting decisions between threshold concepts and quantitative dose–response relationships needs to be improved. Uncertainties in HIAs should be made explicit and put in proper perspective. (3) The application of Geographic Information Systems (GIS) for HIA purposes should be explored, and ways should be sought to combine GIS with standard statistical tools. (4) Whenever possible, opportunities to follow-up and validate HIA methodology should be utilized.

In addition, the meeting identified policy priorities concerning the institutional framework and infrastructure of HIAs. It was stated: (1) An overview of 'current practice of HIA in Europe' would be useful. Such a report would include policies, procedures, methods, institutions, and case studies. (2) The exchange of HIA documents and literature should be facilitated, and the usability of electronic resources should be explored. (3) A series of European HIA workshops should be prepared, for example, several 2-day workshops over the next three years. Strategies for capacity building need to be explored. (4) The production of toolkits should be considered, useful for the efficient preparation of HIAs. The need of consensus-building activities should be explored.

In 2001, the first German national HIA workshop sponsored by federal institutions was held in the context of the German NEHAP. The workshop provided a comprehensive overview of the status quo of HIA in Germany. Health impact assessment is mainly conducted within the framework of EIA. Applied methodologies range from expert judgement via simple modelling and comparison with limit values to sophisticated modelling and assessment. As far as the workshop participants were aware, HIA has not been applied to policy or planning in Germany up to now. Workshop participants agreed on the following objectives for further HIA development in Germany [11].

1. Conducting status quo analysis of HIA in Germany.
2. Creating an organizational framework for further HIA development, including vertical linkages between administrative levels and horizontal linkages between sectors.

3. Fostering collaboration between EIA and HIA experts.

4. Improving the legal basis for coverage of health in planning procedures.

5. To critically discuss (and possibly replace) the term 'Gesundheitsver-träglichkeitsprüfung'.

6. Fostering participation via improved transparency of planning procedures.

7. Creating a framework for HIA developments.

8. Fostering methodological developments, including guidelines, recommendations, and checklists; adaptation of sophisticated methods; quality assurance.

9. Education and training programmes.

10. Measures towards efficient communication and cooperation.

Strategic HIA in Germany

While most of former HIA work revolved around individual projects, it has become increasingly obvious that the policy and planning level deserves at least the same level of attention. In analogy to strategic environmental assessment (SEA), it seems useful and even imperative to investigate how health determinants are being influenced by policies, plans, and programmes. Such HIA of policies, plans, and programmes (i.e., strategic HIA) constitutes a new step. Concerning planning, the following distinction needs to be made: (i) 'general' planning of local, regional, and other development including land use; (ii) sectoral planning, for example, transport, agriculture, provision of water/energy, and so on.

Health impact assessment is now required by state law in several German states (Bundesländer). For example, under the headline of 'Involvement in planning', section 8 of the North Rhine-Westphalian Public Health Service Act of December 1997 requires the local health authorities to be involved in all planning procedures whenever health is an issue [12].

Worldwide there is a tendency towards deregulation in many policy sectors, including liberalization and privatization of drinking water management. Concerns about negative impacts on human health call for prospective HIA on this issue. As a contribution to a more comprehensive HIA of drinking water privatization, an analysis was performed with the cooperation of lögd NRW and the University of Bielefeld in order to contribute quantitative estimates concerning health effects from increased exposures to carcinogens in drinking water. Using data from North Rhine-Westphalia in Germany, probabilistic estimates of excess lifetime cancer risk as well as estimates of additional cancer cases from increased carcinogen exposure levels are presented. The results demonstrate how exposures from contaminations strictly within current legal limit values could cause cancer risks and case loads to increase substantially. The study also serves to demonstrate uncertainty involved in predicting the health impacts of alterations in water quality [13].

Up to now, the experience with strategic HIA in Germany is very limited. It seems reasonable, however, to assume that the basic rationale of project HIA as developed over the last decade should also fit the needs of strategic HIA. An adaptation of the ten-step HIA model to strategic HIA is shown in Table 23.1.

Table 23.1 Comparison of project HIA and strategic HIA, concerning steps 1–7 of the ten-step model

Step	Project HIA	Strategic HIA
1. Subject analysis	Project details incl. (qualitative and quantitative) technical specifications	Contents of policy, plan, or programme
2. Regional analysis	Status quo of the study area(s) potentially affected by the project	Status quo of the study area(s) potentially affected incl. major fluxes of energy, matter, etc.
3. Population analysis	Size, composition, health status, behavioural patterns of population potentially affected (residential and occupational populations of the study area)	Same, plus other aspects incl. life style, socio-economic development, time budgets; also populations indirectly affected via physical and/or informational connections
4. Background analysis	Existing burdens in the physical and social environment, incl. current levels of pollution; existing resources at jeopardy of deteriorization or depletion	Same, plus (i) wider physical surroundings and (ii) other societal sectors potentially affected or potentially acting as a buffer against unwanted implications
5. Prediction of environmental impacts	Expected emissions of chemicals, noise, radiation, microbes, etc., predicted impacts on physical and social environment expected deteriorization or depletion of resources	Same, plus expected impact on fluxes of energy and matter, on socio-economic development, social coherence, and disruption
6. Prediction of health impacts	Expected health impairments and benefits resulting from the predicted changes in the physical and social environment	Same, plus expected impact on 'avoidable deaths and disabilities' in a longer time-range; common metrics, for example, life expectancy, DALYs
7. Summary assessment	Summary assessment of predicted impacts, using environmental and health standards, risk analysis, expert rating, etc.	Same, with a probable shift of focus from specific environmental and health standards towards more comprehensive expert opinion

Due to the limited experience with strategic HIA, it is difficult to identify, with any level of certainty, the differences between project HIA and strategic HIA. Tentatively, such key differences may include the following: (1) Instead of specific emissions and the like, we will typically have to deal with broader changes concerning overall fluxes of energy and matter. Models of 'regional metabolism' exist [14] but have not yet found widespread recognition. In addition, for the purpose of HIA, we need broader concepts of human exposure. (2) Even with project-level HIA, there are close links with monitoring/surveillance of the local situation. A good existing reporting system can both facilitate the preparation of the impact statement and also provide a basis for HIA evaluation in case the project does materialize. For strategic HIA, it seems imperative to consider (long-term) time trends of key parameters and superimpose the expected effect of the policy, plan, or programme onto this trend. (3) Both the true variation and the observer uncertainty will tend to be higher (possibly much higher) in strategic HIA. Due to systemic interactions between components, it can be difficult to predict even the direction of change of relevant parameters; concerning quantification, it may be next to impossible to predict quantities of change with any degree of reliability. (4) Due to less 'obvious' relationships between policies, plans, and programmes to everyday life from the public's perspective, it may be harder to involve representatives of potentially affected population groups.

Germany is currently involved in an EU-funded project 'Policy HIA for the European Union'. Article 152 of the Treaty of Amsterdam [15] made explicit the commitment of the EU to ensure that human health is protected in the definition and implementation of all Community polices and activities. However there is no accepted methodology for assessing the impacts of EU policies on health within the Community, although many organizations are carrying out HIA at regional or member state level. Existing HIA methodologies and methods have been collected, reviewed, and analyzed by project partners from England, Ireland, the Netherlands, and Germany. Based on this analysis and the project partner's skills, knowledge, and experience in the field, a synthesized generic policy HIA methodology was developed. A pilot study using the methodology will be carried out on an EU policy at EU and member state level. The Institute for Public Health North-Rhine Westphalia is responsible for piloting the methodology within Germany. The end product will be applicable to EU and other types of policy.

Strategic HIA is more than the counterpart of project HIA on a higher level. In addition to some minor adaptations, one has to add specific approaches as sketched above. Strategic HIA will need time and resources to evolve properly. Nevertheless, current standard methods and experiences with project-level HIA seem to provide a good basis for facing the new challenge.

Perspectives

Currently the coverage of human health aspects in EIA still tends to be incomplete. Possible reasons include the following: complexity of the task of prospective impact assessment; insufficient provision of specific methods, tools, instruments; inadequate access to data that are both current and reliable; and lack of a systematic evaluation of HIA applications. To date, public health departments often seem to be left out from the circle of participating institutions, especially from the initial scoping step, which is of crucial importance for the entire HIA procedure.

Health impact assessment can be conducted on different levels of complexity, each of which features specific strengths and weaknesses. For future developments, it may be useful to distinguish these levels more carefully in terms of strengths and weaknesses than has been done in the past. For completeness, one could include level 'zero' implying no HIA at all, which means incurring no (immediate) costs but running the risk of missing an important opportunity of prevention and health promotion. The first level of HIA then is an 'ad hoc' assessment as an improvised procedure, mostly based on informally collected opinions. This involves low costs but will tend to be subjective and unreliable. Another level is a predominantly qualitative expert rating(s), based explicitly on the body of established knowledge. This approach will be widely applicable, and moderately time-consuming. It involves limited transparency, however, on how experts reach their conclusions. Under favourable circumstances, a quantitative approach provides a prognosis based on modelling procedures, and an assessment based on explicit standards. In this case, best use is made of existing knowledge, and there is a high level of transparency.

A discussion on HIA quality criteria includes the following candidates: (1) transparency of how the prognosis and the assessment are done; (2) objectivity (as far as achievable); (3) choice of adequate modelling (e.g., agents with versus without threshold limit); (4) empirical basis for the assumptions that enter into the HIA; (5) level of completeness of health outcomes considered; (6) explicit representation of both variation and investigator uncertainty; (7) integration of outcomes.

The current efforts to improve HIA practice in North Rhine-Westphalia, especially on the local level, focus on the development of HIA tools in order to strengthen the local health authorities' role in planning procedures. Currently, the coverage of human health aspects in EIA still tends to be incomplete, and public health departments often do not participate.

One prerequisite for the broader use of exposure assessment in HIA is the provision of default information for key exposure factors, which are often used in exposure assessments such as body weight, time budget, food intake, and the like. Additionally, more advanced techniques like probabilistic analysis need information about the probability distribution for these factors. Compilations of default values and distributions are already available in the United States, but do not necessarily reflect the European or German situation. On the other hand, the German report 'Exposure assessment standards' [16] offers only limited background information on the distribution of exposure factors and reflects the state of the art of one decade ago. An update of this report is currently being prepared by a consortium of universities in Bielefeld, Hamburg, and Bremen, funded by the Federal Environmental Agency. The update offers the opportunity to integrate probability density functions for selected exposure factors into this document. It is planned to also include guidance on methods of probabilistic modelling, an overview of modelling tools, a discussion of the implications of using specific probability density functions with respect to uncertainty analysis, treatment of correlation among variables, and interpretation of the results.

It is encouraging to see how the idea and vision of HIA, in spite of numerous obstacles, currently is gaining momentum on the European level. One of the challenges we face in Germany is to efficiently link the local and regional HIA development with these diverse activities in the international arena.

References

1 California Air Pollution Control Officers Association. *Air Toxics 'Hot Spots' Program—Revised 1992 Risk assessment guidelines.* Sacramento, CA, CAPCOA, 1993.

2 **Wodarg W.** Die Gesundheitsverträglichkeitsprüfung (GVP)—eine präventivmedizinische Aufgabe der Gesundheitsämter. Öffentl. Ges.wesen 1989; **51**:692–697.

3 **Kobusch, A-B, Fehr R, Serwe H-J, and Protoschill-Krebs G.** GVP—eine Schwerpunktaufgabe des öffentlichen Gesundheitsdienstes [*HIA—a key task for the Public Health Service*]. Ges.wesen 1995; **57**:207–213.

4 **Serwe, H-J and Protoschill-Krebs, G.** Gesundheitsverträglichkeitsuntersuchung der Umgehungsstraße B 9n/Krefeld [*HIA of city bypass highway B 9n/Krefeld*]. Ergebnisbericht. Nordrhein-westfälischer Forschungsverbund Public Health, Universität Bielefeld, 1996.

5 Projektgruppe 'GVP im Rahmen der UVP' der AGLMB. Konzept zu Ablauf und Inhalt der Gesundheitsverträglichkeitsprüfung [*Concept for procedure and contents of HIA*]. Bericht gem. Beschluß der Arbeitsgemeinschaft der Leitenden Ministerialbeamten der Länder v. 14./15.4.1994; September 1994.

6 **Nennecke A and Fromm K.** Aufwand und Nutzen von GVP—Ein Resümee aus drei Jahren UVP-/GVP-Praxis in Hamburg [*Cost and benefit of HIA—a summary of three years of EIA/HIA practice in Hamburg*]. In Kobusch A-B, Fehr R, Serwe H-J (Hrsg.): Gesundheitsverträglichkeitsprüfung: Grundlagen—Konzepte—Praxiserfahrungen. pp. 293–308, Baden-Baden, Nomos Verlagsgesellschaft, 1997.

7 **Kobusch A-B, Fehr R, and Serwe H-J.** Gesundheitsverträglichkeitsprüfung: Grundlagen—Konzepte—Praxiserfahrungen [*Health Impact Assessment: basics—concepts—practical experiences*] pp. 318. Baden-Baden, Nomos Verlagsgesellschaft, 1997.

8 **Redecker-Beuke B.** Gesundheitsverträglichkeitsprüfung (GVP): Entwicklung eines Curriculums und Erhebung zum Stand der GVP-Fortbildung in Deutschland [*HIA: Development of a curriculum, and status quo analysis concerning HIA training in Germany*]. Diplomarbeit 1997, Fak. f. Gesundheitswissenschaften, Univ. Bielefeld. Materialien 'Umwelt und Gesundheit', lögd NRW, Bielefeld, Nov. 2000 (viii +97 Seiten).

9 **Machtolf M and Barkowski D.** Status-quo-Analyse zur UVP-Praxis NRW. Abschlußbericht. [*Status quo analysis concerning the EIA practice in North Rhine-Westphalia*]. IFUA-Bericht im Auftrag des lögd. Reihe Materialien Umwelt und Gesundheit, Nr. 13, Bielefeld, 2000.

10 **Fehr R and Lebret E.** Report on the Scientific Meeting Health Impact Assessment of transportation—the regional perspective, Institute of Public Health (loegd) North Rhine-Westphalia, Bielefeld, and Dutch Institute of Public Health and the Environment (RIVM), Bilthoven, in Bielefeld, May 17–18, 1999.

11 **Welteke R and Fehr R.** (eds.) Workshop Gesundheitsverträglichkeitsprüfung/Health Impact Assessment. Berlin, 19–20 November 2001. Proceedings/Tagungsband. Berlin pp. 188, 2002.

12 OEGDG / Gesetz über den öffentlichen Gesundheitsdienst (ÖGDG). Gesetz- und Verordnungsblatt für das Land Nordrhein-Westfalen, Nr.58, 17. Dez. 1997, pp. 431–436.

13 **Fehr R, Mekel O, Lacombe M, and Wolf U.** Towards Health Impact Assessment of drinking-water privatization – the example of waterborne carcinogens in North Rhine-Westphalia (Germany). *Bulletin of the WHO* 2003; **81**(6):408–414.

14 **Baccini P and Bader H-P.** Regionaler Stoffhaushalt. Erfassung, Bewertung und Steuerung [*Regional metabolism. Documentation, evaluation, and control*]. Spektrum Akademischer Verlag, Heidelberg, 1996.

15 European Community: Treaty of Amsterdam amending the Treaty on European Union, the treaties establishing the European Communities and related acts. *Official Journal C* 10 November 1997; **340**:0173—0306.

16 Ausschuß für Umwelthygiene der Arbeitsgemeinschaft der Leitenden Medizinalbeamtinnen und -beamten der Länder (AGLMB). Standards zur Expositionsabschätzung [*Standards for exposure assessment*]. Behörde für Arbeit, Gesundheit und Soziales (BAGS), Hamburg, 1995.

Chapter 24

HIA in Schiphol Airport

Brigit AM Staatsen, Ellis AM Franssen, Carla MAG van Wiechen, Danny Houthuijs, and Erik Lebret

Introduction

Reviews from the Netherlands, Germany, Canada, and the United Kingdom all show that coverage of human health aspects in environmental impact assessment (EIA) tends to be limited, lacking a systematic approach and developed methodology [1–3]. Assessment of health risks is often only qualitative or stops short at comparison of predicted pollution levels with available standards. Often the data needed to quantify risks such as numbers exposed, exposure levels, and exposure–response curves are unavailable. Further, very few health impact assessments (HIA) include follow-up to verify that any predictions made for the option implemented were accurate [4].

This chapter describes the HIA programme for Schiphol Airport, Amsterdam, in which impacts have been quantitated using a combination of health impact calculations, analysis of health registries, and epidemiological field studies [5]. A monitoring programme has been established to check that the predictions made for health impacts of noise management measures and airport expansion are correct. Finally the lessons of this programme for future HIA in the context of an EIA are discussed.

Background

Schiphol is the fourth largest airport in Europe in terms of passenger numbers, freight traffic, and commercial traffic, only London, Paris, and Frankfurt are larger. The airport is situated in a densely populated area on the outskirts of Amsterdam. It originally had four runways and a fifth was added in 2003 (Fig. 24.1).

An EIA published in 1993 assessed the impact of the proposed fifth runway and other options on both the environment and public health [6,7]. The terms of reference for the EIA were based on an advisory report of the State Inspectorate of Public Health and interviews with stakeholders. These terms

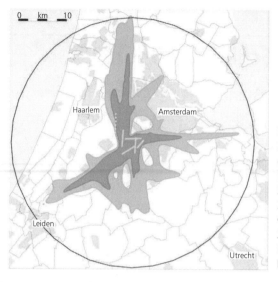

Light grey
20 Ku noise
contour 2001
Dark grey
35 Ku contour
2001
Circle
25 km around airport
Dotted line
New runway 2003

Fig. 24.1 Location of Schiphol Airport.

required a description of current health status using data from health registries and a study of the perceived health risks. They also required proposals for further research and health monitoring.

From the start it was clear that assessment of health impacts in this EIA would be limited by lack of time and unavailability of data. The Ministry of Housing, Spatial Planning and the Environment, the Ministry of Transport, Public Works and Water Management, and the Ministry of Public Health, Welfare and Sport, and the Commission for Environmental Impact Assessment requested the National Institute of Public Health and the Environment (RIVM) to prepare further studies of the health effects of environmental pollution related to air traffic. The Schiphol HIA research programme was therefore established by RIVM working in collaboration with other Dutch research institutes and universities. A Steering Committee with representatives of the ministries oversees the project. Two advisory bodies consisting of policy makers, representatives of local action groups, municipal health services, general practitioners association, the airport, and the local population are periodically consulted. Studies are reviewed by an *ad hoc* advisory committee of experts.

The objectives of the Schiphol HIA research programme are to:

◆ describe the current health status and potential health risks of the population in relation to environmental pollution from Schiphol Airport;

◆ collect information about exposure–response relationships relevant to the airport;

- develop a system to monitor the health status of the population after expansion of the airport, which can be used to inform decision makers.

The Schiphol HIA programme is being conducted in three phases.

Phase I was completed in 1993. It was performed as part of the EIA and consisted of an assessment of the number of people affected, a literature analysis, a qualitative risk assessment, an analysis of health registry data, and a limited survey on risk perception and annoyance. It identified gaps in knowledge and led to proposals for research in the next phase [8].

Phase II was carried out during the period 1995–2002. It consisted of two parts: an analysis of existing health registries and epidemiological field studies. The examination of registries involved semi-ecological studies relating individual health data and aggregated measures for exposure (often at a postal code level). These types of studies have methodological problems due to the limited availability of data on important confounders and must be interpreted with care. However they serve a sentinel function and when differences in disease occurrence are observed they signal that further investigation may be worthwhile. The epidemiological field studies investigated the relationship between aircraft-related exposures and potential health effects.

The various studies undertaken in Phases I and II are listed in Table 24.1.

Phase III started in 2002 and consists of a monitoring study to assess the impacts due to the expansion of the airport.

Selection of health outcomes

Informed by the risk analysis from Phase 1, health outcomes for further study were selected on the basis of biological plausibility, known exposure–response relationships, number of people affected, and public concern. The health outcomes initially selected were cardiovascular and respiratory diseases, sleep disturbance, annoyance, birth weight, cognitive performance, and medication use. This list was then refined by considerations of statistical power and the practicality of measuring the outcomes. Finally it was decided to focus the epidemiological field studies on sleep disturbance, annoyance, respiratory diseases, and cognitive performance. Cardiovascular diseases and birth weight were studied using only health data registries. An additional field study of high blood pressure was considered but not deemed feasible. Medication use, respiratory disease, and sleep disturbance were assessed through a field study as well as with existing data from health registries.

The health outcomes chosen include direct and indirect effects of environmental exposures, pathophysiological function (e.g., bodily complaints, performance, awakenings), well-being (perceived health, risk perception, annoyance,

Table 24.1 Studies included in Phases I and II

Study	Year	Outcomes and method	Analysis and sample
Cardiovascular disease			
1.	1991–1993	Hospital admissions for myocardial infarction, hypertension, ischaemic heart disease, and cerebrovascular disease [9,10]	Analyzed by 4-digit postal code, 1.5 million population in area 55 × 55 km around Schiphol
2.	1996	Postal questionnaire on medicines for cardiovascular disease or raised blood pressure [11]	Related to 1. Noise level for postal code area (6 digit) 2. Distance from airport (proxy for air pollution) 11,812 adults within 25 km radius of Schiphol
Annoyance			
3.	1996	Postal questionnaire on specific (aircraft noise) and non-specific annoyance [11]	As for study 2
4.	1993–2002	Trends in complaints about aircraft noise [12].	Related to number of flights
Sleep disturbance			
5.	1993–1994	Pharmacy data on use of sedatives [13]	Analyzed by noise level for postal code area, (4 digit) 213,524 people served by 32 pharmacies in 55 × 55 km area around Schiphol
6.	1996	Postal questionnaire on sleep disturbance, sleep quality, and use of sedatives and sleeping pills [11]	As for study 2
7.	1998–2002	Panel study of movement during sleep (by actimetry), night-time waking, daytime tiredness (by diary), attention test, perceived health (questionnaire) sleep disturbance, sleep quality [14]	Related to noise events/levels measured in subjects' bedroom and outdoors near their house. 418 subjects for 11 nights living in high and low night-time noise area around Schipol
Respiratory disease			
8.	1993–1994	Pharmacy data on use of medication for asthma [13]	Analyzed by distance from airport (proxy for air pollution) Population as for study 5
9.	1991–1993	Hospital admissions for acute airway infections, upper respiratory symptoms, bronchitis, asthma, and emphysema [9]	As for study 1

Table 24.1 (continued)

Study	Year	Outcomes and method	Analysis and sample
10.	1996	Postal questionnaire—one or more respiratory symptoms (asthma, chronic cough, phlegm, bronchitis), medical treatment for allergies, medication for asthma or allergy [11]	As for study 2
11.	1998	Measurement of pulmonary function, IgE levels in blood, respiratory symptoms recorded by parent questionnaire [15]	Related to PM_{10}, $PM_{2.5}$, NO_2, and VOC measured in ambient and indoor air at schools 2500 primary school children in 30 schools around Schiphol

Neurobehavioural effects

12.	1995	Cognitive and psychomotor function measured with computerized neurobehavioural evaluation system and paper/pencil tests, behaviour and annoyance studied by questionnaire [16]	Related to calculated noise levels near school 159 children (age 8–9) in 2 schools, one in high-noise village and one in low-noise village.
13.	2001–2003	Cognitive and psychomotor function, behaviour and annoyance measured as in 12	Related to measured and calculated noise levels for school postal code area, 730 children (age 8–11) in 33 schools around Schiphol

Birth weight

14.	1989–1993	Birth weight and duration of pregnancy from obstetrics records [17]	Analyzed by postal code area (4 digit), 83,751 babies born in 55 × 55 km area around Schiphol

Perceived health

15.	1996	Postal questionnaire on health complaints and self-rated health [11]	As for study 2

Risk perception and residential satisfaction

16.	1996	Postal questionnaire on safety, concern about health effects, fear of aeroplanes, (dis)satisfaction with and (un)favourable aspects of housing and neighbourhood [11]	As for study 2

disturbance), morbidity, use of medical services (hospital admissions), and medication use.

In many cases, the causal pathways linking environmental exposure to health effect are poorly understood. Some effects are direct and others indirect.

For example, blood pressure may be directly affected by exposure to noise, but may also be indirectly affected through stress caused by noise annoyance. Since the actual pathways are largely unknown, a variety of end points is being studied.

Exposure assessment

Exposure to aircraft noise around Schiphol has been calculated using the National Aerospace Laboratory (NLR) model. For the sleep disturbance study additional measurements of indoor and outdoor noise were taken. In most of the health registry studies, the aggregated noise levels for each wide (4 digit) postal code level, was used as a measure of exposure. For the questionnaire survey however, aircraft noise exposure was calculated for the geometric centre of each smaller (6 digit) postal code area. For most studies the noise level metric used was B65 expressed in Kosten units but a wide range of other metrics of noise level were also used. B65 is a yearly average defined by the maximum noise levels during flights, the total number of flights, and the time at which these flights take place with evening and night flights having more weight than day flights. In calculating B65, aircraft movements with calculated noise levels at the ground of more than 65 dB(A) are included. Metrics based on (then) legally established methods of aircraft noise assessment were included as well as metrics suggested in national and international (EU) discussions on uniform metrics for (aircraft) noise levels.

No suitable data on air pollution and odour caused by air traffic were available so the distance from the airport was used as a proxy measure of exposure to these factors. In the study of respiratory disease in children a combination of measurements and modelling was used.

Health outcomes

Annoyance

Annoyance is one of the chief factors considered in evaluating the expansion of Schiphol Airport. Accurate assessment of the current level of annoyance is therefore extremely important. In Phase I, using modelled noise exposure it was estimated that in 1991 over 100,000 people were severely annoyed by aircraft noise. In 1996 a postal questionnaire survey was carried out, in order to assess the prevalence of annoyance and update the exposure–response relationships. Of the 30,000 people who were approached, 39 per cent responded. A small telephone survey of non-responders showed that selective non-response had probably biased the result. Table 24.2 shows the results from the questionnaire study for annoyance (measured by an 11-point scale with not at all annoying and extremely annoying as end points) due to aircraft and airport-related activities.

Table 24.2 Annoyance in the population aged 18 years and older living within 25 km of the airport and the proportion attributable to aircraft noise [10]

Variable (Highly annoyed[a] due to aircraft)	Percentage highly annoyed or highly sleep disturbed		Absolute number highly annoyed or highly sleep disturbed
	Corrected	Uncorrected	
Highly annoyed by noise			
All study area	18	31	265,000–465,000
High-noise area[b]	36	53	98,000–158,000
Very high noise area[b]	48	65	12,000–15,000
Highly annoyed by odour			
All study area	5	7	80,000–108,000
≤10 km	16	19	47,000–60,000
Highly annoyed by dust, soot, or smoke			
All study area	6	8	100,000–125,000
≤10 km	19	23	57,000–69,000
Highly annoyed by vibrations			
All study area	10	14	150,000–210,000
High-noise area[b]	11	15	60,000–84,000
Very high-noise area[b]	39	45	9000–11,000
Highly sleep disturbed by aircraft noise			
All study area	8	12	120,000–180,000
High night-noise area[b]	33	39	6000–7000

[a] Subjects are defined as highly annoyed or highly sleep disturbed if their score on the 11-point scale (0–10.0) exceeds 7.2.

[b] High-noise area is area in which B65 ≥ 20 Ku; very high-noise area is area in which B65 ≥ 35 Ku; high night-noise area is area where $L_{Aeq, 23-06\ h}$ ≥ 26 dB(A) indoors.

Various estimates have been made to correct the results for the possible bias due to selective non-response. Both the uncorrected and corrected estimates are presented, showing the range of possible values in the results.

Within a range of 25 km around the airport, 18–31 per cent of adults reported serious annoyance by aircraft noise (Table 24.3). In the high-noise zone (with noise levels exceeding the legal limit) 48–65 per cent of adults (12–15 thousand

Table 24.3 Perceived sleep quality, self-rated health and medication use in the population aged 18 years and older living within 25 km of the airport and the proportion attributed to aircraft noise [11]

| Variable | Number | Percentage of people | | Proportion attributed to aircraft noise, in absolute numbers of people ≥ 18 years |
		Reporting the complaint (%)	Proportion attributed to aircraft noise (%)	
Poor perceived sleep quality				
One or more sleep complaints				
All study area	1,520,750	72	—	—
High-noise area[c]	370,280	72	1.4–3.9	5300–14,300
Very high-noise area[c]	23,510	73	3.8–6.1	900–1400
Poor self-rated health				
All study area	1,520,750	20	—	—
High-noise area[c]	370,280	21	−0.4–2.8	−1500[a]–10,000
Very high-noise area[c]	23,510	21	2.3–4.4	500–1000
Use of prescription medication for cardio-vascular diseases or elevated blood pressure				
All study area	1,520,750	15	—	—
High-noise area[c]	370,280	17	0.6–1.4	2100–5200
Very high-noise area[c]	23,510	18	1.7–2.3	400–500
Use of prescription sleeping pills or sedatives[b]				
All study area	1,520,750	8	—	—
High-noise area[c]	370,280	10	1.2–2.2	4500–8100
Very high-noise area[c]	23,510	11	2.6–3.6	600–900

— The estimate is too inexact.

[a] As the wide range in estimate shows, the figures for the 20 Ku zone are particularly approximate. One reason for this is the low precision of the exposure–response relationship in areas with low aircraft noise exposure. The confidence interval on either side of the point estimates is so wide that negative values are possible.

[b] Respondents using medication for cardiovascular diseases, rheumatism, and those working in night shifts are excluded.

[c] High-noise area is area in which B65 ≥ 20 Ku; Very high-noise area is area in which B65 ≥ 35 Ku.

people) report that they were highly annoyed by aircraft noise. In lower-noise exposed zones, the percentage annoyed is lower, but since a large number of people live in them, the absolute numbers annoyed (265,000–465,000) are greater than in the 'high-noise zone'. Multivariate regression analysis showed that noise level was the strongest determinant explaining about 40 per cent of the variation in annoyance. The most important non-acoustical factors affecting annoyance were the respondents' noise sensitivity and their fear of crashes.

The reported aircraft noise annoyance was higher than predicted based on previous Schiphol studies (Fig. 24.2a) or international annoyance surveys (Fig. 24.2b). Possible explanations for the higher than expected annoyance

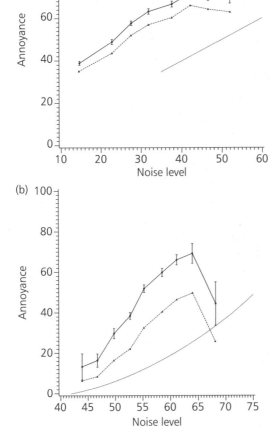

Fig. 24.2 Relationship between annoyance and aircraft noise exposure; (a) expressed in Ku; (b) expressed in L_{dn}.

levels include an increased sensitivity to noise and concern about safety, the actual exposure to noise being higher than the calculated values would indicate, and the influence of the ongoing political and social debate about the expansion of the airport.

About 80,000–108,000 people (5–7 per cent) are highly annoyed by odour from aircraft (Table 24.2). This is also higher than the model calculations indicated in Phase 1.

Sleep disturbance

Sleep disturbance by night-time noise may manifest itself in various ways:

◆ Primary effects like difficulties falling asleep, waking during the night, sleep stage changes, and instantaneous arousal effects during the sleep (temporary increase in blood pressure, release of stress hormones in the blood, increased motility);

◆ Secondary 'after effects' measured the next day: decrease of perceived sleep quality, increased fatigue, and decrease in mood and performance;

◆ Long-term effects on well-being as shown by increased medication use and chronic annoyance. It is not known whether the more or less instantaneous effects will lead to chronic changes or long-term health effects. Recovery mechanisms could prevent the occurrence of further effects.

In the questionnaire survey sleep disturbance was measured in the same way as annoyance. Taking into account non-response bias 8–12 per cent of the adult population (120,000–180,000) living within 25 km of the airport reported severe sleep disturbance from aircraft noise (Table 24.2). The majority of those experiencing sleep disturbance live in areas with noise levels below the legal limit for night-time aircraft noise (26 dB(A) $L_{Aeq, 23-06\,h}$, in the bedroom). In the area with high night-time aircraft noise levels ≥26 dB(A), 33–39 per cent of the population (6000–7000) reported serious sleep disturbance caused by aircraft noise. These numbers are a little lower than earlier predictions in Phase I.

An analysis of pharmacy data on drug dispensing in a circle of 25 km around the airport, showed that the use of sedatives was higher in areas with high aircraft noise exposure [13]. After the occurrence of chronic diseases had been taken into account the use of sedatives was 8 per cent higher in the high-noise area (odds ratio: 1.08 (1.00–1.18)).

The results of the questionnaire survey confirmed that the use of 'sleeping pills or sedatives' was related to aircraft noise. The results, adjusted for age, sex, education level, ethnicity, and degree of urbanization suggest than in high-noise areas 1.2–2.2 per cent of sleeping pills or sedative use could be attributed to aircraft noise (Table 24.3) [11].

Despite the results of these earlier studies and the findings in field studies [18], there was still debate about the noise levels at which sleep disturbance can occur and its long-term effects on the health status of the population. Therefore, an extensive field study was carried out in 400 adults to assess the relationship between night-time aircraft noise exposure and sleep disturbance as well as sleep-related health complaints [14]. Noise levels were measured continuously during the night in the bedroom and outdoors. Aircraft-related noise events were determined by combining noise data with data about aircraft movements. Information about sleep disturbance was collected by actimetry (degree of movement, frequency of waking), diary (sleep quality, medication), and questionnaires (annoyance, health complaints). Mean degree of movement during sleep was associated with the number of sleep and health complaints and self-reported sleep quality. People in locations with low aircraft noise levels were more sensitive to aircraft noise events than people living in locations with high noise levels. The use of sleeping pills and frequency of waking increased with increased exposure to indoor aircraft noise during sleep. No relation was found between night-time noise exposure and reaction time the following day. The prevalence of night-time aircraft noise annoyance and number of health complaints were associated with long-term night-time aircraft noise exposure. The exposure–response relations derived from this study were used to assess the prevalence in the Schiphol population of effects of aircraft noise on sleep. It was estimated that about 7 per cent of the adult population living in an area of 55 × 55 km near Amsterdam Airport has increased movement during the sleep and that for half of these (about 70,000 people) this was associated with aircraft noise exposure [19].

Cardiovascular diseases

In Phase I it was estimated that about 1500 extra cases of hypertension could be expected due to aircraft noise exposure in the 1.6 million adults living around Schiphol Airport. For this and other reasons there could be an increased risk of ischaemic heart diseases, therefore spatial patterns in hospital admission data for cardiovascular diseases were analyzed [9].

Age:sex Standardised Admission Ratios with 95 per cent confidence intervals for cardiovascular diseases were calculated and mapped per 4-digit postal code area for 1991–1993. Spatial patterns were studied using an empirical Bayesian model to reduce random variation and to account for small-area variability and spatial interdependence in the data. There was no consistent spatial pattern that would suggest a relation of cardiovascular diseases with Schiphol Airport. The disease patterns varied per year and differed between men and women [9]. In later analyses with more developed time–space models it was possible to include noise exposure levels in the calculations, but again no significant associations were identified [10]. In the absence of data on important determinants of cardiovascular disease (e.g., socio-economic status, lifestyle) this finding has to be interpreted with care. In most noise reviews the statistical evidence for a causal relation between noise exposure and cardiovascular health risk is considered to be inconclusive [20]. However, a small effect on cardiovascular risk is deemed highly plausible.

The questionnaire survey provided some additional data on the use of 'medicines for cardiovascular diseases or elevated blood pressure' prescribed by a physician. Regression analysis showed that the use of these medicines is related to both aircraft noise exposure and distance to the airport. About 0.6–1.4 per cent of the use of 'medicines for cardiovascular diseases or elevated blood pressure' in areas with an aircraft noise exposure ≥20 Ku could be attributed to aircraft noise (Table 24.3). For areas with an exposure ≥35 Ku this was 1.7–2.3 per cent.

Respiratory disease

Stakeholder interviews and discussion with citizen groups in Phase 1 showed that respiratory disease was an area of substantial concern. The Phase I report concluded that known and modelled air pollution exposure levels in the Schiphol area were similar to levels encountered elsewhere in urban areas and that air traffic emissions accounted for less than 10 per cent of background air pollution levels. On this basis, respiratory health effects from air-traffic related air pollution were considered unlikely. However, there was little information on exposure levels to particulate air pollution (PM_{10}, $PM_{2.5}$). Analysis of pharmacy data revealed an increased prevalence of medication for asthma within a radius of 10 km from the airport as compared to greater distances.

Because of public concern about this report and the findings of another ecological study in the area by the municipal health service an epidemiological investigation of respiratory health in school children was included in Phase II. This study involved 2500 primary school children, age 7–12 years, from 30 schools in different towns. Air pollution measurements were performed at the schools. Respiratory health was assessed with a questionnaire for parents on respiratory symptoms and measurements of lung function and tests for atopy (blood, skin-prick test). There were differences in prevalence of respiratory health problems between the towns, however these were not related to distance from the airport. Primary schools close to busy highways had higher concentrations of air pollution than those situated further from a highway. Levels of NO_2, soot, and benzene decreased with increasing distance from Schiphol Airport. No association was found between measures of air pollution exposure and the prevalence of respiratory symptoms, decreased lung function, or an increased level of IgE in blood [15]. In addition, measurements in 92 houses (high- and low-noise exposed) in the vicinity of the airport showed that sound insulation or changed ventilation behaviour due to noise annoyance from road and air traffic did not result in different indoor levels of contaminants (PM 2.5, aromatic and volatile hydrocarbons), of moulds, or of house dust mite allergens [21].

Psychomotor and cognitive performance effects

Earlier studies of children living in close proximity to airports in Los Angeles and Munich indicate that exposure to (aircraft) noise might result in negative effects on cognitive performance [22]. In Phase I no exposure–response curves for this were available so it was not possible to estimate the number of people around Schiphol Airport in whom cognitive performance might be impaired. Therefore an epidemiological field study was recommended for Phase II.

A pilot study was carried out to test the reliability of selected automated psychomotor and cognitive performance tests, and the feasibility of using them in a school environment [16]. At this stage no definite conclusions can be drawn, but further results will come from the EU-funded RANCH project—Road Traffic and Aircraft Noise Exposure and Children's Cognition and Health. This study is examining the relationship between chronic exposure to aircraft or road traffic noise and impaired cognitive function, health, and noise annoyance in children (9–12 years old), living around three airports (Schiphol Airport, London Heathrow, and Madrid Barajas).

Perception of risks and health

The surveys and stakeholder interviews in Phase I indicated that the presence and expansion of Schiphol Airport were a cause of considerable anxieties. This was confirmed in the more elaborate questionnaire survey in Phase II. Sixteen per cent of the respondents reported that they were very concerned about their safety because of living under the approach route of a large airport, while 64 per cent were not or hardly concerned. More people were concerned about health effects due to air pollution from aircraft (42 per cent) than about health effects from aircraft noise (18 per cent). Those exposed to more aircraft noise were more likely to be concerned about safety. Worries about safety were not only related to exposure to aircraft noise, but also to the frequency with which aircraft are heard and the number of flights passing overhead. Those exposed to higher aircraft noise levels were also more likely to rate their own health as poor. This information on perceived health in relation to aircraft noise exposure can be used as a baseline for future monitoring (Table 24.3).

Phase III—Monitoring programme

In February 2003, a fifth runway at Schiphol Airport became operational, creating space for a 50 per cent increase in the number of flights. Aircraft movements can increase from 400,000–600,000 per year. This expansion and changes in the spatial pattern of aircraft movements will lead to changes in environmental quality, which could affect the health of residents in the vicinity of the airport. At the request of the Dutch government, we designed a programme to monitor environment and health. The primary aim of the programme is to keep a close watch on how the ongoing expansion of the airport impacts on health. The programme should also be able to detect possible changes in environmental quality and changes in both short- and long-term health effects. It should also provide the government with information, needed to make decisions about the future development of air traffic in the Netherlands.

The programme consists of several different studies [23]. In 2002, before the opening of the new runway a postal questionnaire on annoyance, sleep disturbance, self-rated health, and residential satisfaction was sent to 13,000 inhabitants living in a region of 25 km around the airport. These data will provide a reference point against which possible future changes can be evaluated. In a panel study, a smaller group of about 600 subjects selected from the main survey will be followed up annually with a questionnaire. Panel members live in the areas where aircraft noise levels are expected to change. The aim is to assess the trends in effects like annoyance, sleep disturbance, and self-rated health.

The third study will track routinely collected health data at small-area level and monitor spatial patterns of disease around the airport over the long term. Data will be collected on hospital admissions and medication use, focusing on cardiovascular and respiratory diseases, and the use of sleeping pills and sedatives. The wide variation in population of different postal code areas (10–21,000) makes essential the use of spatial smoothing techniques, when estimating and comparing disease rates. In the analyses disease rates will as far as possible be adjusted for the most important confounding factors.

Complaints about aircraft noise provided by the Schiphol Environment Advisory Committee will also be monitored as an indicator of annoyance. Since the 1980s the number of complainants has increased in line with the number of aircraft movements, resulting in approximately 10,000 yearly complainants in 2000.

The frequency of taking measurements and choice of study area depends on the expected spatial and temporal distribution of the environmental changes and health impacts. In discussion of the monitoring programme, policy makers were requested to state how small a difference or change was politically important to detect, and whether they were more worried about missing a possible change or falsely detecting one that was not there (type 1 and type 2 error). These requirements of policy makers were then fed into power calculations for the study design. Interpretation criteria to identify true effects (strong signal) were also developed. A strong signal should fulfil all the following requirements:

- an increase or decrease in effect,
- adequate adjustment for confounders,
- a relation with distance to the airport or increase in exposure,
- consistent with data from other sources.

Discussion and conclusions: lessons learned

The scope of an HIA depends on the situation, the knowledge available, and the importance of potential impacts. Parry and Stevens distinguish between a rapid 'mini' HIA, based on easily available information, and a 'maxi' HIA that consists of an extensive analysis of existing data and literature, collection of new data, and quantification of impacts as well as full participation of stakeholders [4]. In the case of Schiphol, the deeply felt concerns of all parties resulted in a comprehensive maxi approach. However, is such a time-consuming approach required in other situations?

Phase I of the HIA, carried out as part of the EIA in 1993 took about one and a half years. Within this time frame it was possible to do both a qualitative analysis of existing data and collect some new data, allowing quantification of the health impacts of aircraft-related pollution. However, it was not possible to address all concerns of the population. The studies carried out in Phase II confirmed earlier predictions and provided additional exposure–response relationships, which can be used in future HIAs. In addition, the full range of health indicators (pathophysiological changes, use of medical services, morbidity) was studied for some diseases, which helped in assessing the plausibility of observed associations.

For example, an effect on sleep is highly plausible. The results for a full range of sleep disturbance indicators (self-reported sleep disturbance due to aircraft noise, poor sleep quality, the use of 'sleeping pills or sedatives', movements in sleep, health complaints) are all consistent. The extent and frequency of these indicators increase with increasing aircraft noise levels. The results from analysis of pharmacy data are supported by the results of the questionnaire and the actimetry study, in which information on various indicators and major determinants of sleep disturbance was collected at an individual level. The exposure–response relations derived by Passchier–Vermeer were used to estimate the prevalence of effects of aircraft noise on sleep in the adult population living near Schiphol Airport. Since the actimetry study has sufficient power and several shortcomings of earlier studies have been accounted for (e.g., control for confounding and outcome dependency due to repeated measurements, indoor noise measurements), the exposure–response relationships derived from this aircraft noise study can be used for future HIAs. The question about the long-term effects of sleep disturbance on health remains unanswered.

The collection of additional data in Phase II provided the opportunity of validating the health impacts predicted in Phase I. The results from the questionnaire survey show a higher prevalence of annoyance by aircraft noise than suggested by previous research in the Schiphol area and in other countries. The higher figures may be explained by increased sensitivity to noise, concern about safety, and higher than predicted actual exposure to noise. Or they may be explained by the effect of the ongoing political and social debate about the expansion of the airport. The risk assessment in Phase I showed that legal standards for aircraft noise and air pollution were not exceeded. However further analysis and the subsequent studies of phase II indicate that effects occur outside the legally established noise zones. A large number of people living outside these areas are annoyed, disturbed in their sleep, or have health complaints. This illustrates the importance of problem definition of the HIA, i.e. limited to the regulated area, or extended to a wider affected area.

It is difficult to say to what extent the HIA has influenced the decision-making process. It has certainly raised awareness about airport-related health impacts. The results from the noise studies have been taken into account in political discussions at national and EU level on limits and methods for measuring aircraft noise and predicting impacts. The study has also raised fierce discussion whether more widespread but less deleterious effects such as annoyance and sleep disturbance should be dealt with in the same way as other impacts such as effects on cardiovascular health.

Furthermore, this HIA programme has positively influenced the risk communication process. More attention was paid to health concerns in the population and dialogue between policy makers, airport officials, experts, and the public was stimulated. Before these studies were conducted, the discussion concentrated on the lack of data about a long list of possible health effects. After the first EIA, discussions were focused on the significance of the study results. Concern about respiratory health has subsided, since the study on school children showed no association between exposure to air-traffic related air pollution and respiratory symptoms [15].

Difficulties in this type of approach are the lack of information on exposure–response relations, application of aggregated data at a small-area level and the insensitivity of data from available health registries. The inadequacy of information on important determinants such as socio-economic status and lifestyle precludes firm conclusions about the causes of the observed disease patterns. The monitoring programme, which focuses on changes in environmental quality and health, making use of more sophisticated analytical techniques, may be more fruitful. An effort will be made to collect data on important confounders at a small-area level, which also can be used for other future assessments. Discussions about the design and requirement of the monitoring programme raised more awareness in decision makers about the consequences of detecting or failing to detect specific impacts.

In conclusion, the Schiphol HIA has shown a large impact of aircraft-related noise exposure on well-being (annoyance and sleep disturbance). Other health risks to individuals are small, but may produce a substantial population impact since so many people are exposed. The studies were useful for validating earlier predictions and adding to the evidence base for future HIA of air transport. The International Committee of the Dutch Health Council has stated that the integrated approach of the Schiphol HIA should become normal practice in assessing the public health impact of complex developments such as large airports. On the basis of these studies, effective and efficient measures to safeguard public health can be implemented [24].

Acknowledgements

The Health Impact Assessment Schiphol airport is commissioned by three Dutch Ministries: Housing, Spatial Planning and Environment; Transport, Public Works and Water Management; Health, Welfare and Sport.

References

1 **Grontmij** Het belang van de gezondheid in milieu-effectrapportage. De Bilt 1993.

2 **Fehr R.** Environmental health impact assessment: evaluation of a ten step model. *Epidemiology* 1999; 10:618–625;

3 **Passchier W, Knottnerus A, Albering H, and Walda I.** Public health impact of large airports. *Rev. Environ Health* 2000; 15:83–96.

4 **Parry J and Stevens A.** Prospective health impact assessments: pitfalls, problems and possible ways forward. *BMJ* 2001; 323:1177–1182.

5 **Franssen EAM, Staatsen BAM, and Lebret E.** Assessing health consequences in an environmental impact assessment. The case of Amsterdam Airport Schiphol. *Environmental Impact Assessment Review* 2002; 22:633–653.

6 Staatsen BAM, Franssen EAM, Doornbos G, Abbink F, van der Veen A, Heisterkamp SH, and Lebret E. Health impact assessment Schiphol (in Dutch). Report 441520001. Bilthoven: RIVM, 1993.

7 Staatsen BAM, Franssen EAM, and Lebret E. Health impact statement Schiphol airport. Executive summary [in English]. Report 441520003. Bilthoven: RIVM, 1994.

8 Franssen EAM, Staatsen BAM, Vrijkotte TGM, Lebret E, and Passchier-Vermeer W. Noise and public health. Workshop report [in English]. Report 441520004. Bilthoven: RIVM, 1995.

9 Staatsen BAM, Doornbos G, Franssen EAM, Heisterkamp SH, Ameling CB, and Lebret E. Spatial patterns in cardiovascular and respiratory disease rates around Schiphol airport [in Dutch]. Report 441520009. Bilthoven: RIVM, 1998.

10 Heisterkamp SH, Doornbos G, and Nagelkerke NJD. Assessing health impacts of environmental pollution sources using space-time models. *Statistics in Medicine* 2000; **19**:2569–2578.

11 TNO-PG; RIVM. Annoyance, sleep disturbance, health aspects and the perception of the living environment around Schiphol airport, results of a questionnaire survey [in Dutch or English]. RIVM report 441520010; TNO report 98.039. Bilthoven: RIVM, 1998.

12 Wiechen CMAG van, Franssen EAM, Jong RG de, and Lebret E. Aircraft noise exposure from Schiphol: a relation with complainants. *Noise and Health* 2002; 5(17):23–34.

13 Willigenburg APP van, Franssen EAM, Lebret E, and Herings RMC. Medication use as indicator for the effects of environmental pollution (in Dutch). RIVM-report no. 441520006 ISBN 90-393-1036-X, Utrecht Institute for Pharmaceutical Sciences, Utrecht; Bilthoven: RIVM, 1996.

14 Passchier-Vermeer W, Miedema HME, Vos H, Steenbekkers HMJ, Houthuijs D, and Reijneveld SA. Sleep disturbance and aircraft noise (in Dutch). RIVM report 441520019, 2002.

15 Vliet PHN van, Aarts FJH, Jansen NAH, Brunekreef B, Fischer PH, and Wiechen CMAG van. Respiratory diseases in children living in the vicinity of Schiphol airport [in Dutch]. RIVM report 441520014. Bilthoven: RIVM, 1999.

16 Emmen HH, Staatsen BAM, and Deijen JB. A feasibility study of the application of neurobehavioural tests for studying the effects of aircraft noise on primary school children living in the vicinity of Schiphol Airport, the Netherlands [in Dutch]. Report 441520007. Bilthoven: RIVM, 1997.

17 Franssen EAM, Ameling CA, and Lebret E. Variation in birth weight around Schiphol Airport, the Netherlands [in Dutch]. Report 441520008. Bilthoven: RIVM, 1997.

18 Ollerhead JB, Jones CJ, Cadoux RE et al. *Report of a Field Study on Aircraft Noise and Sleep Disturbance.* London: Civil Aviation Authority, 1992.

19 RIVM: Environmental Balance. Milieubalans 2003. Alphen a/d Rijn: Samson bv, 2003 (in Dutch). www.rivm.nl.

20 Babisch W. Epidemiological studies of cardiovascular effects of traffic noise. In Proceedings of the 7th International Congress on Noise as a Public Health Problem. Noise Effects '98. Sydney, Australia, 1998.

21 Strien RT van, DOuwes J, and Brunekreef B. The influence of noise insulation and ventilation behaviour in dwellings around Schiphol Airport on indoor air quality (in Dutch). EOH report 2000–490. RIVM report 441520016. Bilthoven: RIVM, 2000.

22 **Hygge S.** Recent developments in noise and performance. In Proceedings of the 7th International Congress on Noise as a Public Health Problem. Noise Effects '98. Sydney, Australia, 1998.

23 **Lebret E, Houthuijs DJM, and CMAG van Wiechen.** Monitoring environmental quality and health status around Schiphol Airport (in Dutch). Report 441520018. Bilthoven: RIVM, 2002.

24 Health Council of the Netherlands. Public health impact of large airports. Report 1999/14E. The Hague: Health Council of the Netherland, 1999.

Chapter 25

The Finningley Airport HIA: a case study

Muna I Abdel Aziz, John Radford, and
John McCabe

Introduction

Doncaster is a district in South Yorkshire, England, covering an area of 224
square miles. The population in 2001 was about 290,000 [1]. It forms part of
the South Yorkshire Coalfields Health Action Zone (HAZ) along with Barnsley
and Rotherham. Economic depression from closure of local coalfields has had a
demonstrable influence on ill-health in the HAZ [2]. Coal mining had been the
traditional form of employment for many villagers. Unemployment in Doncaster
was 6.3 per cent in April 2000, compared to the 4.0 per cent national average.

In November 1999, a planning application was submitted to Doncaster
Metropolitan Borough Council (DMBC) to develop a former Royal Air Force
base into a large commercial airport, with over 2 million passengers and
62,000 tonnes of freight annually by 2014 [3,4].

Responding to initial consultation, the Director of Public Health (DPH) for
Doncaster Health Authority (HA) suggested the need for a health impact
assessment (HIA). This requirement for HIA was incorporated into the local
planning guidance for the site [4]. A deadline of September 2000 for the HIA
was agreed on, and also that the decision on the airport would be deferred
until the HIA was complete.

This was the first time in the United Kingdom for an HIA to be undertaken
at the initial stage of a planning application for an airport. Previously, HIAs
had been conducted for public inquiries on proposed expansion of established airports; Manchester Airport's second runway [5] and Terminal Five at
Heathrow [6].

Outline of the planning process

Finningley Airport was the largest planning application in Doncaster for
decades.

The Developers are Peel Holdings, a large property and transport company, who own Liverpool Airport, the Manchester Ship Canal Company, the Trafford Shopping Centre in Manchester, and a 50 per cent stake in Sheffield City Airport.

The Planners were Doncaster Metropolitan Borough Council planning officers up to and including the Executive Director. They ensured the planning process was managed according to legislation, and considered the environmental impact assessment (EIA) submitted by developers. They prepared the documentation and negotiated with developers the planning conditions and Section 106 (S106) agreement.

The (Planning Act) S106 agreement is the statutory regulatory framework for airport operations. This is a negotiated legal agreement between planners and developers on criteria for environmental monitoring and mitigation measures proposed. The HIA was required by the local planning guidance for the site, so a decision on the application could not be made without it.

The primary decision-makers were the Development Control Panel of DMBC; a committee of Councillors who deliberated on the application. They supported the airport (subject to planning conditions and the S106) but could not approve it. It was called in for public inquiry by the Secretary of State (who following consideration of the Inquiry Inspector's Report approved the airport in April 2003).

The HIA process

The objectives of the HIA were to provide planners and the Development Control Panel with information on positive and negative impacts of the proposed airport on the health of local populations.

Partnership structures

A steering group was set up, chaired by the DPH for Doncaster HA, with representatives from DMBC, Doncaster East Primary Care Group, the local parish council, and the neighbouring North Nottinghamshire HA. Due to conflicts of interest, DMBC planners and airport developers (Peel Holdings) were not members of the steering group, but were invited to attend at critical points in the process. Good working relations were established and information shared throughout.

A working group was set up to share out tasks and make use of expertise available locally. This was initially drawn from Doncaster HA and DMBC. Because of a lack of technical expertise regarding environment and health, the Institute for Environment and Health (IEH) at Leicester University, was commissioned to assist with the HIA.

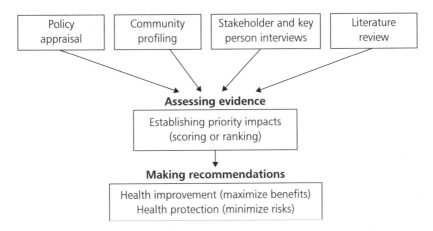

Fig. 25.1 Methods of the Finningley HIA.

Scope

The steering group decided to limit the HIA to the population of Doncaster (the district boundary for DMBC and the HA), but recognized that results were generalizable to neighbouring districts (hence the representative from North Nottinghamshire HA).

Methods

The HIA adapted the Merseyside model for HIA, by undertaking the four activities simultaneously in work-streams to save time (Fig. 25.1):

Policy appraisal of planning application documents

The application included statutory environmental, transport, and economic impact assessments [4]. From these, IEH extracted potential health impacts and mitigation measures onto a data extraction form (Fig. 25.2).

Profiling local communities

Public Health Information specialists used routine statistics (Census and Office of National Statistics data) to profile health and deprivation in Doncaster. Unemployment, coronary heart disease, respiratory disease, and cancers were mapped using geographical information systems (GIS) to convey information easily. Some technical data were obtained from the application documents (flight paths and noise contours). The Public Health GIS Unit at Sheffield University used an extract of the Exeter system to estimate numbers of residents affected by aircraft noise and those within 50–100 m of the busiest roads and railways (where >30 per cent increase in traffic was predicted).

Potential/Probable/Definite health impact(+ve / −ve):			Stated	Identified
Source (e.g., pollutant):				
Activity (factors bringing about impact):				
Scope (neighbourhood, local, regional, national):				
Who (persons affected; by age, sex, social status, health status):				
Duration/timing (how long and when?):				
Evidence (evidence base—stated/identified):				
Sensitivity analysis:				
Comments:				

Mitigation/enhancement measure:			Stated	Identified
	Potential/ Probable/ Definite	Quantified/ Estimated/ Speculated		
Source/action (of mitigation/enhancement):				
Scope (local, regional, national):				
Who (persons affected; by age, sex, social status, health status):				
Duration (how long and when?):				
Evidence (evidence base stated/identified):				
Sensitivity analysis:				
Comments:				

Fig. 25.2 Policy appraisal form.

Stakeholder and key person interviews

This activity focused on four villages directly adjacent to the flight path, and two others further away (purposively selected because of their location in relation to the airport). A researcher from the HA Community Involvement Unit and co-facilitator from DMBC interviewed 16 key informants (general practitioners, schoolteachers, youth leaders, and local businesses), and held six focus group discussions with 42 residents. The responses to consultation received by DMBC were also reviewed.

Literature review

Literature searches identified the health determinants to be included in the HIA (Table 25.1). Literature on health and social impacts of airports and regeneration/transport projects were identified. Researchers from IEH summarized these documents and provided reviews on health effects of noise, pollution, and employment. Data extraction forms similar to the policy appraisal forms were used.

Selected results from the HIA

Detailed findings were presented in the HIA report (46 pages) and technical annexes (236 pages) [7]. Selected results are presented here as examples of the information gathered by each of the four HIA activities.

Policy appraisal of planning application documents

A number of air pollutants of potential importance were identified. These stem primarily from the combustion of fossil fuels. Predictions for CO, SO_2, and hydrocarbons indicate that levels in 2014 are likely to fall well within currently accepted limits. However, NO_2 levels may begin to approach (but not breach) the WHO guidelines [8] and National Air Quality Standards [9].

Profiling local communities

The proposed development is situated within an area having unemployment rates consistently higher than the national average (Fig. 25.3). While the area immediately surrounding the airfield is not as deprived as other parts of South Yorkshire, a large part of the South Yorkshire Coalfields HAZ is within 30 min drive of the airfield and could benefit directly from employment, provided suitable training and transport infrastructure are in place. An estimated 4.2 million people live within an hour's drive of the airfield; an area likely to benefit from regeneration.

The number of persons who will be affected by significant aircraft noise is few due to the relatively isolated situation of the airfield (Fig. 25.4). Residents

Table 25.1 Determinants of health and potential health impacts associated with airports

Determinants	Source(s)	Potential health impacts
Air pollution	Aircraft Road and rail traffic Fuel handling Combustion Demolition Construction	Cancer Cardiovascular disease Respiratory disease Allergies Annoyance
Land and water pollution	Airport construction Fuel handling Fuel dumping Contaminated land De-icing/washing	(As above)
Noise	Road traffic Aircraft Rail Ground operations Construction	Annoyance Anxiety and stress Sleep disturbance Hearing loss Low birth weight Distributed cognition, communication, and motivation
Vibration	Road traffic Aircraft	Sleep disturbance Annoyance Occupational health
Odour	Fuel Combustion Sewage	Annoyance Anxiety and stress Reduced quality of life
Accident/ Fire/explosion Risk	Aircraft crashes Fuel handling Traffic accidents Aircraft vortex	Injury Death Risk perception Emergency services (demand)
Communicable diseases	International travel	Imported diseases Travel health
Policing and crime	International travel Terrorism	International crime Smuggling/drugs Terrorist incidents
Road traffic (road networks)	Volume of traffic Congestion	Anxiety and stress Access to the area
Public Transport	Change of routes	Annoyance
	Cycle lanes	Increased fitness
Employment	Direct airport-related Construction Local businesses	Creation of jobs Economic regeneration Social inclusion
	Job displacement	Reduces job opportunities
Tourism and travel	Air travel Transport links	Employment Economic regeneration

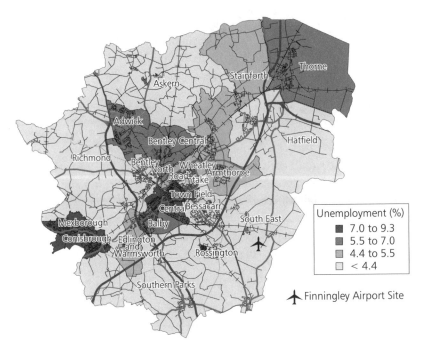

Fig. 25.3 Doncaster wards with the April 2000 unemployment rates.

Fig. 25.4 Aircraft noise from daytime flying predicted in 2014 (57 dB 16 h outer contour and 69 dB 16 h inner contour).

numbering 693 were estimated to be potentially affected by daytime noise and 1885 from night-flying in 2014. As they live in the area that exceeds the negotiated limits properties will be soundproofed. Two schools are located close to (but outside) this statutory mitigation area.

Stakeholder and key person interviews

Most local respondents felt that increased volume of road traffic would impact negatively on safety particularly of children, elderly, and the disabled. As well, the heavy traffic would contribute to less opportunity for recreation and sports such as cycling, horse-riding, jogging, and walking. It was also felt that heavy traffic could put off carers and domiciliary helpers from visiting those in need living near the airport. A link road between the motorway and the airport that bypasses the main villages was suggested as a way around the problem that could alleviate traffic congestion.

Literature review

There are a few locations around the proposed airport that will exceed the WHO guidance value for noise (55 dB 16 h average; general environmental goal for outdoor noise in residential areas [10]). Potentially, the main public health impacts of exposure to noise are community annoyance, anxiety, sleep disturbance, and effects on child health. There is some evidence [10] that these may in turn lead to secondary health impacts like cardiovascular disease, immune system effects, cognitive dysfunction (reduced memory, performance, and social behaviour), and respiratory illness.

Establishing priority impacts

On 4 September 2000, the working group met in a workshop session to consolidate findings from the four activities, and prioritize impacts so as to make recommendations. However, the group could not rank impacts in any particular priority order because:

- The framework did not allow for spatial distribution of impacts (negative impacts from noise and air pollution were concentrated around the airport while positive impacts from regeneration potentially included the whole of South Yorkshire).

- A marginal impact for most people could be significant for another more vulnerable person (the public health or the individual perspective).

- The group could not weigh negative against positive impacts as these would not cancel out.

◆ The group could not decide whether to rank impacts with or without mitigation in place, for example, noise impacts with or without sound-proofing. This was an issue because mitigation for environmental impacts was proposed, but not sufficiently detailed in the application.

So, the group decided not to rank impacts. Instead, common themes were identified (noise, pollution, and employment). All other impacts were listed in a separate section. Individual reports of the four activities were annexed to the report to incorporate all information gathered by the HIA.

Conclusions and recommendations of the HIA

Employment and regeneration opportunities are the main positive impacts that could lead to improvements in the health of the Doncaster population. The main negative impacts are from noise and pollution affecting local residents. The balance of effects will be a positive benefit to the health of Doncaster through the creation of jobs and regeneration of the area, provided effective amelioration measures are implemented to minimise negative impacts. [7]

The proviso about amelioration in the conclusion enabled the steering group to make an overall judgement of relative weights of positive and negative impacts without ranking; as noise and air pollution would be minimized below statutory levels.

Recommendations were made to maximize the positive benefits and minimize negative impacts. These addressed the following areas:

◆ Contributing to the S106 planning agreement negotiations between planners and developers as details of environmental mitigation are agreed (Recommendation 1).

◆ Setting up an independent Airport Health Impact Group (AHIG) for continued HIA if the airport is approved (Recommendation 2).

◆ Specific recommendations on employment, noise management, a Green Transport Plan, and consideration of a motorway link road that takes traffic away from local villages (Recommendations 3–6).

◆ Review of public services infrastructure if the airport goes ahead (Recommendation 7).

Influence of the HIA on the planning decision

Councillors had agreed to delay their review of the planning application until the HIA was complete. A progress report was presented to the Development Control Panel in July 2000 and the HIA report was published on 30 September 2000. Doncaster HA, which had called for the HIA during the initial consultation, accepted the conclusions and recommendations of the report and fully supported the planning application.

The health evidence supporting the airport was presented to the Development Control Panel in October 2000. In January 2001, the Development Control Panel recommended that the airport be approved, subject to S106 agreement.

Throughout 2001, the steering group continued to liase with planners and developers to maintain the health contribution to airport planning. There were two main thrusts to this contribution; establishing an independent AHIG for continued HIA (Section 5.1), and implementing specific HIA recommendations (Section 5.2).

A full planning inquiry was held between October 2001 and March 2002. Two lobby groups and a consortium of airports opposed the Finningley application. The HIA report was included as an inquiry document. It also provided a basis for the DPH to prepare a separate Proof of Evidence [11] describing the health contribution to the S106, and Supplementary Evidence [12] countering evidence submitted by an opposing lobby.

The AHIG and continued HIA

The HIA steering and working groups felt that HIA activities should continue beyond the time constraints of the initial HIA. It was important to provide a health perspective as details of the S106 planning agreement were concluded, and thereafter. The AHIG was established to monitor health impacts in parallel with airport operations, validate the initial HIA, and respond to unexpected health impacts (if they arose).

Ensuring health is considered throughout the lifetime of the airport

The HIA steering group made its last submission to the Development Control Panel on 26 January 2001 again supporting the application. It wrote 'It is encouraging that health has been a core concern of planning services, DMBC, in parallel with its statutory duties. The HIA Steering Group has been kept updated on the process of negotiating the Section 106 agreements. As these become finalised the Steering Group will continue to make detailed comments.' The steering group suggested that the health section of the S106 planning agreement should be strengthened by specifying the remit for the AHIG.

The critical role of the AHIG

This was the first time in the United Kingdom that an AHIG had been incorporated into the regulatory framework for an airport. This came about because the HIA was itself part of the planning process. Through the S106 planning agreement, the AHIG would be operational throughout the lifetime of the airport; a unique opportunity to improve and protect health.

One of the challenging, but critical, notions to justify in the remit for the AHIG was the WHO definition of health [13]. This was to ensure that all matters relating to well-being would be considered (including information from other airport committees, health monitoring, and review of complaints). This would also facilitate the consideration of non-statutory WHO guidelines, some of which are more stringent than statutory limits [10,11].

Implementing specific HIA recommendations

The steering group was reassured that statutory guidelines for noise and air pollution would not be breached, but sought even greater health protection where possible.

Noise.   The DMBC HIA representative was the noise expert and helped negotiate a detailed agreement with developers on soundproofing, noise monitoring, and night-flying quotas that took account of health. These include noise restrictions, bans on the noisiest flights, and an airport environmental management scheme covering the monitoring of railway and traffic noise. In one of the S106 meetings, was raised the issue of soundproofing for schools just outside the 57 dB (16 h average) [4] statutory mitigation area for noise as noise may exceed 35 dB (the WHO recommended maximum in classrooms) [10]. It was agreed that this was an appropriate area for the AHIG to monitor.

Air pollution.   Most of the air pollution was predicted to come from road traffic, not from planes. At the time of S106 negotiations, the Green Transport plan had not been developed to a stage where the steering group could comment, but DMBC modelling concluded that the statutory limits would not be exceeded. Early scoping reports had also concluded that the local road network could cope with the predicted increase in traffic. However, the HIA argued from the health perspective (community severance, accidents, and quality of life) that a motorway link/bypass road should be considered. Accordingly, a link road is being actively considered by DMBC. Although it falls outside the planning application, airport developers have pledged to meet part of the costs.

Training and employment.   The HIA urged developers to consider targeted opportunities for the more deprived communities in Doncaster. They had commissioned a survey of skills and training in South Yorkshire, and subsequently, Doncaster College secured new premises on the airport site.

Evaluation of the HIA

The HIA has added value to the planning process in the following ways:

Making health-based decisions whether to support the application.   The HIA played in an important role in supporting the airport, so that Doncaster and the entire region should benefit from the employment and regeneration predicted.

Translating environmental impacts into health terms.   Without an HIA, planners could not fully judge from environmental assessments what the effects on people in Doncaster would be. Baseline health profiles for future monitoring have been drawn up. The outputs have been healthy airport planning with regard to noise abatement and air pollution.

A systematic approach to involving communities.   Championing their views in the planning process. Local communities suggested the motorway link/bypass road, which is being actively considered.

Inputs into the S106 agreement.   This was the first time for health advocates to sit at the table in discussions of S106, but was accepted by planners and developers in recognition of the HIA work already done. Direct outputs are a detailed and stringent noise abatement agreement, and the first AHIG in the United Kingdom to be part of the S106.

Building capacity to conduct HIA.   Contributors to the HIA felt that the HIA had been relevant and comprehensive, with a good working relationship established between DMBC and the HA. The main output was a proposal to main-stream HIA in Doncaster.

Lessons for future HIAs

Finningley Airport HIA was an intensive process as it had to be completed within five months. The total time committed by working and steering group members (excluding commissioned experts) was 348 person-days. Assuming an average cost of £150–200 per day (including overheads), this work cost £52,200–69,600 in staff time. The monetary costs of commissioning external expertise and disseminating the report were over £17,000, shared equally by Doncaster Health Authority and DMBC.

Finningley HIA was a robust and comprehensive process. It enabled health and local communities views to be considered explicitly in the planning process. It demonstrated the value of having an HIA built into the planning process at an early stage. It informed decision making and considered more stringent WHO limits than existing UK statutory levels.

In identifying priority impacts for recommendation, the HIA steering group could not rank or score impacts—largely due to the uncertainty of proposed mitigation at that stage. The approach taken by the group was to minimize negative impacts and maximize benefits.

Recommendations were made that would not have otherwise been considered. These influenced the negotiation of the (Planning Act) Section 106 Heads of Terms between planners and developers. The AHIG would continue HIA throughout the lifetime of the airport—monitoring health and initiating action where necessary.

References

1 Office for National Statistics. 1996-based sub-national population projections. In Radford J. The annual report of the Director of Public Health, Doncaster Health Authority, 1999.

2 **Dunn J.** HAZ Profile, South Yorkshire Coalfields Health Action Zone. South Yorkshire Coalfields HAZ Profile Steering Group, 2000.

3 Finningley Airport Planning Brief. Doncaster Metropolitan Borough Council, 1999.

4 Finningley Airport Planning Application. Airport Planning and Development Ltd, Leeds, October 1999.

5 **Will S, Ardern K, Spencely M, and Watkins A.** A prospective health impact assessment of the proposed development of a second runway at Manchester International Airport. Proof of evidence of Stockport Health Commission (Manchester Airport Second Main Runway Inquiry). Manchester and Stockport Health Commissions, 1994.

6 Case study. Heathrow Airport Terminal 5. In *Health Council of the Netherlands: Public Health Impact of Large Airports*. The Hague: Health Council of the Netherlands, 1999; publication no.1999/14E: 150–156.

7 **Abdel Aziz MI, Radford J, and McCabe J.** Health Impact Assessment, Finningley Airport. Doncaster Health Authority, Doncaster Metropolitan Borough Council, MRC Institute for Environment and Health, University of Sheffield Public Health GIS Unit, September 2000.

8 WHO. Update and revision of the Air Quality Guidelines for Europe. Meeting of the Working Group 'Classical' Air Pollutants, Bilthoven, the Netherlands, 11–14 October 1994. (EUR/ICP/EHAZ 94 05/MT 12). WHO Regional Office for Europe, Copenhagen, 1995.

9 Expert Panel on Air Quality Standards. *Nitrogen Dioxide*. London: The Stationery Office, 1996.

10 **Berglund B, Lindvall T, and Schwela DH** (eds.). *Guidelines for Community noise*. (Outcome of the WHO Expert Task Force Meeting, April 1999). Geneva: World Health Organisation, 2000.

11 **Radford J.** Proof of Evidence on health impacts. Finningley Airport Public Inquiry, DMB/30, 2001.

12 **Radford J.** Supplementary Proof of Evidence. Finningley Airport Public Inquiry, DMB/32, 2001.

13 Preamble to the Constitution of the World Health Organization as adopted by the International Health Conference, New York, 19–22 June, 1946; signed on 22 July 1946 by the representatives of 61 States (Official Records of the World Health Organization, no. 2, p. 100) and entered into force on 7 April 1948.

Chapter 26

HIA and urban regeneration: the Ferrier estate, England

Ruth Barnes

Introduction

Since the election of the New Labour government in the United Kingdom in 1997, urban regeneration has been at the heart of government policy. Regeneration is seen as a mechanism for tackling social and economic disadvantage, inequality, and social exclusion through initiatives such as the New Deal for Communities, Healthy Living Centres, and the Single Regeneration Budget (SRB).

The SRB was first introduced a few years earlier, in 1994, bringing together programmes from several government departments and consolidating the resources available for regeneration. It provides funding for initiatives by local regeneration partnerships to enhance the quality of life of people in disadvantaged areas by reducing the gap between those areas and others, and between population groups. SRB programmes commonly include a range of objectives relating to employment and education, sustainable economic growth, the physical environment, housing, social exclusion, crime and community safety, and the enhancement of quality of life through improved health, cultural, and sports opportunities [1].

Single Regeneration Budget partnerships, like other current regeneration initiatives, are expected to involve a diverse range of local organizations in the management of their schemes and there is an emphasis on the involvement of local communities, which is expected to increase local accountability, improve service standards, and lead to sustainable development [2,3].

In a health impact assessment (HIA) of a major SRB programme to regenerate the Ferrier estate in Greenwich (London), community involvement was a key component in bringing together qualitative as well as quantitative evidence and in ensuring that the recommendations were geared towards the current population's needs.

The Ferrier estate

The Ferrier estate in Greenwich was built in the late 1960s for letting to higher income tenants. Since then the area has deteriorated. Across Greenwich, the Under Privileged Area (UPA) score rose from 24 in 1981 to 37 in 1991. However the figures for the Ferrier are even bleaker, with UPA scores of 23 in 1981 (similar to Greenwich as a whole) and 50 in 1991, indicating a much more marked decline than that seen elsewhere in the locality [4].

Today the estate is home to around 6800 people and it represents an area of severe disadvantage. The population is young, a high proportion of residents are from minority ethnic groups—many without English as their first language—and there are significant numbers of refugees. Average income levels are low, unemployment is high, and fear of crime is a significant concern for local people. Health-related data indicate that there are high levels of ill-health that are so often concomitant with indicators of significant social and economic deprivation.

In 1999 the South Greenwich Partnership attracted £23 million of government finance for an extensive regeneration strategy which includes, as a main priority, radically transforming the Ferrier's housing stock [5]. Given the scale of the developments and the established links between health and housing, the potential health impacts are sizeable.

The HIA of the Ferrier estate

An in-depth HIA of the proposals for the Ferrier was undertaken in 2000 at a time when the land use options—refurbishment, demolition, and rebuilding or a combination of the two—were being considered by the local authority and SRB Partnership Board. The HIA was timed for a strategic planning stage where its recommendations could be most effective not only in informing the options for the future development of the estate but also in assessing the health impacts of the transitional phase of the changes.

Each option will result in a considerable reduction in the number of housing units on the Ferrier, involving the movement of some people to other locations. Taking this into account, the main focus of the HIA was the health and well-being of current residents and health impacts on them rather than issues relating to people who may move into the area as a result of the regeneration.

The broad aims of the HIA were

- to assess the potential positive and negative health impacts of possible changes in housing and land use;

- to make recommendations to enhance the predicted positive impacts and minimize the negative ones, including long-term effects and shorter-term impacts of the transitional phase; and

♦ to highlight the impact of any proposed development on health in-equalities.

It was envisaged that the work would also have a longer-term influence on policy development locally by raising awareness and understanding of health, establishing the principle of undertaking HIA, and making public health improvement a routine part of policy development.

Methodological framework

The Merseyside Guidelines for HIA [6] were used as a starting point for the development of the methodology. At the outset, a multi-agency steering group was established to oversee the development of the HIA and a small working group formed to carry out the day-to-day work. A brief review of the literature on health, housing, and regeneration was made to inform the HIA process and to provide an evidence base to facilitate the assessment of potential health impacts and the appraisal of the land use options. The overall policy context was also reviewed, setting the work within the wider local and national contexts. A community profile was collated from routinely available data sources including the Office for National Statistics (ONS) small-area statistics based on Census data from 1981 and 1991, data derived from the ONS annual death and birth extracts and from the ONS Public Health Common Data Set, and housing and unemployment data from the local authority and the London Research Centre. This review of routine statistics highlighted a number of important gaps in the available data and informed the development of a questionnaire, which was subsequently administered by the SRB Partnership to all residents of the estate [7].

In identifying the key health issues and the potential health impacts of the land use options and in developing the recommendations, the participation and active involvement of local residents and professionals from a wide range of organizations working in the area was crucial. Initially, 'key stakeholders' were identified through the steering group and included, for example, representatives from the local authority's regeneration, housing, education and environmental health departments, local health care providers such as general practitioners and health visitors, members of the SRB Partnership Board, the police and the South Greenwich Forum (an umbrella group for local community organizations). In a series of structured interviews with individuals and groups, the stakeholders were asked to help identify

♦ current health issues that they perceived as important;

♦ long-term potential health impacts (positive and negative) of the refurbishment and demolition options for the estate; and

♦ shorter-term transitional health impacts that might be expected during the period of change.

Community representatives were also consulted informally. Members of the working group attended the Tenants' Forum and worked alongside the Independent Residents' Advisers to be part of an integrated consultation process with the residents, to involve residents in the HIA and to ensure its widest possible ownership. This added important anecdotal and illustrative detail to the more structured information, which was collected during the interviews and also helped to raise awareness of the HIA and of health issues in general amongst residents.

The key issues and areas of concern which emerged from the interviews were

- poverty, unemployment, education and training;
- recreation and leisure;
- social networks and belonging to the local community;
- housing;
- management by the council;
- the local environment; and
- crime and the fear of crime.

The qualitative information on key issues and potential health impacts was subsequently explored in more detail along with quantitative evidence from the community profile and the residents' survey.

As a result, a number of recommendations were made relating to

- *support to existing communities* to ensure that positive social and community networks were safeguarded, for example, by allowing residents to self-define family, friends, and communities and, where they were to be moved, housing them in close proximity, particularly where extended family networks provide essential care and support;

- *management of the redevelopment process* including, for example, guidelines for clear policies to deal with issues such as decanting during refurbishment or permanent relocation to other areas and measures to minimize the shorter-term health risks associated with the period of building works;

- *asbestos*, including the updating of survey records and full risk assessment where the location of asbestos was unclear before any refurbishment or demolition work started, and independent monitoring of asbestos levels during the period of demolition or refurbishment;

- *sustainability*, for example, by minimizing disruption to residents, using local labour for the regeneration where possible and focusing training and employment schemes on routes to wider labour markets and the creation of jobs in the immediate locality; and

◆ *health inequalities*, particularly relating to poverty and income, women and young children, elderly people, and people from black and other minority ethnic groups. This included a recommendation to consider commissioning a cohort study following a sample of the current population and a sample from a control population over the next 10–20 years with the aim of ascertaining how residents of disadvantaged areas benefit or lose out from regeneration programmes in the longer term [8].

The advantages of community involvement and participation

For the HIA process, involving the local community had a number of distinct benefits, including

◆ gaining a broader perspective and insight;

◆ obtaining something closer to a complete picture than that shown by the quantitative evidence alone; and

◆ offering opportunities to find tailored solutions to some of the issues that emerged.

Whilst the qualitative evidence did not produce any major surprises, by analyzing it alongside the quantitative data it was possible to interpret the 'hard facts' in the light of 'real life' experiences and consequently to build up a much broader and more complete picture which was invaluable in assessing the potential health impacts of the regeneration and in developing the recommendations.

For example, Census data show that the estate has a significant minority of people who have lived there for over ten years together with larger numbers whose length of residence is much shorter [9]. What the figures mask, however, is the complexity of the situation, particularly in relation to social networks and residents' perceptions of the estate as a place to live. The interviews with key stakeholders revealed that, for a small core of residents, there are good community and social networks and a sense of belonging. There are generations of people who have grown up there, some with their extended families, and networks have developed and people have invested time and money in their homes. The social networks they have developed over the years are crucial to their quality of life and it is important that they are housed close together. This was particularly relevant for older people who make up only a small proportion of the population so that, had the HIA relied only on routine statistics, their needs may have been neglected.

As a whole, however, the community is not cohesive. There are groups differentiated by geographical location (the estate is based around a series of squares, each with its own distinctive character) or by ethnic group. The statistics show that almost one in five residents belong to a minority ethnic group but only through the involvement of these communities could light be shed on their needs. Whilst many people from minority ethnic groups, such as those from Eritrea, Vietnam, and Somalia, are well established in the area and have good social support structures, others, such as newly arrived refugees, can be isolated and marginalized. Their problems are compounded by poor interpreting services and there were widespread reports of racial harassment and victimization, which do not appear in any routine statistics. For this group, which has a large turnover, there were few incentives to make friends locally or to take part in community activities as most were looking for a chance to move out at the earliest opportunity. Involving some of them in the HIA helped to elucidate some of their perceptions about the underlying causes of social isolation and the lack of social cohesion.

The issue of crime also illustrates the benefits of combining quantitative evidence and qualitative information gleaned from the community. Crime statistics show that whilst domestic crime, burglary, vehicle crime, personal assaults, and criminal damage are long-standing and growing problems on the estate, the overall crime rate is no higher than in other comparable areas. Understandably, perceptions of crime amongst local residents posed a significant threat to their psychological health and well-being. Children who were interviewed commented on the problems of vandalism, including arson, graffiti, car crime, and petty theft. Others voiced concerns about drug-related crime and violence. Many felt intimidated by the large groups of young people who congregate on the stairwells at night. It was suggested that the recorded crime figures may be misleading as many residents, particularly those who see the estate as a temporary home, have no vested interest in crime reduction and do not wish to draw attention to themselves by reporting crime.

For the HIA process, then, the benefits of involving the community and combining the qualitative and quantitative evidence are clear but, apart from the strengthening of the HIA's recommendations, the benefits for the community itself are less tangible. However, anecdotal evidence suggests that there were advantages, including

- having a voice;
- expressing choice and influencing policy development;
- providing opportunities for involvement with neighbours; and
- providing opportunities for developing new skills.

The interviewees and group work participants said they appreciated being consulted and felt that the HIA gave them an opportunity to express their views in a way that had not been possible during the formal consultations with the SRB Partnership. This was because their involvement was on a one-to-one basis or in a peer group rather than in formal meetings led by local authority representatives.

Obtaining community involvement and participation

Set against the benefits are several difficulties in securing community involvement and in making sense of the qualitative evidence it provides. The difficulties included:

◆ identifying and accessing all sections of the community;

◆ getting a balanced view;

◆ finding ways of effective two-way communication;

◆ obtaining time and other resources;

◆ dealing with conflicting interests; and

◆ managing expectations.

At the start of the HIA the key stakeholders were identified by steering group members or by other contacts made through them. The local residents in this initial list were members of the Tenants' Forum and already engaged in the SRB Partnership's consultation. Interviews with these people produced some valuable evidence but it quickly became apparent that they were representative of only a small proportion of the population, being long-standing residents with well-developed social networks and a vested interest in the area's future. As a result there was concern that the voices of others, those who see the estate as a 'last resort' or temporary measure before moving on, were not being heard as these people were, on the whole, disengaged from the consultation processes conducted by the SRB Partnership.

One way of overcoming this was to work alongside the Independent Residents' Advisers, who had good contacts across the whole population and facilitated the involvement of residents who might not normally have come forward. In addition, all the local community groups were contacted and group sessions held with those wishing to participate including, for example, the Refugee Women's Group and older people meeting at the Elderly Resource Centre. HIA working group members also attended a primary school class, at the invitation of their head teacher, to solicit the children's views on health issues on the estate. This was an extension of a health-related project they had already been working on and led to further work, in an art class, on health-related issues.

As a final check on balance and representation, the interview findings and residents' survey results were compared. The survey had been carried out with interpreters where necessary. Many of the interpreters were drawn from the local community, having been offered training as appropriate, and their help was also key to the participation and involvement of some interviewees from minority ethnic groups.

The SRB Partnership had put a great deal of effort into consulting local residents but it emerged from the HIA interviews that many people still felt excluded from the process, either because of language or cultural barriers or for other reasons. This was reflected in the disparate views about the management of the estate. Some interviewees felt that the Ferrier was well managed and that there was a high level of resident involvement whilst others felt that their voices were not heard.

As a result the HIA was able to make specific recommendations about the involvement of residents in the change process and suggest ways in which the SRB Partnership could develop more effective mechanisms for two-way communication. Recommendations were also made about managing conflicting interests, for example, between older residents in supported housing who wished to remain there and those who saw demolition as the preferred option, and the HIA highlighted the need to ensure that unrealistic expectations were not created. Many people commented, for example, on how they had felt let down by 'the authorities' in the past and questioned the extent to which the findings of the HIA would be acted upon in the future.

Using the HIA to inform policy and programme development

The HIA was appraised and discussed in detail with the HIA steering group, the SRB Partnership Board, the Council, and residents. It formed a major input into planning for the estate's regeneration and the SRB Partnership has acted on many of the recommendations, particularly those relating to managing the process of change.

Since the HIA was completed, the decision has been taken to demolish and rebuild large swathes of the estate. Local residents, many of whom were not involved in the consultation process prior to their involvement in the HIA, have been remarkably active in helping shape the plans and have used the findings of the HIA to support their arguments.

Conclusions

It is unquestionable that broad community involvement in the HIA strengthened its recommendations and enabled them to be tailored closely to the needs of the current population, particularly for groups that may have been neglected if they had not been engaged in the HIA process. The main benefit of their involvement is perhaps that their voices were heard and their views and perceptions taken into account in a way that would not have happened if the HIA had relied primarily on routinely available statistics or involved only those people who were easily identifiable and most vocal.

There is also some evidence of the HIA having provided wider benefits, raising awareness of health issues, and engaging 'hard to reach' groups in the SRB Partnership's consultation process. The HIA is reported to have encouraged the involvement of some of these groups and they have used its findings to advocate their views.

Nevertheless, the difficulties of securing participation, particularly in terms of time and other resources, and the dangers of raising unrealistic expectations which cannot subsequently be met, cannot be underestimated. This is not a reason for failing to involve communities in HIA but it does highlight the need for careful planning and anticipation of some of the biases that may arise if their involvement is only partial. In the case of the Ferrier it was fortuitous that there was already some local capacity for this, most notably in the Independent Residents' Advisers and the established local community groups.

Three years on the signs are positive but only time—and a retrospective evaluation of processes and outcomes—will tell whether the right balance was struck.

References

1 Ward K. The single regeneration budget and the issue of local flexibility. *Regional Studies* 1997; **31**:78–81.

2 White L. The role of systems research and operational research in community involvement: a case study of a health action zone. *Systems Research and Behavioural Science* 2003; **20**:33–145.

3 Taylor M. *Unleashing the Potential: Bringing Residents to the Centre of Regeneration.* York: Joseph Rowntree Foundation, 1995.

4 Eltham Locality Needs Assessment. Greenwich: London Borough of Greenwich, 1999.

5 Single Regeneration Budget Delivery Plan. Greenwich: South Greenwich SRB Partnership, 1999.

6 Scott-Samuel A, Birley M, and Ardern K. *The Merseyside Guidelines for Health Impact Assessment.* Liverpool: Liverpool Public Health Observatory, 1998.

7 A survey of Ferrier estate residents carried out on behalf of the London Borough of Greenwich. Greenwich: First Call, 2000.

8 Barnes R and Macarthur K. The Ferrier estate: health impact assessment. Available on www.hiagateway.org.uk

9 Office for National Statistics, Census of Population 1991. Data derived from small area statistics.

Chapter 27

Impact Assessment in Canada: an evolutionary process

Roy E Kwiatkowski

Canada was one of the first countries in the world to develop a standardized national approach to health impact assessment (HIA). This chapter will describe the HIA framework and its evolution over the past decade.

Background

> Canadians are increasingly worried about the quality of the environment in which they live. Much of this concern stems from anxiety about the possible danger to human health. . . .
>
> Excerpt form *Canada's Green Plan* (1990)

Development projects are carried out by people for people; most often for economic gain. However, economic activity depends on a healthy environment. Economically and environmentally, Canadians are enormously wealthy by comparison to much of the world. Almost half of Canada is forested, representing about 10 per cent of the world's temperate and boreal forests. Canada's rivers and lakes provide approximately 8 per cent of the global freshwater supply. One-quarter of the earth's remaining wetlands are located in Canada. It is bordered by three oceans, has a 224,000 km coastline, and the second largest continental shelf in the world.

Canada clearly needs economic development to ensure a secure future. Indeed, the health of Canadians depends on a prosperous economy so that not only are the basic prerequisites for health provided, but so too is the high standard of living, since health is clearly related to wealth [1]. It is in everyone's best interest to protect the environment, locally, nationally, and globally. Canada's environmental resource wealth has driven its economy and its economy depends heavily on sustaining Canada's diverse and rich ecosystems. Resource-based industries are essential components of the economy. One in every three working Canadians is employed in five main resource-based industries: agriculture, forestry, fishing, mining, and energy production [2].

Over the last 25 years, environmental impact assessment (EIA) has evolved within Canada, moving from a policy of general but voluntary application to a legislated process designed to identify, analyze, and evaluate the environmental effects of a development project in an open and transparent manner. Since the mid-1990s, the scope of the EIA process in Canada has slowly but progressively broadened to include cumulative effects, follow-up monitoring and related health, and other social considerations.

In Canada, constitutional responsibility for the environment and health and environmental assessment is shared between the federal and provincial governments and as a result all ten provinces, as well as the federal government, have environmental assessment legislation in place. The three Canadian territories are also in the process of developing their own processes (rather than depending on the federal EIA process). Irrespective of which jurisdiction one works in, the EIA process serves a number of objectives. Early identification of potential environmental problems reduces risks, legal liabilities, and costly retrofits for the proponent of the development project. It provides the public an opportunity to learn about the project and comment on its perceived impacts. For government decision-makers it provides an opportunity for all stakeholders (proponents, non-government organizations, individuals, other government departments, and interested stakeholders) to participate in an open and transparent decision-making process.

Even though the EIA process is legislated, activities under EIA are remarkable dynamic. Concepts and approaches change as new information (obtained from follow-up monitoring) or concerns expressed by any stakeholder group, are brought forward. One such event is the changing approach to HIA within the EIA process. A decade of independent reviews on HIA implementation nationally and internationally [3–8] indicated that the health aspects within an EIA are inconsistently or only partially addressed; there is a clear need for a more systematic approach. Responding to this situation the Federal, Provincial, Territorial Committee on Environmental and Occupational Health (FPTCEOH), which has membership from the Departments of Health, Environment and Labour from the federal, provincial, and territorial jurisdictions within Canada, established a Task Force in 1995.

The task force on HIA

This task force was asked to develop:

- a definition of HIA acceptable to all jurisdictions;
- a population health framework appropriate to HIA;
- guidance and training material for HIA,

- strategies to increase awareness about HIA, EIA, and the linkages between human health and the environment.

One of the first activities it carried out was the development of a number of guiding principles, which where to be incorporated into any Canadian HIA framework, guidance material, and training activities. These included:

- the WHOs 1967 ('*a state of complete physical, mental and social well-being and not merely the absence of disease or infirmity*' [9]) and 1984 ('*the extent to which an individual or a group is able, on the one hand, to realize aspirations and to satisfy needs, and on the other, to change or cope with the environment*') definitions of health that are to be incorporated into the Canadian HIA framework;

- environmental and human health are inextricably interlinked and therefore, HIA is an integral part of EIA;

- a cornerstone of HIA is the recognition of the need for public participation in the definition and scoping of human health concerns and in decision making;

- HIA is required throughout the life cycle of the project (planning, construction, operation, decommissioning, and follow-up monitoring) and takes into consideration occupational health and safety;

- development of a scientific approach to HIA will focus efforts and diminish resource requirements, providing a fair, effective, and efficient process of information gathering for decision makers and the public; and,

- educational tools are required to promote or increase awareness of environmental/human health assessment, risk assessment, and communication, and the linkages among environmental, social, economic, cultural, and human health effects.

HIA, as envisaged by the Task Force, demanded a high level of multidisciplinary health expertise and the integration of social, economic, environmental, and health components. As a first step the Task Force reviewed a number of conceptual frameworks. The most notable were by: Evans and Stoddart [10], who developed a model that linked an individual's well-being to their physical environment (environmental impact); their level of prosperity (economic impact); their social environment (social impact) and health factors (health impact); the healthy community model proposed by Hancock [11] that linked community sustainability and well-being by focusing on community, environment, and economy, with a central focus on health; and, the Determinants of Health model proposed by the Federal, Provincial, and Territorial Advisory Committee on Population Health (FPTACPH) [12].

The determinants of health

As the FPTACPH Determinants of Health [12] model became a cornerstone to the Canadian HIA framework, a brief description of the nine determinants of health is provided here:

1. *Income and social status.* Growing evidence indicates that income and social status is the most important determinant of health in Canada. People perceive themselves as being healthier the higher their socio-economic status and the higher their income level. This may be surprising considering that Canada has a health system that provides equal access for all Canadians, regardless of their income. Yet studies in provinces and cities throughout Canada consistently indicate that there is not only a difference between people in the highest and lowest income scale, but that people at each step on the income scale are healthier than those on the step below. Furthermore, many studies demonstrate that the more equitable the distribution of wealth, the healthier the population, regardless of the amount spent on health care.

2. *Education.* For a variety of reasons, health status improves with an increasing level of education. Education improves opportunities for employment, income, job security, and job satisfaction and equips people with knowledge and skills necessary for problem solving. People with higher levels of education also have more control over their work environment and are better able to access and understand information to help them stay healthy.

3. *Employment and working conditions.* Employment and unemployment contribute significantly to health status, with unemployed experiencing significantly more psychological distress, anxiety, health problems, and hospitalization than the employed. Within the employed population, however, other factors that negatively affect health include stress-related demands of the job, workplace injuries, and occupational illnesses.

4. *Physical environment.* Our basic physical necessities are derived from the natural environment: air, water, food, and shelter. Therefore, human health is fundamentally dependent on the natural environment. Exposure to man-made environmental hazards is the cause of a large and growing proportion of illnesses, injuries, and premature deaths. Diseases linked to environmental hazards include diabetes, immune depression, cancers, and respiratory diseases. Factors in our human-built environment such as housing, workplace, and community safety have equally important influences on health.

5. *Biological and genetic endowment.* The organic make-up of the body, the functioning of various body systems, and the process of development and aging serve as fundamental determinants of health. Biological differences between the sexes and the traits and roles that society ascribes to females and males form a complex relationship between individual experience and the development and functioning of key body systems. At the same time, genetic endowment predisposes certain individuals to particular diseases or health problems.

6. *Social support networks.* The support that families, friends, and communities provide contributes to improved health. Social support networks can help people cope with daily stresses and solve their problems. Studies have shown that the more social contact people have the lower their premature death rates.

7. *Personal health practices and skills.* Healthy lifestyle and behavioural choices increase opportunities for improved health. A balanced diet and regular exercise have been shown to provide substantial health benefits while the use of tobacco, recreational drugs and excessive consumption of alcohol are linked to many health problems.

8. *Healthy child development.* Mounting evidence indicates that prenatal and early childhood experiences have a powerful influence on subsequent health, well-being, coping skills, and competence. Not only are infants with low weights at birth more susceptible to infancy deaths, neurological defects, congenital abnormalities, and retarded development, they also experience negative effects later in life, which can include premature deaths. Of further interest, a strong correlation exists between a mother's level of income and the baby's birth weight; mothers at each step up the income scale have babies with higher birth weights, on average, than those on the step below.

9. *Health services.* Health care services contribute to health status, particularly when they are designed to maintain and promote health and prevent disease. Services such as prenatal care, immunization, early detection, and those that serve to educate children and adults about health risks and choices all serve to improve health.

Clearly, the health and social well-being of individuals, families, and communities are subject to many factors, few of which can be conveniently placed within specialized disciplines or government agencies. Their influences on the human environment defy simple relationships and equations. Factors such as those described above, are intricately related and include many, often intangible variables and constants defining the health and well-being of people and communities.

Evolution of HIA in Canada

Within Canada, approximately 5000 development projects undergo an EIA annually and therefore any HIA framework developed by the Task Force would have to be robust enough to handle a variety of different environmental health issues. The Task Force quickly realized that HIA was not the responsibility of a single government department or agency; as it included environmental, social, economic, occupational, and population health components. As well, the Task Force rejected the concept of developing a separate legislated HIA process and promoted the integration of HIA within the existing legislated federal or provincial EIA processes. However, this integration could only occur if guidance material on HIA was prepared, promoted, utilized, and freely available.

The Canadian Handbook on Health Impact Assessment [13] was written in 1996 with the overall objective of acting as a world class resource, promoting the integration of HIA into EIA. The Handbook (presently in three volumes) describes HIA as a new, holistic approach to impact assessment, which clearly places human beings as the centre of considerations about development, while seeking to ensure the durability of the ecosystem of which they are an integral part.

The framework outlined in the Canadian HIA Handbook has been used successfully over the last several years to assist decision makers to guide and direct policy, programme, and project development so that these initiatives are protective of human health and meet the goals of sustainable development. The Handbook is a living document, developed through consultations with concerned stakeholders (government agencies, industry, academia, aboriginal, non-government organizations, consultants, and the general public). Its main strength is the clear recognition that the relationship between humans and the natural environment generates many complex feedback mechanisms— changes in the environment result in changes to the community dependent on the environment; similarly changes in the community can cause changes to the environment. These changes may be positive, reinforcing feedbacks, or they may be negative feedbacks. A proper assessment of these feedbacks and development of mitigation measures to minimize or negate the negative feedbacks requires a team of experts in HIA, as well as in EIA.

The HIA framework continues to evolve within Canada. After two rounds of cross-country consultations the Handbook is being rewritten to reflect the comments received and will be released in 2004 on the World Wide Web. As well, greater efforts linking Social impact Assessment (SIA) methodology and approaches with HIA methods is underway [14]. Two organizations within Canada, the office of Environmental Health Assessment Services, part of Health

Canada; and, the WHO/PAHO Collaborating Centre on Environmental and Occupational Health Impact Assessment and Surveillance, Centre Hospitaler Universitarie de Québec have formed a partnership [15] to work cooperatively with institutions nationally and internationally to exchange information and experience, providing expertise when required for addressing specific HIA problems and implementing training at the local, regional, national, and international level.

Conclusions

While economic development is a crucial factor in health improvements and well-being, it can also generate adverse effects on health and the environment. Over the last quarter century, thinking about environment–health linkages has evolved towards a much more global, more ecological approach. Similarly, natural resource management thinking has also progressed and now includes environmental, social, and health factors, along with economic parameters. Both fields have seen a move to a more integrated approach to management— whether of health, economics, or of the environment. Linking the environment and human health offers a unique opportunity to address challenges that are facing both the developed and developing worlds. Our current ability to carry out effective EIA is constrained by our understanding of the environment/ecosystem we are attempting to protect. Similarly, in HIA, our ability to predict, assess, understand and mitigate the impacts of development projects on quality of life, human health, and the well-being of an individual or a community, is constrained by our understanding of the complex interactions between humans and their environment. Creating changes in a community as a result of a project, programme, or policy, without learning from, or knowing what the impact of those changes were, can generate uncertainties within the community leading to a loss of control over and deterioration of the quality of life and health of the community. Whether beneficial or negative, it is important to understand, assess, and respond to changes and if possible, prevent or enhance them as required. Communities might notice a marked decrease in their quality of life and health, yet be incapable of determining when or from what processes these changes emerged. On the other hand, their quality of life may have improved, as a result of a project, programme, or policy, yet without the knowledge of just where and when these improvements began, enhancing such changes or duplicating them in the future or in other communities may prove difficult or impossible, and attempts to do so may be counterproductive.

Including HIA in EIA is a cost-effective strategy that integrates the health effects of projects, programs and policies in decision making at the same time as environmental, economic, social, and political issues. This approach is fully consistent with the recommendations made in Agenda 21, notably in Chapter 8, which deals with integrated decision-making.

References

1 Environment Canada. *A Framework for Discussion on the Environment.* Minister of Supply and Services Canada, 1990. ISBN 0-662-57411-7.

2 Health and Welfare Canada. *A Vital Link: Health and the Environment in Canada.* Ottawa, Canada: Canada Communications Group, Publishing, 1992. ISBN 0-660-14350-X,

3 Martin JE. Environmental health impact assessment. *Environmental Impact Assessment Review* 1986; **6**:7–48.

4 Giroult E. WHO interest in environmental health impact assessment. In Wathern P (ed.) *Environmental Impact Assessment: Theory and Practice.* London, Unwin Hyman, 1988.

5 Turnbull RTH (ed.). *Environmental and Health Impact Assessment of Development Projects: A Handbook for Practitioners.* London and New York: Elsevier Applied Science, 1992.

6 Arguiaga MC, Canter LW, and Nelson DI. Integration of health impact considerations in environmental impact studies. *Impact Assessment* 1994; **12**:175–197.

7 Sloff R. Consultants Report on the Commonwealth Secretariat Expert Group Meeting on Health Assessment as Part of Environmental Assessment. Aberdeen, Scotland, 1–3 February 1995. Available from the Commonwealth Secretariat. London, United Kingdom.

8 Davies K and Sadler B. *Environmental Assessment and Human Health: Perspectives, Approaches and Future Directions: A Background Report for the International Study of the Effectiveness of Environmental Assessment.* Minister of Supply and Services Canada, p. 51, 1997. ISBN:0-660-17063-9.

9 World Health Organization (WHO). The Constitution of the World Health Organization. *World Health Organization Chronicles* 1967; **1**:29.

10 Evans RG and Stoddart GL. Producing Health, Consuming Health Care. Population Health Program Working Paper No. 6. Toronto: Canadian Institute for Advanced Research, 1990.

11 Hancock T. Developing healthy communities: a five year project report from the Community Health Development Centre. *Canadian Journal of Public Health* 1988; **79**: 416–419.

12 Federal, Provincial, Territorial Advisory Committee on Population Health. *Strategies for Population Health: Investing in the Health of Canadians.* Minister of Supply and Services Canada, 1994. ISBN 0-662-22833-2.

13 *Canadian Handbook on Health Impact Assessment* obtainable as a series of English PDF files from http://www.hc-sc.gc.ca/ehas/ or a series of French PDF files from http://www.hc-sc.gc.ca/behm/. It is currently being revised.

14 Rattle R and Kwiatkowski RE. Defining Boundaries: Health Impact Assessment and Social Impact Assessment. In Becker HA and Vanclay F (ed.) *The International Handbook of Social Impact Assessment.* Edward Elgar publishers, 2003 ISBN 1-84064-935-6.

15 Kwiatkowski RE and Gosselin P. Promoting human impact assessment within the environmental assessment process: Canada's work in progress. *International Journal of Health Promotion and Education* 2001; **8**:17–20.

Chapter 28

HIA and waste disposal

Ian Matthews

Introduction

There is increasing demand for more and different waste disposal facilities and it is frequently suggested that proposals for these facilities should be the subject of an health impact assessment (HIA). Industry, commerce, and households in England and Wales produce approximately 106 million tonnes of waste each year and most of this waste is landfilled. In some parts of the country especially near urban centres there is no land available for waste disposal. Landfill is a major source of methane (a greenhouse gas contributing to climate change). For these reasons as well as for issues of sustainability the European governments agreed on the EU Landfill Directive [1], which sets targets for the reduction of biodegradable municipal waste sent to landfill. The EC legislation (Framework Directive on Waste 75/442/EEC and 91/692/EEC) also required governments to draw up plans, which would:

- Prevent or reduce waste production and its harmfulness,
- Recover waste by means of recycling, re-use, or reclamation,
- Use waste as a source of energy,
- Ensure that waste is recovered or disposed off without endangering human health.

The UK government proposes by 2010 to reduce biodegradable waste landfilled to 75 per cent of that in 1995 [2]. The Environment Agency (EA) is responsible for enforcing and regulating waste management facilities by means of licence and permit conditions. These require operators to control or prevent releases of substances so that public exposures are acceptable when judged against UK or international standards. In future, the National Protection Agency may assume relevant epidemiological and surveillance functions in the United Kingdom and the EA may develop a National Exposure Registry.

Quantification of health impacts of waste disposal

When assessing the health impacts of any waste management facility it is important to realize that there are no risk free ways of managing waste, and any alternative method of managing the waste will also have health impacts.

Health impacts of proposals may be assessed by considering the level of hazard, the pathways to a receptor, and the likelihood of that receptor being adversely affected by the hazard (the 'sources–pathway–receptor' methodology). Information that is required includes: the type and quantity of emissions from the waste; the location, nature, and size of populations that could be affected by those emissions; how they could be exposed to the toxic substance; and the impact on their health of this exposure. The first three parameters are very dependent upon local decisions on the type and siting of each facility. The impact of exposure of toxic chemicals on health is only known for a very limited number of chemicals. Therefore the ability to carry out a quantitative HIA is severely constrained by the very limited scientific evidence that is available.

The risks to health are assessed using available or estimated emissions data, pathway modelling, and human dose estimates. The dose estimates are then related to standards or risk estimates based on prior research. A number of the more significant potential hazards associated with landfills, incinerators, composting, and recycling are outlined in the following sections. It should, however, be noted that understanding of the interaction between the environment and health is constantly evolving and, in many instances, current scientific knowledge is insufficient. In general it is helpful to quantify impacts on health but where the data are insufficient, the best assessment should be made accepting that some degree of uncertainty will be unavoidable.

Studies on the impact on health of waste management usually have great difficulty in furnishing the epidemiological and biological evidence required to prove that a specific chemical emission has caused an observed health effect. Epidemiological evidence must demonstrate an association between health effects and exposure levels to that chemical, taking account of confounding factors. Biological evidence, involving *in vivo* and *in vitro* experiments must offer a plausible explanation of how that exposure leads to the health effects. Nevertheless, it is possible to undertake HIA on the types of waste management facilities without fully knowing the chemicals involved.

Different methods are used to assess the health impacts of existing waste management facilities. One approach involves measurement of self-reported illness but these data suffer from participation and recall bias. Another approach compares disease incidence between exposed and comparison populations, either making use of existing disease registries (e.g., cancer or congenital malformations) or special surveillance systems. Increasingly, biological tests, which measure organ damage or dysfunction, are employed in exposed and control populations.

Landfill—possible hazards

Landfilling for disposal of municipal solid waste will continue to be the major waste stream for many years, despite the increasing role of recycling, composting, and incineration. A complex sequence of chemical and biological processes produces liquid and gaseous emissions from the parent waste. As water passes through a landfill, contaminants are leached from the solid waste and may reach the groundwater, although in recent years leachate containment designs have reduced this problem. Much organic waste is converted to gaseous products, termed landfill gas (LFG) comprising approximately 40–60 per cent methane, approximately 30 per cent of carbon dioxide (CO_2), and traces of gases such as hydrogen sulphide (H_2S). Trace concentrations of toxic volatile organic compounds (VOC) are also observed in LFG including halogenated aliphatics, heterocyclic compounds, aromatics, and ketones [3,4].

Each landfill site is unique with respect to age, quantity, and type of waste contained, local meteorology, hydrogeology, and engineering control of leachate and LFG. Toxic pollutant emissions can be minimized through optimization of biodegradation, leachate and gas collection, and treatment.

The US Agency for Toxic Substance and Disease Registry (ATSDR) have characterized emissions from hazardous waste sites [5–7] and are studying the hundred or so trace chemicals, which may be measured in leachate or LFG at such sites [8]. People in the vicinity of hazardous waste sites may also be exposed to chemical mixtures [9,10].

Airborne exposure may lead to people inhaling constituents of LFG or emissions from LFG flares or particulate matter. It is in general unlikely that any abstraction points for public drinking water supplies will be near the site but leachate contamination of nearby private water supplies is a possibility. If offsite soil is contaminated by atmospheric deposition or surface water then exposure may also occur by skin contact, ingestion of soil by casual hand to mouth contact, or by eating crops grown in the soil. Exposure of children is of particular relevance since young children are more likely to inadvertently ingest dust adhering to their hands.

In the United States, results of public health assessments conducted at 167 waste sites during 1993–1995 showed that about 1.5 million people had been exposed to site-specific contaminants. Complete or potentially complete exposure pathways involving 56 substances were identified at 10 per cent or more of the sites. Of these substances, 19 are either known or anticipated human carcinogens and 9 are associated with reproductive or endocrine-disrupting effects.

Landfill—possible health impacts

In the United States, the ATSDR has responsibility for conducting epidemiological studies, undertaking surveillance, and setting up disease and exposure registries to determine the relationship between exposures to hazardous substances and adverse effects upon public health. It has established a National Exposure Registry and has sub-registries on dioxins, benzene, and trichloroethene. It has developed a priority list for hazardous substances. The ATSDR has also produced a list of seven priority health conditions [11], which might possibly be caused by waste sites. These are birth defects and reproductive disorders, cancers (selected sites), immune function disorders, kidney dysfunction, liver dysfunction, lung and respiratory diseases, and neurotoxic disorders. The particular vulnerability of children and pregnant women is a consideration.

The epidemiological evidence of health effects associated with exposure to substances from landfill sites has been the subject of a number of recent reviews [12,13]. Most studies of landfill sites have focused on hazardous waste sites rather than household or domestic waste sites and many of these focus on sites with relatively high emissions.

Public concern in the vicinity of landfills has prompted a number of studies of particular sites [14–18]. These are prone to recall bias and are limited in statistical power due to the size of population residing in the vicinity of a single site. To increase statistical power, several multi-site studies have been undertaken, where sites have been selected independent of community concerns or reported disease clusters. A collaborative European study (EUROHAZCON) examined the association of non-chromosomal congenital anomalies with 21 hazardous waste landfill sites. A 'proximate' zone of 3-km radius from the site, within which it was assumed that most exposure to chemical contaminants would occur, was compared to a zone of radius 3–7 km from the site. Increased risk of non-chromosomal [19] and chromosomal [20] anomalies for residents living within 3 km of the sites was reported. In Great Britain, those living within 2 km of 9565 landfill sites operational at some time between 1982 and 1997 were found to have a marginally higher relative risk (1.01) for all birth anomalies than those residing more than 2 km from all known landfill sites [21]. Multi-site studies in the United States have produced both positive [22] and negative [23,24] results. The lack of information on alternative pollution sources hampers the interpretation of multi-site studies.

Most multi-site studies have concentrated upon congenital malformations, but increased bladder cancers and leukaemias have been reported in women residing in areas likely to be exposed to landfill gas [25]. There have also been a large number of health surveys, which have relied upon residents reporting symptoms through questionnaires [26]. These may be subject to reporting bias but nevertheless indicate the impact that concerns can have on health.

Incinerators

Modern, well-managed incinerators can be an effective means of reducing and disposing of waste materials so that any potential health risk is minimized. The main residue from the process is bottom ash that can be disposed of to landfill or can be used as an aggregate substitute in road building or construction. The other main residue is fly ash that includes the remains of materials (such as lime or activated carbon) used in the various stages of cleaning the gases vented to atmosphere from the chimney stack. Fly ash disposal is a tightly controlled process since the material is usually a hazardous waste containing contaminants in a concentrated form. The products of the combustion process may contain hazardous or toxic pollutants and emissions will add to background pollution levels. As a result, there is often considerable public concern over the possible health impacts of incinerators processing hazardous, clinical, or municipal waste.

The substances of principal concern with regard to emissions from waste incineration are set out in the Waste Incineration Directive. Dioxins and furans, polycyclic aromatic hydrocarbons, and heavy metals (arsenic, nickel, cadmium, and chromium) are perhaps of greatest concern. It has been hypothesized that exposure to dioxins and furans (either directly via inhalation or indirectly via the food chain) are major causes of cancer in communities around incinerators. Whilst older incinerators were often significant sources of dioxins and furans in the local environment, modern incinerators are significantly cleaner.

Although incinerators generate considerable public concern, there have been surprisingly few published epidemiological studies that examine the health of communities living close to incinerators. The majority of published studies concentrate on the effects of exposure to emissions from the older generation of incinerators, which were phased out in the United Kingdom after the introduction of stricter emission controls implemented through the Integrated Pollution Control (IPC) regime. The level of exposure from such facilities would have been substantially higher than would be expected from modern or future incinerators.

Several epidemiological studies have suggested a possible association between incinerator emissions and cancer. The possibility of a cancer cluster, particularly of cancer of the larynx, near to the Charnock Richard incinerator prompted a detailed study, which included the other nine UK incinerators licensed to burn waste solvents and oils [27]. This study concluded that the apparent cluster of cancer cases was unlikely to be due to the incinerator.

Despite reports of cancer clusters, no consistent or convincing evidence of a link with incineration has been published. In the United Kingdom, a large epidemiological study by Elliot and colleagues of the Small Area Health Statistics Unit (SASHU) examined an aggregate population of 14 million people living within 7.5 km of 72 municipal solid waste incinerators. This included all incineration plants irrespective of age up to 1987. Initial analysis suggested a possible association with stomach, colorectal, and liver cancers [28], but once confounding by socio-economic and other factors was taken into account no excess of cancers around the incinerators was demonstrated. As a result, the Department of Health's Committee on Carcinogenicity published a statement in March 2000 evaluating the evidence linking cancer with proximity to municipal solid waste incinerators in the United Kingdom [29]. The Committee concluded that 'any potential risk of cancer due to residency (for periods in excess of ten years) near to municipal solid waste incinerators was exceedingly low and probably, not measurable by the most modern techniques'.

Several studies have examined possible adverse effects on respiratory health among people living near incinerators [30] and failed to show any excess of acute chronic respiratory symptoms.

Composting

Composting is a complex aerobic microbiological process by which the organic fraction of municipal solid waste and other organic wastes are converted into compost products. The United Kingdom currently landfills 27 million tonnes a year of municipal waste and 60 per cent of this is biodegradable. Composting organic materials produces biological aerosols (bioaerosols) consisting of actinomycetes, bacteria, fungi, protozoa, and organic constituents of microbial and plant origin [31–34]. Bioaerosol concentrations existing in the ambient environment (i.e., background levels) will vary with geographical location and with season. The species components within the total concentration will also vary according to geography and season.

The EA reported [35] a monitoring programme of key environmental emissions from various types of composting facilities treating various types of compostable waste. On many occasions the concentrations of bioaerosol measured both upwind and downwind of the sites exceeded 1000 cfu/m^3 bacteria, 300 cfu/m^3 gram-negative bacteria, and 1000 cfu/m^{-3} fungi.

Bioaerosols produced by composting have the potential to produce adverse health effects such as aspergillosis, hypersensitivity pneumonitis, and exacerbation of asthma [36–39]. There is also potential for disease if pathogens survive the composting process and are present in bioaerosols. Faecal contamination of raw material is highest when it incorporates large quantities of urban wastewater sludge or farm wastes. However, household wastes may contain human and domestic animal faeces and municipal solid wastes consist of about 1 per cent by weight disposable nappies of which about one-third are soiled with faeces. Such faecal material may be contaminated with potentially pathogenic bacteria (e.g., Salmonellae), prototozoa (e.g., Cryptosporidium parvum, Giardia lamblia), worms (e.g., *Toxocara* spp.), or enteric viruses (Hepatitis A, Poliovirus, Coxsackie) [40].

The risk to health for an individual exposed to bioaerosol from composting operations depends upon the concentrations in air of different components of the bioaerosol as well as personal exposure and prior health status. However occupational health and individual case reports demonstrate the potential for health risk in uncontrolled settings. *Aspergillus fumigatus* is an opportunistic pathogen in that it colonizes and infects individuals who are immunocompromised. Hypersensitivity pneumonitis (extrinsic allergic alveolitis) may result from repeated inhalation of and sensitization to a wide variety of organic aerosols including bacteria and fungi. It is completely reversible if antigen exposure ceases but continued exposure commonly leads to progressive interstitial fibrosis [41]. Inhalation of specific allergens is a well-recognized cause of exacerbations of asthma. Asthma may be caused by allergens of microbial or plant origin but the amounts of airborne allergens that sensitize and incite asthmatic or allergic episodes cannot be defined given the wide variation in host sensitivity.

Waste collection–transfer–recycling

In future waste collection authorities will be under pressure to meet recycling targets, which may necessitate fortnightly collection intervals and compromises on waste containers for kerb-side collection. Putrescent matter held in unsuitable domestic situations for too long may encourage infestation. Most published research on waste transfer or sorting sites has focused on occupational health and internal air quality. A wide-ranging review, undertaken in Denmark [42], reported respiratory symptoms, gastrointestinal symptoms, and irritation of the eyes and skin in those exposed. A Finnish study [43] highlighted that concentrations of micro-organisms and VOCs around a waste transfer site were higher than around landfill sites but a Canadian study [44] concluded that microbial air quality outdoors 100 M downwind was not affected by operations.

Road traffic generated by waste movements has potential health impacts and includes the use of private vehicles using recycling/civic amenity sites. Many significant air pollutants are primarily generated by road vehicles (e.g., PM_{10}, benzene, and NO_2).

Anxiety and distress

HIA must take account of the health effects arising from public anxiety about health impacts of waste management facilities (be they actual or perceived). Several studies have reported data on psychiatric symptoms amongst residents living close to a waste disposal site. Only five of these studies included samples of unexposed residents as a comparison group. There was some evidence to support the hypothesis that residents exposed to hazardous waste facilities exhibit greater levels of psychiatric morbidity than residents who are not exposed to such sites. However it seems likely that at least some of this association might be explained by response bias, measurement bias, and confounding. Local incinerators appear to generate the greatest public concern but people also display anxiety about living close to landfills incorporating hazardous waste. Psychiatric disorder is common, disabling, and burdensome and any excess associated with waste disposal needs to be accurately quantified as a matter of urgency. Psychiatric morbidity amongst residents living close to hazardous waste sites might be improved through transparent and accurate communication of the health risks involved, with the aim of alleviating the heightened yet understandable concern in the exposed population. A well-run HIA process will do this at the same time as making more quantitative analyses of health risks.

The way forward

The review of the different waste management options demonstrates that all produce emissions that have the potential to harm health. It is impossible to say that a strategy maximizing recycling and composting and minimizing incineration and landfill will reduce local health impacts. The areas where better evidence to support HIA of waste strategies most immediately needed are:

+ more sophisticated spatial epidemiology of health outcomes married to dispersion modelling of emissions;

♦ more investigation of the role of confounding factors in determining psychological morbidity of individuals living close to waste facilities and evaluation of interventions directed to preventing psychological morbidity.

Local impacts on health from proposed waste management facilities must be considered by the local decision-making processes. In the United Kingdom, new facilities require planning permission for land use from the local authority and an operating license from the EA, who may also impose operating conditions. HIA can contribute to these decision-making processes and may be submitted as evidence to the statutory authorities. The Chemical Hazard Management and Research Centre, University of Birmingham, has identified the principles that should underpin local health authority input and suggests key components of a public health assessment for IPPC applications [45]. They will need to consider not only the general risks associated with each type of facility but also how characteristics of the local communities and geography could affect the health impacts of the proposal. The potential for health improvement relating to local employment and the economic consequences of alternative waste disposal may need to be considered.

References

1 European Commission Council Directive 99/31/EC on the landfill of waste.

2 Waste Strategy 2000 for England and Wales. Part I HMSO, Norwich, 2000. ISBN 0-10-146932-2.

3 Brosseau J and Heitz M. Trace gas compound emissions from municipal landfill sanitary sites. *Atmospheric Environment* 1994; **28**:285–293.

4 El-Fadel M, Findikakis AN, and Leckie JO. Environmental impacts of solid waste landfilling. *Journal of Environmental Management* 1997; **50**:1–25.

5 Johnson BL. Hazardous waste: human health effects. *Toxicology and Industrial Health* 1997; **13**:121–143.

6 De Rosa CT, Stevens Y, and Johnson BL. Role of risk assessment in public health practice. *Toxicology and Industrial Health* 1998; **14**:389–412.

7 Amler RW and Lybarger JA. Research program for neurotoxic disorders and other adverse health outcomes at hazardous chemical sites in the United States of America. *Environmental Research* 1993; **61**:279–284.

8 De Rosa CT, Johnson BL, Fay M, Hansen H, and Mumtaz MM. Public health implications of hazardous waste sites. Findings, assessment and research. *Food and Chemical Toxicology* 1996; **34**:1131–1138.

9 Hansen H, De Rosa CT, Pohl H, Fay M, and Mumtaz MM. Public health challenges posed by chemical mixtures. *Environmental Health Perspectives* 1998; **106**:1271–1280.

10 Etkina EI and Etkina IA. Chemical mixtures exposure and children's health. *Chemosphere* 1995; **31**:2463–2474.

11 Agency for Toxic Disease Registry (ATSDR). Hazardous—waste sites: priority health conditions and research strategies—United States. *Morbidity and Mortality Weekly Report* 1992; **41**:72–74.

12 Vrijheid M. Health effects of residence near hazardous waste landfill sites: a review of epidemiologic literature. *Environmental Health Perspectives* 2000; **108**:101–112.

13 Johnson BL. A review of the effects of hazardous wastes on reproductive health. *American Journal of Obstetrics and Gynaecology* 1999; **181**(Suppl 1):S12–S16.

14 Kharrazi M, Von Behren J, Smith M, Lomas T, Armstrong M, Broadwin R, Blake E, McLaughlin B, Worstell G, and Goldman L. A community based study of adverse pregnancy outcomes near a large hazardous waste landfill in California. *Toxicology and Industrial Health* 1997; **12**:229–310.

15 Berry M and Bove F. Birth weight reduction associated with residence near a hazardous waste landfill. *Environmental Health Perspectives* 1997; **105**:856–861.

16 Najem GR, Strunck T, and Feverman M. Health effects of a superfund hazardous chemical waste disposal site. *American Journal of Preventive Medicine* 1994; **10**:151–155.

17 Kilburn KH. Neurotoxicity from airborne chemicals around a superfund site. *Environmental Research Section A* 1999; **81**:92–99.

18 Williams A and Jalaludin B. Cancer incidence and mortality around a hazardous waste depot. *Australian and New Zealand Journal of Public Health* 1998; **22**:342–346.

19 Dolk H, Vrijheid M, Armstrong B, Abramsky L, Bianchi F, Garne E, Nelen V, Robert E, Scott JES, Stone D, and Tenconi R. Risk of congenital anomalies near hazardous waste landfill sites in Europe: the EUROHAZCON study. *Lancet* 1998; **352**:423–427.

20 Vrijheid M, Dolk H, Armstrong B, Abramshy L, Bianchi F, Fazarinc I, Garne E, Ide R, Nelen V, Robert E, Scott JES, Stone D, and Tenconi R. Chromosomal congenital anomalies and residence near hazardous waste landfill sites. *Lancet* 2002; **359**:320–322.

21 Elliott P, Briggs D, Morris S. de Hoogh C, Hurt C, Jensen TK, Maitland I, Richardson S, Wakefield J, and Jarup L. Risk of adverse birth outcomes in populations living near landfill sites. *BMJ* 2001; **323**:363–368.

22 Geschwind SA, Stolwijk JAJ, Bracken M, Fitzgerald E, Stark A, Olsen C, and Melius J. Risk of congenital malformations associated with proximity to hazardous waste sites. *American Journal of Epidemiology* 1992; **135**:1196–1207.

23 Marshall EG, Gensburg LJ, Deres DA, Geary NS, and Cayo MR. Maternal Residential exposure to hazardous wastes and risk of central nervous system and musculoskeletal birth defects. *Archives of Environmental Health* 1997; **52**:416–425.

24 Sosniak WA, Kaye WE, and Gomez TM. Data linkage to explore the risk of low birth weight associated with maternal proximity to hazardous waste sites from the national priorities list. *Archives of Environmental Health* 1994; **49**:251–255.

25 Lewis EL, Kallenbach LR, Geary NS, Melius JM, Ju CL, Orr MF, and Forand SP, Investigation of cancer incidence and residence near 38 Landfills with soil gas migration conditions: New York State, 1980–1989. ASTDR/HS-98-93 Agency for Toxic Substances and Disease Registry, 1998.

26 Vrijheid M. Health effects of residence near hazardous waste landfill sites: a review of epidemiologic literature. *Environmental Health Perspectives* 2000; **108**:101–112.

27 Elliott P, Hills M, Bereford J, Kleinschmidt I, Jooley D, Pattenden S, Rodriques L, Westlake A, and Rose G. Incidence of cancer of the larynx and lung near incinerators of waste solvents and oils in Great Britain. *Lancet* 1992; **339**:854–858.

28 Elliott P, Shaddick G, Kleinschmidt I, Jolley D, Walls P, Beresford J, and Grundy C. Cancer incidence near municipal solid waste incinerators in Great Britain. *British Journal of Cancer* 1996; **73**:702–710.

29 Committee on Carcinogenicity. Cancer incidence near municipal solid waste incineration in Great Britain. Statement COC/00/S/001, Department of Health, 2000.

30 Hu S-W, Hazucha M, and Shy CM. Waste incineration and pulmonary function: An epidemiologic study of six communities. *J Air Waste Manage Assoc* 2001; **51**:1185–1194.

31 Millner PD, Olenchock SA, Epstein E, Rylander R, Haines J, Walker J, Ooi BL, Horne E, and Maritato M. Bioaerosols associated with composting facilities. *Composting Science & Utilization* 1994; **2**:8–57.

32 Gilbert J and Ward C. Monitoring bioaerosols at composting facilities. *BioCycle* 1998; **39**:75–76.

33 Van der Werf P. Bioaerosols at a Canadian composting facility. *BioCycle* 1996; **37**:78–83.

34 Fischer G, Muller T, Ostrowski J, and Dott W. Mycotoxins of *Aspergillus fumigatus* in pure culture and in native bioaerosols from compost facilities. *Chemosphere* 1999; **38**:1745–1755.

35 Environment Agency. *Monitoring the Environmental Impact of Waste Composting Plants.* R&D Technical Report P428. Bristol: Environment Agency, 2001. ISBN: 1 857 05683 3.

36 Epstein E. Human pathogens: hazards, controls and precautions in compost. *Compost Utilization in Horticultural Cropping Systems* 2001; **361**:361–380.

37 Browne ML, Ju CL, Recer GM, Kallenbach LR, Melius JM, and Horn EG. A Prospective study of health symptoms and *Aspergillus fumigatus* spore counts near a grass and leaf composting facility. *Compost Science & Utilization* 2001; **9**:241–249.

38 Bünger J, Antlauf-Lammers M, Schulz TG, Westphal GA, Mülle MM, Ruhnau P, and Hallier E. Health complaints and immunological markers of exposure to bioaerosols among biowaste collectors and compost workers. *Occup. Environ Med* 2000; **57**:458–464.

39 Douwes J, Wouters I, Dubbeld H, van Zwieten L, Steerenberg P, Doekes G, and Heederik D. Upper airway inflammation assessed by nasal lavage in compost workers: a relation with bio-aerosol exposure. *American Journal of Industrial Medicine* 2000; **37**:459–468.

40 Pahren HR and Clark CS. Micro-organisms in municipal solid waste and public health implications. *Crit. Rev. Environ. Contam.* 1987; **17**:187–228.

41 Rose CS. Hypersensitivity pneumonitis 1867–1884. In Murray JF and Nadel JA (eds.), *Textbook of Respiratory Medicine.* London: WB Saunders Company, 2000.

42 Poulsen OM, Breum NO, Ebbehøj M, Hansen AM, Ivens UI, van Lelieveld D, Malmros P, Matthiasen L, Nielsen BH, Nielsen EM, Schibye B, Skov T, Stenbaek EI, and Wilkins KC. Sorting and recycling of domestic waste. Review of occupational health problems and their possible causes. *The Science of the Total Environment* 1995; **168**:33–56.

43 Kiviranta H, Tuomainen A, Reiman M, Laitinen S, Nevalainen A, and Liesivuori J. Exposure to airborne microorganisms and volatile organic compounds in different types of waste handing. *Ann. Agric Environ Med.* 1999; **6**:39–44.

44 Lavoie J and Guertin S. Evaluation of health and safety risks in municipal solid waste recycling plants. *Journal of the Air & Waste Management Association* 2001; **51**:352–360.

45 Chemical Hazard Management and Research Centre. *IPPC: A Practical guide for Health Authorities.* Birmingham: University of Birmingham, 2000.

Chapter 29

HIA and fears of toxicity: health risk assessment of a control programme for the white-spotted tussock moth in New Zealand

Donald M Campbell

Background

On 17 April 1996, a resident of Kohimarama, Auckland, first identified caterpillars of the white-spotted tussock moth (*Orgyia thyellina*). The species is native to Japan, Korea, Taiwan, China, and the Russian Far East but does not occur naturally in New Zealand. In response to the potential threat to our native and planted forests, horticultural crops, amenity trees, and gardens, a programme to eradicate the pest commenced in October 1996, using aerial and ground application of the biological pesticide Foray 48B [1]. The campaign, named Operation Ever Green, was implemented by the then New Zealand Ministry of Forestry.

Foray 48B is a commercial formulation containing, as the active ingredient, the bacterial species *Bacillus thuringiensis* var. *kurstaki* (*Btk*) and water, traces of essential elements, minerals, or salts plus a number of 'inert' components such as thickening, sticking, and wetting agents. The inerts are so named because they are considered not to contribute directly to the pesticide activity of the formulation; it does not reflect necessarily their toxic potentials. The *Btk* is a bacterium that occurs naturally in soil, water, foliage, and air throughout the world, including New Zealand. It is a biological insecticide that is activated in the uniquely alkaline gut contents of the caterpillar, causing gut paralysis and feeding cessation. It breaks down relatively quickly in the environment through exposure to UV light and other micro-organisms. For over 30 years large quantities of *Btk* have been used throughout North America for the control of the gypsy moth (*Lymnatria dispar*), a relative of the tussock moth. Formulations and patterns of *Btk* use differ according to local conditions and moth species [2].

Table 29.1 Exposure to *Btk* during the 1996–1997 campaign by means of delivery

	DC6 plane	Helicopter	Ground spraying
Type	Aerial mist—*Btk*	Aerial mist—*Btk*	Ground-level mist blower or micronair—*Btk*
Volume of *Btk*	130,000 L	28,090 L	
Duration	5 Oct 1996–9 Dec 1996 (9 flights)	5 Oct 1996–17 Apr 1997 (23 flights)	2 Oct 1996–17 Apr 1997 (21 rounds)
Location	Eastern suburbs	Subarea of Mission Bay, Kohimarama, and Meadowbank	Some properties within subarea
Population	81,389	5640	Est. 1269
Residential properties	c30,000	2395	539

A health risk assessment preceded the aerial spraying in 1996/1997 and incorporated a plan for risk management and communication [3]. This did not cover the then unanticipated use of helicopter and ground spraying. A health risk assessment was carried out in early 1997 of ground spraying [4]. The area included in the 1996–1997 aerial eradication with *Btk* was predominantly urban residential, with a resident population of 91,389 (1996 Census; Table 29.1). A smaller 'infested' area, with a population of 5640, underwent a longer duration aerial spray programme supplemented with ground spraying of 539 residential properties in the immediate vicinity of detected caterpillars, egg cases, or moths.

A synthetic pheromone (chemical sexual attractant for catching male moths) was developed and used in an extensive pest surveillance programme during 1997/1998—so called 'moth trapping'.

In 1997 a second health risk assessment was performed preparatory to possible continued control measures [5]. The Ministry was considering a number of options for further pest management activities. Those elements considered to require health risk assessment were:

- Nine aerial sprays between October 1 and early December 1997, using *Btk* in the formulation Foray 48B at 5L/hectare over approximately 300 hectares of eastern Auckland

- BK117 helicopters as the delivery platforms

- The use of Micronair sprayers

- The placing of white-spotted tussock moth pheromone in up to 10,000 traps at a density of 100 traps/hectare.

No further applications of *Btk* were used in the area and eradication of the white-spotted tussock moth was declared in June 1998.

This chapter focuses on the second health risk assessment.

HIA process and outcomes

The framework of risk assessment used was based on the method described by the WHO [6]. Four steps are accepted as being essential [7]:

- *Hazard identification.* Identifying the agents responsible for the health problem, its adverse effects, the population exposed, particularly susceptible groups, and conditions or routes of exposure.

- *Dose/response assessment.* Describing the potential health effects of the hazard at different levels of exposure.

- *Exposure assessment.* Estimating the magnitude, duration and frequency of exposure, and number of people exposed by different routes.

- *Risk characterization.* Combining dose/response and exposure assessment to quantify the risk level in a specific population. The end result is a qualitative and, where possible, quantitative statement about the health effects expected and the proportion and number of affected people in a target population, together with estimated of the uncertainties involved.

A number of steps were taken to identify hazards of the possible future programme that might pose a risk to the population, to define those likely to be exposed to these hazards, to identify possible health effects arising from these hazards, and to identify any subgroups that might be particularly susceptible. These were:

- *Community inspection.* Local authorities were consulted to identify any public areas, facilities, institutions, or industries or any proposed activities or events planned for the next 12 months that might present a risk of unusual exposure from the possible future programme.

- *Demography.* 1996 census data was used to describe the population in the area of the proposed activity. Key indicators of health status were estimated for the eastern suburbs and surrounding areas.

- *Scientific publications.* Standard literature searches were undertaken to identify publications on the health effects of the hazards identified. In light of the emotive nature of the topic Internet searches were also performed.

- *Community consultation.* The Ministry of Forestry undertook extensive community consultation. The health risk assessors attended public meetings and also met with a residents group, Society Targeting Overuse of Pesticides (STOP), to identify their concerns.

◆ *Commissioned review*. An expert toxicological review of the evidence was commissioned. The toxicologist produced a 'worst case' scenario of exposure to Foray 4B for more detailed appraisal of risk.

The second assessment had the benefit of being able to examine health effects reported during and since the 1996–1997 spraying programme. Monitoring of the health effects undertaken were:

◆ A review of health complaints

◆ Health reports

◆ Laboratory reports

◆ Hospital discharge reports

◆ Road traffic accident reports

◆ Occupational health reports

◆ Population questionnaires

A review of health complaints of attendees of an accident and emergency centre in the spray zone during the programme undertaken by Public Health Protection, Auckland Healthcare, produced no evidence of increased attendance during the initial month of campaign (October 1996) compared to the previous month or corresponding month in 1995. Health reports were received by Ministry of Forestry on a free telephone service (Bugline), initiated on commencement of the campaign, and notifications and self-reports received by the Medical Officer of Health were reviewed. Two hundred and seventy eight individuals reported a total of 682 specific symptoms. Respiratory symptoms were reported by 40 per cent (asthma 12 per cent), skin 30 per cent, eye 31 per cent, and 'general' 28 per cent. Concerns about possible future health effects were articulated by 15 per cent. Medical laboratory reports from community and hospital laboratories serving the area were reviewed. *Bacillus species* commonly occur as contaminants in bacteriology cultures so the proportion of *Bacillus* sp. that is *B. thuringiensis* was unknown. Growths of the organism from an eye and a skin swab were reported during the spraying time. One growth from a blood culture occurred though its clinical significance was not clear. Concerns were raised over an alleged excess of premature births and low birth weights occurring in the spray zone. Hospital discharge reports were reviewed but too few of the pregnancies conceived during the spraying time had come to term for epidemiological study at the time of the assessment. No excesses of postulated possible associated infections (corneal ulcers, respiratory and gastro intestinal infections, and meningococcal disease) were identified.

The potential for accidents to pedestrians or vehicle occupants through driver distraction from low flying aircraft was postulated. There was no evidence from road traffic accident reports of anything but normal year-on-year variation during the spray period. Researchers or field workers employed by Operation Ever Green may have had higher levels of exposure than the general public. No reports of occupational health concerns were received.

A questionnaire survey of residents in the original spray zone was carried out using random telephone samples. Ninety per cent expressed concerns about the white-spotted tussock moth and its effect on the environment. Eight per cent of the total sample reported a member of their household to have been adversely affected by the spray used with a slightly higher percentage (10 per cent) in target area households. Irritation of eyes, throat and sinuses, headaches, breathing difficulties, and fatigue accounted for 91 per cent of the conditions described.

Quantifiable risks of specific diseases were not identified in association with the programme.

The health risk assessment considered the suggested spray programme, including its method of delivery, the use of synthetic pheromone traps and the white-spotted tussock moth [3]. No evidence was found that using Btk or the other ingredients of Foray 48B would pose a significant health risk to this population though the available toxicological database was limited. However, as with the past programme, it was expected that there would be some complaints of minor eye, skin, and upper respiratory tract irritation. Teratogenicity and mutagenicity studies were not available. The risks from exposure to aircraft noise or accidents and road accidents were assessed as extremely small though nuisance effects were unquantifiable. The assessed risk from exposure to white-spotted tussock moth pheromone traps was exceedingly small. The risks to health from exposure to an infestation of the white-spotted tussock moth were identified as small but real, with children and those working in gardens most likely to be exposed. These are principally eye irritation, skin lesions, and respiratory reactions. There is a demonstrable dose–response effect related to egg masses/acre [8].

Recommendations

Recommendations were made for risk management and communication. These covered public health advice, operational advice to the Ministry of Forestry, surveillance, public health management, and research.

A policy of prudent avoidance to minimize exposure was suggested. This was proposed in light of the concern in the community and the lack of epidemiological information, despite the assessment that the proposed programme would not cause significant health effects. Specific advice was given for schools, on food hygiene, to users of roof drinking water, on gardening, and on swimming pool maintenance. Special guidance was given for groups of potentially susceptible people, identified as those with asthma or skin disease, pregnant women, and immunosuppressed individuals.

It was recommended that spraying be targeted as far as possible to avoid times when children might be walking to and from school or of major outdoor events.

Recommendations on laboratory work and proposals for examining the acute and future health experience of the exposed population were made. As well as physical disease, it was proposed that the psychological status of the community should be monitored.

In light of the concerns expressed over the previous programme and the perceived lack of independent health advice, recommendations over a Helpline and expert review panel were made.

Gaps in knowledge identified in the health assessment process were listed.

Discussion

It is impossible to conclusively prove safety. Experiments can demonstrate if there are harmful effects from an exposure and the levels at which these occur. However, the absence of some effect under particular exposure conditions may not prove safety because:

- the range of other effects and exposure conditions that could be tested;
- not demonstrating an effect does not mean no effect; and
- associations can be shown conclusively between exposure and disease but not with absence of disease.

In microbiological studies *Btk* has been shown not to be associated with infective illnesses or changes in balance of cutaneous micro-organisms. However, studies have not been performed specifically among individuals with a range of illnesses to ascertain whether they have not had infections with *Btk*. Such studies have not been performed for any other infective agents.

The toxicological studies of the chemical components of the spray are based on animal models. There are difficulties in relating these to humans, especially in this unique programme with multiple exposures of possibly varying durations in a relatively short time-period.

Epidemiological enquiries are observational rather than experimental, making it complex to accurately assess exposure to an agent or compound. Even if exposure can be estimated, it may be difficult to determine a single source and therefore attribute effects to it and to distinguish these from other exposure sources (e.g., food stuffs). Studies examine health effects on populations of variable sizes. While statistical analysis may not demonstrate a statistically significant effect in the entire group researched, there may be individuals within the collection who have experienced the effect under examination. This highlights the difference between statistical, clinical, and public health significances. Bradford Hill enunciated epidemiological criteria with which to judge whether an association may be causal [9].

A frequent criticism is that health risk assessments are too technical. Those who will be, or have been, affected by the impact are not involved in gauging its effects. Ideally the community concerned should be full and equal participants in the process. This has the benefit of a greater chance of voluntary acceptance of risks by an informed community. Studies of risk perception suggest that voluntary risks are more acceptable than involuntary ones [10]. Ethical guidelines for risk communication have not been defined but measures for converting an 'arrogation of wisdom' into a 'stewardship of wisdom' have been proposed [11]. Public attitudes to chemical hazards may diverge considerably (in either direction) from probabilistic risk assessments made by engineers and scientists [12].

Strengths and weakness of work

The principal strength of this work was its ability to focus on reasonably definable hazards—a biological agent and its associated chemicals, the means of their delivery and exposure, and an identifiable geographic area with a relatively stable population. This was not a novel use of *Btk* though aspects of the proposed programme were unique. In addition there had been the experience of the 1996–1997 spray programme in the same area to draw on.

This was a multidisciplinary assessment carried out by individuals from a range of backgrounds—microbiology, public health medicine, and toxicology who were able to utilize a reasonably extensive literature. The complexity of public health issues rarely, if ever, allow an assessment from the perspective of any one single discipline.

The foremost weakness was that of being a standalone exercise rather than part of an integrated risk management project. The overall driver of the eradication programme was the Ministry of Forestry who commissioned a health risk assessment as part of their overall risk assessment. Thus the public consultation was driven by them and covered wider issues than health. This resulted in health positions being articulated by a small number of individuals only, rather than the overall community viewpoint being available and expressed. Community feedback on the health risk assessment was done via STOP and the local and national media. The residents of the eastern suburbs were thus deprived of full ownership of the assessment.

The health information used in the assessment, based principally on hospital death and discharge and primary consultation data, would not identify the types of non-specific symptoms described. Spurgeon *et al.* [13] have commented on the importance of the psychosocial as well as the physical pathways to symptoms. A comprehensive health surveillance programme was not established at the commencement of the 1996–1997 spray programme. Self-reports of symptoms were received by either Operation Ever Green or by Auckland Medical Officers of Health. However investigation and expert panel validation was not available until after the 1997–1998 health risk assessment; similarly with health surveillance using records of sentinel medical practice. No adverse health patterns were found from either of these data sets [14]. It would have been useful to have had this evidence at the time of this assessment.

Impact on decision-making process

This health risk assessment was not subjected to the test of the decision-making process as ongoing surveillance in the area for the moth and its larvae proved negative. No further applications of *Btk* were used and the white-spotted tussock moth was declared eradicated in June 1998. However in May 1999 the painted apple moth (*Teia anartoides*) was discovered in Glendene, west Auckland. This pest is unique to Australia and is considered a potential threat to New Zealand's native trees. Again aerial spraying using Foray 48B was identified as part of the eradication programme. The Auckland District Health Board carried out a further health risk assessment [15]. The only new findings were recognition of the Foray 48B's distinctive odour, which many people might find unpleasant, and that odours can cause health symptoms. The recommendations for risk management were essentially unchanged. As the programme was exempt under regulations from needing consent under New Zealand's Resource Management Act, it required the joint authorizations of the government Ministers of Conservation, Forestry and Health. This health assessment, those done for the white-tussock moth, the health surveillance report, and an EIA, were the determining influences on the advice to the ministers from their Technical Advisory Group and an independent review team to proceed with eradication using aerial spraying, again in a residential area.

Neither this health impact assessment (HIA) process nor its predecessor has been subjected to formal evaluation. The draft final report was subjected to peer review both by experts in the field of health risk assessment from within New Zealand and overseas. The interested community groups, including

STOP, were reasonably accepting of the findings at the time and media interest rapidly vanished. Unfortunately a proposal to examine the community's knowledge, attitudes, and perceptions on the topic was not considered appropriate for funding from the Operation Ever Green budget, Forestry preferring to utilize their own public opinion surveys. A report on the Operation Ever Green Health Surveillance was published in 2001 [13]. It incorporated documentation and investigation of self-reported health concerns. The most frequently reported single concern between October 1996 and June 1999 was 'fear of unspecified future disease', followed by headache and respiratory symptoms such as sore throats. No adverse patterns were found by health surveillance using sentinel general medical practitioners. No adverse patterns were found. Review of health data from suitable sources revealed no significant findings. There was no consistent difference between birth weights, gestational age, or pattern of birth defects between those exposed and not exposed to spraying.

A voluntary register of 3144/5640 (55.7 per cent) of normally resident individuals has been compiled and placed in the National Archives as a master list for any suitably approved future health studies on this population.

This and the other health impact assessments of Foray 48B plus other inputs to the decision-making processes have become the subject of legal opinions as challenges have been mounted to the painted apple moth spray programme by concerned individuals, community groups, and politicians. They have not yet been subjected to the definitive evaluation process—the courts.

Resources required

The need for this assessment occurred in New Zealand's largest conurbation of approximately 1.3 million population, which is served by one regional public health service—Public Health Protection Services, Auckland District Health Board. Its staff has extensive knowledge of the geographical, economic, political, and social influences, as well as health influences on the distinct areas that make up greater Auckland. Because of its size, this organization is multidisciplinary including Public Health Medicine Specialists with special interests in environmental health issues, supported by doctors in training and public health scientists with microbiological and chemistry backgrounds. Due to some external funding from the Ministry of Forestry, expert toxicology and external peer review services could be utilized. This assessment was the major task for the Environmental Health Team over a six-week period though involvement preceded and continued after the submission of the report. The ability to access worldwide literature and expertise rapidly by electronic means greatly facilitated the speed of carrying out the assessment.

Lessons learned

Health risk assessment is not an isolated process. The purpose is to contribute to the protection and maintenance of the public's health through the prevention or management of avoidable health hazards. It should be undertaken in the full public gaze. The ideal may not always be achievable but should be an explicit aim. It fits into a framework that incorporates risk perception, risk management, stakeholder engagement, risk communication, and community consultation [16]. Most of this health risk assessment and the other associated assessments were done at a technical level with the overall risk management being managed by another organization. This meant that the particular concerns of residents might not have been addressed fully in the health risk assessment. Communication is a two-way process and the inability of the community to contribute to the design and dissemination of the work denied them ownership. They would have had unique insights into how the proposal might affect their lives, their community, and their health-related behaviour.

It would have been better if the health risk assessors and the community had had a more defined input to the overall risk assessment process and a model such as that proposed by Chess and Hance had been part of the overall risk assessment and management programme [17]. However the assessors and managers had to be sensitive to distinctions between risk assessment and risk management. The assessors generated a credible, objective, realistic, and scientifically balanced analysis. They presented information on each component of the risk assessment. The confidence in each assessment was explained, by delineating strengths, uncertainties, and assumptions.

Risk assessments have limitations. These include information gaps, poor exposure information, the limitations of toxicological and epidemiological research, and the complex causality of many health conditions.

It was not predicted that the assessment would continue to have active currency over such a protracted period. The work was performed in the expectation that it was related to the hopefully short-lived white-tussock moth hazard. However it has been used also in relation to another—the ongoing painted apple moth infestation. These two programmes have operated in very different local political environments. The former had reasonable public support and no opposition from local body or national politicians while the latter has generated much local, legal, and even national controversy.

The means of implementing some of the recommendations to reduce possible exposure of vulnerable groups were not fully described. This has again led to controversy in the painted apple moth programme. The HIA process has to recognize the implementation phase. This includes how any conditions will be enforced.

It would have been useful to have information on the health experience of the potentially exposed population prior to the proposed programme. Most of the predicted health effects were likely to be either self-managed or to present in primary care. There was no suitable routine surveillance system that could be used to determine whether there had been any significant health outcomes. Consequently specially developed health impact auditing systems had to be implemented to provide early warning of any undesirable changes in people's health that were not predicted by the HIA. Insufficient resources were committed by the programme manager to this part of the risk management process, as is being demonstrated by the inability to answer some of the health questions being raised during the current painted apple moth programme. No information is available on the psychological effects of this programme on the community, which would have been an invaluable baseline given the painted apple moth controversy.

References

1 Gibbs N. *Environmental Impact Assessment of Aerial Spraying Btk (Bacillus thuringiensis var, kurstaki) in New Zealand to Eradicate Tussock Moth (Orgyia thyellina).* Wellington: Ministry of Forestry, 1996.

2 Otvos IS and Vanderveen S. *Environmental Report and Current Status of Bacillus thuringiensis var, kurstaki Use for Control of Forest and Agricultural Insect Pests.* Victoria, BC: Forestry Canada and British Columbia Ministry of Forestry, 1993.

3 Public Health Protection Service and Jenner Consultants. *Health Risk Assessment of Btk (Bacillus thuringiensis* var. *kurstaki) Spraying in Auckland's Eastern Suburbs to Eradicate the White-Spotted Tussock Moth (Orgyia thyellina).* Auckland: Auckland Healthcare Services, 1996.

4 Public Health Protection Service. *Health Risk Assessment of Ground-Spraying With MIMIC 70W (tebufenozide) on Selected Properties in Auckland's Eastern Suburbs to Eradicate White-Spotted Tussock Moth (Orgyia thyellina).* Auckland: Auckland Healthcare Services, 1997.

5 Public Health Protection Service. *Health Risk Assessment of the Proposed 1997–1998 Control Programme for the White-Spotted Tussock Moth in the Eastern Suburb of Auckland.* Auckland: Auckland Healthcare Services, 1997.

6 World Health Organisation. *Assessment and Management of Environmental Health Hazards.* Mimeograph WHO/PEP/89.6. Geneva: WHO, 1989.

7 Public Health Commission. *A Guide to Health Impact Assessment. Guidelines for Public Health Services.* Wellington: Public Health Commission, 1995.

8 Durkin PR, Fanarillo PA, Campbell R, *et al. Human Health Risk Assessment for the Gypsy Moth Control and Eradication Program.* Radnor, Pa: US Dept. of Agriculture, Forest Service and North-Eastern State and Private Forestry, 1995.

9 Hill AB. The environment and disease: association or causation? *Proceedings of the Royal Society of Medicine* 1965; **58**:295–300.

10 Starr S. Social benefit versus technological risk. *Science* 1969; **165**:1232.

11 **Jamieson D**. Scientific uncertainty: how do we know when to communicate research findings to the public? *The Science of the Total Environment* 1996; **184**:103–108.

12 **Lee TR**. Public attitudes towards chemical hazards. *The Science of the Total Environment* 1986; **51**:125–147.

13 **Spurgeon A, Gompertz D, and Harrington JM**. Modifiers of non-specific symptoms in occupational and environmental symptoms. *Occupational and Environmental Epidemiology* 1996; **53**:361–366.

14 Operation Ever Green Health Surveillance. Report to the Ministry of Agriculture and Forestry. Aer'aqua Medicine Ltd. Auckland, 2001.

15 Public Health Protection Service. *Health Risk Assessment of the 2002 Aerial Spray Programme for the Painted Apple Moth in Some Western Suburbs of Auckland.* Auckland: Auckland District Health Board, 2002.

16 enHealth. *Environmental Health Risk Assessment; Guidelines for Assessing Human Health Risk from Environmental Hazards.* Canberra: Dept of Health and Ageing and enHealth Council, 2002.

17 **Chess C and Hance BJ**. *Communicating with the Public; 10 Questions Environmental Risk Managers Should Ask.* Rutgers, New Jersey: Center for Environmental Communication, The State University of New Jersey, 1994.

Chapter 30

The HIA of crime prevention

Alex Hirschfield

Background

Crime is a major issue in terms of its impact on individuals, communities, and the state. In exploring the links between crime and health it is important to distinguish between the health impacts of being a victim of crime (especially on more than one occasion), those of being an offender or at risk of offending, and those of being afraid of becoming a victim (fear of crime). Fear of crime may also affect those who are concerned about the safety of relatives, friends, and neighbours. This introduces the notion of 'vicarious fear of crime' that may also be an important influence on health outcomes.

Deleterious health impacts can be generated by different types of crime and by related problems, such as disorder, that are not strictly criminal offences. These include:

- Crime against the person (robbery and theft);
- Violent crime (assault and wounding, domestic violence);
- Sexual offences (rape, indecent assault);
- Acquisitive property crime (domestic burglary, theft of/from vehicles);
- Anti-social behaviour (e.g., neighbour disputes, vandalism, racial harassment).

Impact of crime on health

There is a growing body of research on the health impacts of different types of crime. A considerable number of studies have been undertaken in recent years, notably in the United States, although much of this has focused on violent crime. Criminal injury, although only a small proportion of all recorded crime, may result in physical injuries, including fractures, bruises, and wounds to limbs and to the face and head, and infection with sexually transmitted diseases. Psychological impacts, including Post-Traumatic Stress Disorder (PTSD) can be serious and long-lasting [1,2].

There is evidence that the nature of the crime experienced influences the severity of symptoms experienced by the victim. Several studies show that rape victims are more symptomatic (or have longer recovery periods) than assault victims, that assault victims (sexual or physical) are more symptomatic than robbery victims, and that violent crime victims (assault or robbery) are more symptomatic than property crime victims [3–8].

In the United Kingdom, research in this field is less well-developed, however over the last few years this trend has been reversed. Most notable is the Public Health Alliance's (PHA) report 'Framing the debate', which looked at the impacts of crime on public health [9]. This study explored the complex relationships between crime, the fear of crime, and health using a range of methods including questionnaires dispensed at general practitioners' surgeries, in-depth interviews with health and criminal justice practitioners, and focus groups with community organizations. Both the experience of victimization and anxiety or fear of crime were shown to impact upon health through 'symptoms' such as stress, sleeping difficulties, loss of appetite, depression, loss of confidence, and health harming 'coping mechanisms' (e.g., smoking, alcohol). Similar relationships described as 'detrimental emotional impacts' have been identified in the British Crime Survey [10].

The PHA research also revealed that crime has a negative impact on the behaviour both of victims of crime and non-victims. These behaviour changes, particularly avoidance behaviour (e.g., staying in after dark, avoiding certain areas, travelling by different means), were common to all respondents. Particular defence mechanisms were often different for different groups, for example, young people felt safer in a group of friends, a minority indicated that carrying a weapon increased their sense of personal security.

Repeat victims of crime (a single type of crime perpetrated more than once against the same individual) are more likely to be adversely affected than victims of a single incident [11], as are multiple victims (victims of more than one type of crime) [12]. Similarly, multiple victims are more likely to be affected by a subsequent crime [11].

'Fear of crime' can profoundly affect the quality of individuals' lives by causing mental distress and social exclusion. It is not necessarily the result of previous victimization and those most in fear of crime are not necessarily those most vulnerable [13].

Impact of crime prevention on health

Since crime itself can result in many known health consequences, crime policy deserves attention from health impact assessors. A health impact assessment (HIA) of a crime prevention scheme is of added interest because of its focus on the changes in health and health determinants associated with preventive actions, rather than merely outcomes in terms of crimes that have been avoided.

Significantly, despite the substantial and growing body of research outlining the impact that crime has upon health, there is very little hard evidence of the health impacts of crime prevention. One existing evaluation of a crime prevention initiative targeted at older people in Plymouth explicitly mentions the reduction of fear of crime and positive consequences for people's quality of life [14]. Only a few HIA case studies so far have attended to this subject. An HIA case study of burglary prevention and youth diversion on Merseyside was undertaken as part of a larger Department of Health-funded research and development project [15,16]; related HIAs have assessed local community safety projects [17,18].

The role of HIA in this context is essentially one that focuses upon the impact of policies, programmes, and interventions aimed at preventing or reducing crime. However, crime reduction policy is wide-ranging covering a plethora of problems and approaches. It includes strategies aimed at making offending more difficult by blocking off opportunities to commit crime, programmes that seek to divert people at risk from offending and pursuing 'criminal careers', and fear-reduction strategies that attempt to reassure the public and reduce their fear of crime. The types of policy intervention that characterize each of these approaches are illustrated in Table 30.1.

Research carried out by the author at the University of Liverpool has concentrated on applying HIA to burglary reduction initiatives [15,16]. Such situational crime prevention, or indeed any of the other policy approaches in Table 30.1, will vary in their effectiveness to impact upon crime. They may also generate positive or negative 'spillover' effects such as crime displacement or diffusion of benefit (i.e., where crime reduces in areas not directly subject

Table 30.1 A simple typology of crime reduction policy

Fear reduction (reassurance, police presence)	Situational crime prevention (blocking off opportunities)	Criminality and anti-social behaviour (lifestyle, behavioural change)
High visibility policing	Target hardening (bolts, locks, alley gates)	Intensive supervision of offenders
Police on the beat	CCTV, Alarms	Drugs treatment and testing orders
Neighbourhood wardens	Property marking	Youth diversion programmes
Homewatch	Steering locks	
Street lighting	Defensible space architecture (designing out crime)	
Residents' associations	Disruption of stolen goods markets Police crackdowns	

to an intervention). How these strategies perform will also influence the nature and scale of any health impacts attributable to them.

HIA of crime prevention: some examples

Two examples of research carried out by the author, for the Department of Health, can be used to identify some of the health impacts that emerge and to illustrate some of the issues that arise in applying HIA to crime prevention. The first example is an HIA of a Target Hardening (TH) programme in Liverpool, the second is an HIA of a national crime prevention policy, the Reducing Burglary Initiative (RBI).

A TH programme

The TH programme aimed to reduce the incidence of repeat burglary to domestic properties by securing 6000 domestic dwellings located in deprived neighbourhoods. The project targeted both vulnerable properties in burglary 'hot spot areas and householders, for example, older people, people on benefits, women, lone parents, repeat burglary victims, and involved installing security measures such as new door and window locks free of charge. A comprehensive, largely retrospective HIA was undertaken and included a documentary review, community profiling, and semi-structured interviews (face-to-face and telephone) with stakeholders (project workers and other workers associated with the project). Forty victims of crime, whose homes had been protected, were also interviewed.

The health impacts were predominantly positive. Greater home security (e.g., window locks, door bolts, alarms) was found to have prevented subsequent burglary, thereby preventing the trauma and associated health impacts of being a victim of crime. There was also evidence to indicate that the programme had reduced the psychological distress of those burglary victims who had been protected. For example, a sizeable majority of those suffering sickness/dizziness, feelings of stress, depression, and panic attacks following a burglary not only claimed that their condition improved following the installation of security measures in their homes, but also attributed the change directly to the crime prevention measures and the greater peace of mind that these provided. There was also evidence from key informants that the programme initially stimulated community spirit and increased local social support networks within the neighbourhood.

As part of the TH programme, assisted households were given a general home safety risk assessment that potentially could be used as a basis of referring those at risk (particularly older people and families with small children) to other services and agencies (e.g., fitting of smoke alarms by the fire brigade). However, a clear message to emerge from this HIA was a need to 'Link the Thinking' between agencies following a domestic burglary. This might include referrals by crime prevention agencies to family doctors when patients with existing health problems are victimized or when victims suffer acute psychological distress from the burglary. It might also include referral to social services for vulnerable people in need of care and support. Key informants pointed to the additional caring responsibilities that families of victims often have to undertake following a burglary.

Negative health impacts were mainly thought to arise as a result of the displacement of crime and the fear and trauma associated with it into new areas or through 'crime switch', whereby, offenders choose new targets and thereby create new victims of crime. Informants also noted a heightened awareness and anxiety about crime experienced by neighbours not targeted by the project (i.e., households on the 'wrong side' of the target area boundary), as well as by the victims' families.

The RBI

The RBI was a national crime prevention programme aimed at reducing domestic burglary in areas of high crime through inter-agency collaboration and innovative complementary strategies. These included prevention of initial and repeat burglaries through the use of CCTV, high visibility policing, the targeting of offenders, and the involvement of local communities in crime prevention through neighbourhood watch and formation of residents' associations.

A rapid HIA was undertaken focusing on six RBI projects in the north of England. RBI projects were selected on the basis of their activities on the ground (i.e., their mix of strategies or 'interventions') and their stage of implementation. This approach enabled health impacts and their possible mitigation/enhancement to be explored for each type of intervention. Given their differing stages of implementation, three projects were assessed retrospectively, two concurrently and one prospectively using the approach set out in the Merseyside Guidelines [19]. The objectives for the HIA included:

- Introducing to the crime prevention community a workable methodology for assessing the health impacts of crime prevention;
- Building a firm foundation for the HIA of other crime prevention;

♦ Encouraging closer working between health, social services and housing, the police, and emergency services through identifying the links between victimization, crime prevention, and health;

♦ Demonstrating the extent to which there are solid public health grounds for preventing and reducing crime.

The RBI was identified as potentially impacting upon health through (a) changes in the nature and extent of burglary; (b) changes in perceptions and levels of fear; and (c) the project implementation process. A number of these findings echoed those from the TH programme, but in addition the RBI assessment highlighted how the process of implementing crime prevention measures could also potentially impact upon health. For example, interventions whose implementation requires greater participation of communities (e.g., the fitting of gates to alleys behind terraced properties, 'alleygating') may foster and encourage social interaction, neighbourliness, and build social cohesion. On the other hand, implementation failure (i.e., the inability to carry out the intended interventions) and/or theory failure (i.e., the misdiagnosis of the crime problem and perceived solutions) may raise the fear of crime. This occurs when the promise of action increases awareness that there is a crime problem in the area. If this is not followed by the reassurance of swift and visible action on the ground or detectable impacts on burglary, this can translate into heightened fear and anxiety. Such cases are likely to generate mostly negative health effects.

If one assumes that the RBI will be successful in achieving its aim, a range of potential health determinant and health impacts would be expected to result from the reduction in burglary. The benefits from burglary reduction within RBI target areas included those that arise from prevention of an initial or repeat burglary and a lowering of levels of fear plus a number of additional positive health impacts. These included:

♦ Feelings of safety in own homes allowing residents to sleep better;

♦ Peace of mind at leaving property unattended to go to place of employment/ fulfil employment commitments, take exercise, engage in leisure activities, visit family and friends (i.e., pursue healthy social connections);

♦ More confidence in leaving the house, making it easier for people to arrange visits to facilities and services that they need or would like;

♦ A reduced likelihood of older people finding it necessary to move into residential accommodation on account of burglaries being prevented or through a reduction in the fear of crime [14]. Hence, the opportunity to continue an independent life in one's usual surroundings is improved;

- Reductions in the fear of burglary and other types of crime such as car crime or violent crime as a result of physical interventions such as CCTV, gating, and improved street lighting leading to increased mobility during those times of the day or evening when residents would previously have feared for their safety;
- Reduction in the consumption of medication.

The principal negative health (determinant) impacts from burglary reduction within RBI project target areas were identified as:

- Increased fear of crime and perceived vulnerability where publicity about projects is out of proportion to the crime prevention measures that are delivered (even if the latter are effective in reducing burglary).
- Displacement of the crime to other areas, particularly those surrounding the geographically restricted target areas of the RBI.
- Change in the nature of crime, for example, in Liverpool there was a switch from burglary to theft from vehicles.

The study made recommendations both for improving HIA methodology and for the design, implementation, monitoring, and evaluation of crime prevention programmes. In terms of methodology, the report stressed the need to conduct more research into methods for assessing the health impacts of national policies characterized by considerable heterogeneity. The Merseyside Guidelines work well at the local level but were not designed to undertake HIAs of national strategies.

The health impacts of crime prevention depend so much on the effectiveness of interventions in reducing victimization and the fear of crime. Health impact assessments really need to take into account alternative scenarios of the implementation process and outcomes of crime prevention policies when identifying the nature and direction of likely health impacts. One idea that emerged from this study was the notion of 'positive health impacts foregone'. This is defined as health impacts that should be realized but that are lost because interventions fail to impact upon crime on account of poor planning, poor targeting, inefficient management, or other forms of policy implementation failure.

A common vein running through those projects that have generated the greatest positive health impacts seems to be the committed involvement of the community. Crime prevention strategies that work with the residents rather than for them maximize their health benefits through the empowerment of the community. Spin-off effects include the formation of residents associations and homewatch schemes, which, in turn, benefit the community's cohesiveness and strengthen mutual support. A good example of the maximization of benefits of an intervention can be found in Liverpool's alley-gating project, where recovering drug users manufacture the gates and health relevant spin-offs such as community involvement, job creation, and liaison with supporting agencies are strengthened.

The most successful projects in terms of generating potential health benefits seem to be those that have succeeded in establishing and utilizing pre-existing links and networks between different agencies and players relevant to health. This has allowed very creative spin-offs to occur, for example, the referral by Victim Support or the police of identified vulnerable individuals to social and voluntary agencies. Partnerships bidding for crime prevention funding should be encouraged and supported to establish creative links with other agencies far beyond the criminal justice system, so that positive health impacts from crime prevention initiatives can be maximized.

References

1 Cohen MA and Miller TR. The cost of mental health care for victims of crime. *Journal of Interpersonal Violence* 1998; **13**:93–110.

2 Norris F and Kaniasty K. Psychological distress following criminal victimization in the general population—cross-sectional, longitudinal and prospective analyses. *Journal of Consulting and Clinical Psychology* 1994; **62**:111–123.

3 Davis R and Friedman L. The emotional aftermath of rape and violence. In Figley C. (ed.), *Trauma and its Wake: The Study and Treatment of Post Traumatic Stress Disorder*. New York: Brunner Mazel, 1985.

4 Lurigio A and Resick P. Healing the psychological wounds of criminal victimisation: predicting post crime stress and recovery. In Lurigio A, Skogan W, and Davis R (ed.), *Victims of Crime: Problems, Policies and Programs*. CA: Sage, Newbury, 1990.

5 Resick P. Psychological effects of victimization: implications for the criminal justice system. *Crime and Delinquency* 1987; **33**:468–478.

6 Skogan W. The impact of victimisation on fear. *Crime and Delinquency* 1987; **33**:135–154.

7 Wirtz P and Harrell A. Assaultative versus non-assualtative victimization: a profile analysis of psychological response. *Journal of Interpersonal Violence* 1987; **2**:264–277.

8 Wirtz P and Harrell A. Victim and crime characteristics, coping responses and short- and long-term recovery from victimization. *Journal of Clinical Psychology* 1987; **55**:866–871.

9 McCabe A and Raine J. *Framing the Debate: The Impact of Crime on Public Health*. Birmingham: Public Health Alliance, 1997.

10 Simmons J. *Crime in England and Wales 2001/2002*. London: Crime and Criminal Justice Unit, Home Office, 2002.

11 Matthews R. Crime and its consequences in England and Wales. *Annals AAPSS*, 1995; **539** (May), 169–182.

12 Hirschfield A and Douglass J. Exploring links between Crime, Disorder and Health: A Wirral Case Study. Final Report submitted to Wirral Health Authority, 2001.

13 Evans D and Fletcher M. Fear of crime: testing alternative hypotheses. *Applied Geography* 2000; **20**:395–411.

14 Mawby RI. Providing a secure home for older residents: evaluation of an initiative in Plymouth. *The Howard Journal* 1999; **38**:313–327.

15 Hirschfield A, Boardman S, Douglass J, and Strobl J. The Health Impact Assessment of the Home Office Reducing Burglary Initiative. Final Report submitted to the Department of Health, March 2001.

16 Hirschfield A, Abrahams D, Barnes R, Hendley J, and Scott-Samuel A. Health Impact Assessment: Measuring the Effect of Public Policy on Variations in Health. Final Report. Department of Health, Health Inequalities Research Programme, August 2001.

17 Winters L and Scott-Samuel A. Health Impact Assessment of Community Safety Projects—Huyton SRB. Observatory Report Series No. 38. Liverpool Public Health Observatory, 1997.

18 Fleeman N. Health Impact Assessment of the Southport Drug Prevention Initiative. Observatory Report Series No. 39. Liverpool Public Health Observatory, 1997.

19 Scott-Samuel A, Birley M, and Ardern K. *The Merseyside Guidelines for Health Impact Assessment.* Liverpool: The Merseyside Health Impact Assessment Steering Group, University of Liverpool, 1998.

Chapter 31

Expanding the number of places for medical student training in England: an assessment of the impacts

Jonathan Mathers and Jayne Parry

Introduction

This chapter describes a piece of work undertaken by the Health Impact Assessment Research Unit (HIARU) based in the Department of Public Health & Epidemiology at the University of Birmingham in England. It describes the key elements of the national policy to be assessed, outlines the various impact areas that were identified, reflects on the rationale underpinning the conduct of the assessment, and considers the implications that this had for the selection of methods for assessment and the dissemination of findings.

Background

The National Health Service (NHS) Plan sets the context for modernization of the UK NHS [1]. A key element of the plan is the need to increase the number of doctors working in the health service. In order to meet this need a rapid expansion in the intake to UK medical schools is currently underway. In the United Kingdom, medical student training takes place both on the university campus and in a range of primary and secondary care clinical settings, in particular general practices and hospitals. The proposal to expand the numbers of medical students necessitates a parallel increase in clinical teaching capacity that cannot be accommodated by existing teaching hospitals and general practices. There must therefore be an increase in capacity outside of the traditional clinical teaching base, including the introduction of training places in hospitals and general practices that have not previously offered placements to students.

As part of the national expansion of student numbers, the University of Birmingham Medical School (UBMS) will increase its undergraduate intake from 218 in 1999 to 332 in 2003. To accommodate this substantial increase in student numbers, basic clinical teaching will expand from its traditional base within hospitals and general practices in the city of Birmingham to placements in other general practices, and in acute and community hospitals elsewhere in the West Midlands region. Specifically, a series of general practices that have not previously had contact with undergraduate students will become teaching general practices, and two 'non-teaching' hospitals will begin to offer student placements. This expansion will initially take place in the 'Black Country', an area to the northwest of Birmingham comprising the towns of Dudley, Sandwell, Walsall, and Wolverhampton.

Why was an HIA conducted?

At this point it is worth outlining why the HIARU decided to conduct an HIA of this policy action. The UBMS wanted to know how the implementation of the student expansion policy would impact on the local health care environment and economy, and so asked the HIARU to explore how best the policy could be evaluated. The HIARU believed that the way in which the national policy was being implemented locally in Birmingham—that is, some hospitals and general practices being encouraged to take on students whereas others were being left as 'non-teaching' units—offered a unique opportunity to establish a quasi-experimental study to evaluate the impacts of a major change in the organization and delivery of secondary care service. The results of this study might usefully inform future NHS decision-making. However there was little information available in the published literature (see below) on the likely impacts of this policy. To guide the evaluation further information was required about the various 'pathways' by which the policy might bring about change. It was therefore decided to gather 'expert knowledge' from a series of key informants involved in the policy's implementation.

Assessing the impacts

An exploration was needed of the impacts on the staff, students, and patients who work, train, and are treated in the hospitals and general practices to be converted to 'teaching' status, and also to assess the potential impacts on the wider population. Methods common to many HIA models were adopted for assessing the impacts of the expansion of student numbers and it was decided to undertake both a literature review and interviews with key informants and stakeholders [2]. The latter included interviews with staff working in hospitals and general practices that (a) already train medical students, (b) were about to begin to offer placements to medical students, and (c) that would not be offering placements to medical students.

It was decided not to speak to patients or to members of the local community directly but instead to rely on the views of professionals (Directors of Public Health and Community Health Council Chief Officers), who represent the views of these groups. The reasons for this were primarily practical: the time and resource limitations of the HIA did not allow a larger study sample that could have included a suitable number of patients and students. Therefore the available resources were targeted on the group of respondents who we felt would have knowledge of the roles and organizations that will be impacted upon by the policy. The possible limitations of this approach are discussed later in this chapter.

The first tranche of medical students took up their placements in the 'new' teaching hospitals and general practices in September 2001. The literature review and the key informant interviews were undertaken during the preceding summer and were completed by the time implementation began.

Literature review

A rapid review of the literature (published up to and including 1998) found no systematic reviews, trials, or pre–post comparisons looking at the impacts of a non-teaching hospital changing to a teaching hospital [3]. Some studies that were identified concentrated on differences in patient outcomes and treatment modalities between teaching and non-teaching hospitals but did not address the organizational issues that are likely to be fundamental to the success or otherwise of student expansion. There was some literature available that described patient perceptions of involvement in medical student teaching and on the perspectives of students undertaking placements in established teaching hospitals and in hospitals where undergraduate teaching had not traditionally taken place (e.g., [4–9]).

More information was available to guide impact predictions from student placements in primary care (e.g., [10–18]), but the generalizability of these findings from primary care to the secondary care context was uncertain. It was concluded that although the literature review suggested areas where impacts might be experienced, in themselves the studies identified could not be used to infer what would happen when students were placed in clinical settings where teaching had not previously taken place. In view of this it was decided to seek a breadth of opinion from a variety of health care professionals, likely to be affected directly or indirectly by the proposed organizational changes, in order to scope the potential impacts.

Interviews with health care professionals

The interviews with staff in hospitals and general practices adopted a similar design and method, and these have been described elsewhere [19,20]. Briefly, interviews were semi-structured with pre-agreed prompts for specific content areas. The interview sample was purposively designed to include a multidisciplinary group of interviewees likely to be affected by the expansion of medical school training places. Those asked to take part were the Regional Director of Public Health, the Directors of Public Health, and the Community Health Council Chief Officers (as representatives of patients and of the wider community) and a range of health care and managerial professionals working in hospitals and general practices. In hospitals those participating included chief executives, medical directors, general managers of surgery and medicine, consultant staff, junior doctors, and nursing staff. From general practices GPs, practice managers, nursing staff, health care assistants, and administrative staff were included. The hospitals and general practices were chosen to reflect current and future variation in involvement with medical student teaching.

Table 31.1 shows impacts anticipated by the interviewees working in the primary and secondary health care environments. Five main impact areas were identified: effects on students, staff, patients, the organization, and the wider (ex-hospital) environment. While stakeholders were broadly enthusiastic about the implications of the policy action, there were some areas where there was clear disagreement with regard to the nature and scale of possible impacts (e.g., whether patients would welcome involvement in teaching). Additionally, many of the 'positive' impacts were dependent on sufficient funds being made available to implement and support changes—for example, positive impacts such as an increase in morale or an improvement in the ability to recruit and retain staff might be reversed if insufficient resources are provided to support the development of teaching duties.

Evaluation of impacts

Such uncertainty in predicting effects of major policies is not unusual and has to some extent undermined the credibility of HIA [21]. If an HIA is to influence the way in which a policy is operationalized then it is reasonable for those involved in developing and implementing that policy to have some assurances that the changes they are being asked to make are based on accurate predictions of effect. In the context of this HIA the authors have begun to explore whether the predictions made by the stakeholders were correct. This is being done using a variety of techniques including a survey of Year 5 students,

Table 31.1 Impact areas identified by staff working in hospitals and general practices

Impact area	Themes identified within the impact area	Comment
Effects on students: the effect on medical students who undertake new placements in the Black Country hospitals and general practices	Influence of the Black Country as a teaching 'setting'	The Black Country is an ethnically diverse and deprived area—students undertaking placements in this setting may be anticipated to gain a greater understanding of cultural differences and values, and an appreciation the impact of socio-economic status on health
	Quality of teaching given to undergraduates	Existing teaching hospitals were felt to be more specialized than other hospitals. Placing students in less specialized hospitals may provide them with a greater opportunity for students to see 'run of the mill' medicine and to gain experience of practical procedures
Effects on teachers: the effect on those personnel undertaking new or additional teaching duties in the Black Country hospitals and general practices	Continuing medical education	Being involved in teaching students may be a stimulus for staff to 'keep up to date' with good practice
	Morale and motivation	Teaching is enjoyable for staff and adds variety to everyday work
	Workload	Teaching may result in heavier workloads especially if adequate time and resources are not provided
Effect on patients: the effects visible to patients attending the Black Country hospitals and general practices for medical care	Doctor—patient relationship	Most patients enjoy taking part in teaching and students are an additional source of information for patients. Some patients may be uncomfortable with additional examinations and consultations
	Information sharing	Patients may tell students information about their illness that they are unwilling to disclose to a qualified doctor. Having a student sit in with a doctor during a consultation may allow patient to learn more about their illness as the doctor explains things to the student
Organizational effects: the potential effects of the Black Country hospitals and general practices	Recruitment and retention of health care professionals	Being termed a 'teaching' hospital may impact positively on recruitment and retention of staff. Students who enjoy their placements may choose to come back and work in the hospital or general practice when they have qualified. Staff who joined the hospital when it did not have teaching responsibilities may resent being required to take on this additional duty

Table 31.1 (continued)

Impact area	Themes identified within the impact area	Comment
	Resource allocation	Hospitals may gain additional resources for providing teaching places. If they attract new staff with an interest in research, additional research funds may also be developed. Some concerns that money might be directed towards the 'new' teaching units at the expense of established teaching units
	Service provision and sub-specialization	If additional staff are recruited there may be the option to widen the interest base and increase the range of services provided
	Research activity	See 'resource allocation'
Wider implications: the potential impacts outside of health care settings	Job creation	Little if any impact likely on the wider population outside the hospital. Student placements and subsequent organizational changes unlikely to impact significantly on the local economy. New jobs most likely to be for medical staff who will choose to commute to, rather than live in, the Black Country area, and not for local people

a series of focus groups with Year 3 students, who have recently undertaken their first placements in hospitals, and a survey and series of interviews of consultant doctors responsible for teaching medical students in different hospitals [22]. Although no work with patient groups is planned in the immediate future the results of previous studies may be utilized in this regard [6,7].

However, although such projects may provide some insight into the accuracy of the predictions they are limited in that they are not able to offer an exploration of the counterfactual; that is, would these changes on staff morale, teaching quality, and so forth have emerged even if the student expansion policy had not been implemented? It would be desirable to establish a quasi-experimental trial with 'case' and 'control' hospitals and general practices in order to explore changes in key outcome measures relating to the impacts identified in the HIA although time and resources may constrain this. Table 31.2 sets out some sources of 'routinely collected' data that might possibly be used to measure change in the five outcome areas. The authors acknowledge the many limitations inherent in such data sources and recognize that more refined data collection for outcome measurement may require the development of bespoke survey tools.

Table 31.2 Examples of routine data sources that might enable the measurement of outcomes relevant to impact areas identified within the five domains in hospitals

Domain	Impact area	Examples of indicator areas	Source of data
Patients	Doctor–patient relationship	Number of complaints against trust	Routine data held by Hospital Trust
	Quality of care	Patient satisfaction National performance indicators	Hospital patient survey Dept. Health
Staff	Continuing medical education/ evidence-based medicine	Treatment protocols, audit activity, participation in research	Data held by Hospital Trust
	Morale and motivation	Sick leave, staff satisfaction, length of time in post	Data held by Hospital Trust + data from West Midlands Occupational Health Project
	Workload	Sessional/teaching commitments	Data held by Hospital Trust
Students	Range of disease/ conditions encountered	Number of patient-days per specified condition (selected from undergraduate prospectus) per student	Hospital episode statistics data + Undergraduate curricula
	Quality of teaching	Results of university quality assurance exercise on teaching standards.	Data held by University and General Medical Council
		Student performance in examinations	Data held by Medical School
		Student satisfaction	Student survey/focus groups discussions undertaken as part of routine in-house quality assurance processes
Institution	Recruitment and retention of health care professionals	No applicants per post, length of time in post, ratio of applicants–appointees per post	Data held by Hospital Trust
	Resources allocation	Trust capacity and performance data	Data held by Hospital Trust

Table 31.2 (continued)

Domain	Impact area	Examples of indicator areas	Source of data
	Service provision including sub-specialization	Procedures and services offered, number of consultants per specialty	Data held by Hospital Trust + Local Management Workforce Advisory Group + Hospital episode statistics
	Research activity	Research proposals, grant income, peer-reviewed publications	Local research ethics committees applications
Wider implicat-ions	Job creation	Number of new jobs created within the Hospital Trust	Data held by Hospital Trust

Discussion

In this study, techniques advocated in HIA were used to develop an outcome framework for the evaluation of a national policy. This rationale has implications for the conduct of the HIA. Thus, for example, although the purpose was to predict impacts, we emphasized a more 'scientific' approach to data collection than is usually adopted in HIAs. This has included explicit sampling frameworks and methods as well as standardized tape-recorded semi-structured interviewing with formal dual analysis. In addition, the approach was non-participatory: for example, only a limited number of stakeholders were invited to contribute to the identification of impacts, these stakeholders were 'professional experts' and their contribution was framed within the context of a semi-structured interview, the content of which was decided upon by the research team. Medical students are undertaking many of the techniques described in the section on 'evaluation' as part of their undergraduate training in public health and epidemiology and as such this component of the HIA might be considered to have a 'participatory' element. However in the actual assessment of impacts neither students nor patients were invited to participate. Given that 'participation' is stated to be a central value of HIA [2] the absence of students and patients from the assessment may appear to be a key weakness. However, as indicated elsewhere, we believe that in situations where time and resources are constrained, discussions with professionals only may offer the most efficient mechanism whereby the majority of impacts can be identified [22].

The timing of the HIA in relation to the implementation of the policy also reflects our purpose—that is, to identify outcomes to evaluate the policy post-implementation rather than to change the nature and implementation of the policy itself. It may be argued that if the purpose of HIA is to attain the objective set out in the Gothenberg Paper then any HIA must have two fundamental characteristics: it must be undertaken prospectively before the policy is finalized and implemented, and the results of the HIA must feed into the decision-making process in order that the intervention can be modified prior to implementation. Against these criteria, the work we have described here is not an HIA: although the work commenced before the first wave of student expansion was in place, the timing of the assessment was too late to influence the policy development at either the national or local level. For example, the hospitals that were to provide the additional student placements had already been decided upon and changes to their infrastructure (staff appointments, refurbishment of areas for student teaching) were well underway. That being said, the findings of the HIA have been reported back to the Education Committee of the UBMS, and to the organizers of the Year 3 medical student training placements. However this feedback has not formally been tied into the policy development and implementation process, and it is unclear what, if any, effect it has had on decision makers. In typical academic fashion, our work has probably been more successfully disseminated to other research colleagues via conferences and peer-reviewed journals than to the policy makers whose work we evaluate [19–23].

The work described in this chapter was initiated by the need to identify outcome measures for a policy evaluation. In the context of HIA however, a 'screening' process has been advocated as the optimal method for selecting policies/programmes on which to perform detailed assessment. It is unlikely that the student expansion programme would have been the subject of an HIA if standard screening criteria had been applied: the limited scale of the capital investment and minor modification of existing service delivery inherent in the student expansion programme in effect has resulted in a relatively small-scale organizational change, the impacts of which do not naturally relate to those typically addressed in an HIA (i.e., determinants of health such as housing, social cohesion, or the physical environment). But this is not to say that other health service changes—for example, a 'new build' hospital or merger of two or more existing units—should be 'exempt' from HIA. Here, the balance of impacts will shift more to those affecting the determinants of health of the wider population—for example, changes in employment opportunities both during construction and operationalization of the new unit, changing patterns of demand and use of the local transport infrastructure—than simple organizational change affecting a limited number of employees and patients.

Finally, in setting out to use an HIA of a policy action to inform the subsequent evaluation of that policy, we have implicitly acknowledged that HIA should act as a provider of evidence. If predicted impacts are not validated then the findings of any HIA can make only a limited contribution to the development of an evidence base to inform future assessments. For some areas, for example, transport and housing, a substantial amount of work is now being undertaken to develop the relevant evidence base, and future HIAs of policies concerned with these issues may begin to predict impacts with some degree of certainty. However for other policy areas such as the one described here, the evidence base is very limited. In these situations, HIAs may act as a valuable provider of evidence—but only if predicted impacts are validated. Herein lies the problem: although the monitoring of health and determinants of health has been advocated as a means of evaluating HIAs, this approach does not assess the counterfactual. Formal, more rigorous evaluations are required and these require the availability of time and resources, which is not always possible. This is not to say that HIA should be abandoned, nor that no policy should be implemented without robust evidence of (positive) effect. Such an approach would be to invite policy nihilism given paucity of evidence for many complex initiatives. But the need to demonstrate the 'accuracy' of HIA in terms of predicting policy impacts on health is a key element in establishing the credibility of HIA as an aide to the decision-making process. How this can be done within existing resource constraints is an issue that the HIA community needs to address urgently.

References

1 The NHS plan. *A Plan for Investment. A Plan for Reform.* London: Stationery Office, 2000.

2 European Centre for Health Policy and World Health Organization Regional Office for Europe. *Health Impact Assessment: Main Concepts and Suggested Approach.* Brussels: WHO, 1999.

3 What are the effects of teaching hospital status on patient outcomes? Aggressive research intelligence facility (ARIF). April 1998. http://www.bham.ac.uk/arif/teach.htm

4 Johnson B and Boohan M. Basic clinical skills: don't leave teaching to the teaching hospitals. *Medical Education* 2000; **34**:692–699.

5 Lynoe N, Sandlund M, Westberg K, and Duchek M. Informed consent in clinical training—patient experiences and motives for participating. *Medical Education* 1998; **32**:465–471.

6 Lehmann LS, Brancati FL, Chen M-C, Rotter D, and Dobs AS. The effect of bedside case presentations on patients' perceptions of their medical care. *New England Journal of Medicine* 1997; **336**:1150–1155.

7 Wakeford RE. Undergraduate students' experience in 'peripheral' and 'teaching' hospital compared. *Annals Royal College Surgeons of England* 1983; **65**:374–377.

8 Glen JM. AMEE medical education Guide No. 23 (Part 1): curriculum, environment, climate. Quality and change in medical education—a unifying perspective. *Medical Teacher* 2001; **23**:337–344.

9 Glen JM. AMEE medical education Guide No. 23 (Part 2): curriculum, environment, climate. Quality and change in medical education—a unifying perspective. *Medical Teacher*, 2001; **23**:445–454.

10 Illife S. All that is solid melts into air—the implications of community-based under-graduate medical education. *British Journal of General Practice* 1992; **42**:390–393.

11 Hampshire AJ. Providing early clinical experience in primary care. *Medical Education* 1998; **32**:495–501.

12 Cooke F, Galasko G, and Ramrakha V. Medical students in general practice: how do patients feel? *British Journal of General Practice* 1996; **46**:361–362.

13 Jones S, Oswald N, Date J, and Hinds D. Attitudes of patients to medical student participation: general practice consultations on the Cambridge community-based clinical course. *Medical Education* 1996; **30**:14–17.

14 O'Flynn N, Spencer J, and Jones R. Does teaching during a general practice consultation affect patient care? *British Journal of General Practice* 1999; **49**:7–9.

15 Murray E and Modell M. Community-based teaching: the challenges. *British Journal of General Practice* 1999; **49**:395–398.

16 O'Sullivan M, Martin J, and Murray E. Students' perceptions of the relative advantages and disadvantages of community-based and hospital-based teaching: a qualitative study. *Medical Education* 2000; **34**:648–655.

17 Gray J and Fine B. General practitioner teaching in the community: a study of their teaching experience and interest in undergraduate teaching in the future. *British Journal of General Practice* 1997; **47**:623–626.

18 Wilson A, Fraser R, McKinley RK, Preston-Whyte, and Wynn A. Undergraduate teaching in the community: can general practice deliver? *British Journal of General Practice* 1996; **46**:457–460.

19 Mathers JM, Parry JM, Lewis SJ, and Greenfield S. What impact will the 'conversion' of two district general hospitals into 'teaching hospitals' have on doctors, medical students and patients?—views from the field. *Medical Education* 2003; **37**:223–232.

20 Mathers JM, Parry JM, Lewis SJ, and Greenfield S. What impact will the development of an increased number of teaching general practices in a deprived and ethnically-diverse community have on patients, their doctors and medical students? 2003; (submitted).

21 Parry JM and Stevens AJ. Prospective health impact assessment: problems, pitfalls and possible ways forward. *British Medical Journal* 2001; **323**:1177–1182.

22 Parry JM, Mathers JM, Al-Fares A, Mohammad M, Nandakumar M, and Tsivos D. Hostile teaching hospitals and friendly district general hospitals: final year students' views on clinical attachment locations. *Medical Education* 2002; **36**:1131–1142.

23 Parry JM and Greenfield S. Community-based teaching; killing the goose that laid the golden egg? *Medical Education* 2001; **35**:722–723.

Chapter 32

HIA in developing countries

Martin Birley

The health impacts of development projects in rural areas of developing countries are far larger than in European market economies. Rural communities are often living in subsistence conditions with very little infrastructure support and with little insurance to offset times of adversity. In addition to problems of mental, social, and spiritual well-being, there is often a high rate of communicable disease. The most common communicable diseases are usually malaria, respiratory infection including tuberculosis, gastrointestinal infection, and HIV/AIDS.

History

The history of tropical communicable disease control provides insights into why there has long been an academic interest in the health impacts of development projects, although this has not been matched by substantive action.

The discovery that mosquitoes transmit malaria in the 1890s provided an understanding that the design of the physical environment affects the risk of disease. This knowledge was used by the Indian colonial service, for example, to modify road construction in Bengal and to choose the site for New Delhi. It contributed to the successful building of the Panama Canal. Environmental modification following survey became a standard feature of new projects in India, Malaysia, Indonesia, Africa, Latin America, and southern Europe. The introduction of the powerful insecticide DDT in the late 1940s changed this approach. Local spraying with residual insecticides proved to be far cheaper and far more effective than environmental modification and has saved millions of lives.

One consequence of the DDT era was that new development projects no longer had to take account of malaria control. Malariologists lost their influence over the design, operation, or maintenance of new projects. So agriculture,

The views expressed are those of the author. Accounts of experience in particular countries are not intended as criticism of specific governments or individuals. The problems illustrated are universal.

mining, transport, housing, energy, and many other projects contributed to the malaria risk while relying on chemicals for mitigation. The project proponents were usually unaware of the risks they were creating. Attempts to communicate with them failed. They were not required to listen.

The twentieth century was also a time of dam construction and some 40 million people were involuntarily displaced [1]. Big dams provide a vital source of cheap renewable energy. They stand as a monument to our ability to undertake complex physical engineering projects and our inability to safeguard social systems and human health. For example, Kariba Dam on the Zambezi River displaced tens of thousands of Gwembe Tonga without compensation from the relatively fertile valley floor to the relatively infertile hills [2]. The process of involuntary displacement continues today with recent examples from the Senegal River basin, China, India, and the Middle East. The principal of compensation is now firmly established, but the reality may be somewhat different.

In Africa, South America, the Middle East, and China, big dams provide breeding sites for the aquatic snail that is the intermediate host of schistosomiasis, or bilharzia [3]. This chronic debilitating disease can cause long-term damage to the internal organs. The snail host thrives in the shallow margins of water impoundments. Infection occurs when people enter the water to bathe, wash, collect drinking water, fish, or get into boats. The permanent impoundment of water in an otherwise arid region can lead to a large increase in prevalence rates. There is a cheap curative drug, but it is still too expensive to be stocked regularly in many rural health clinics. The schistosomiasis risk prompted many dam proponents to undertake health impact assessment (HIA). The assessments tended to be narrow in focus and it is debatable to what extent the recommendations led to effective action.

The tropical medicine community has seemed powerless to influence development decisions outside the health sector. Their debates lie firmly 'within the fence' of the health sector and take little account of other sectors; the research priorities include molecular biology, drug trials, health sector reform, vaccination, and insecticides. This research has produced cheap, robust, effective interventions such as insecticide treated bednets, oral rehydration salts, and pharmaceuticals. It is highly valued, biomedical, and essentially reactive. By contrast, HIA is poorly valued but holistic and prospective. The central problem has remained: development continues to transform the physical and social environment without health safeguards and the political will to change this has been lacking.

During the last quarter of the twentieth century, the environmental movement changed the political agenda. It became standard practice for big development projects to receive a prospective environmental impact assessment (EIA). But that assessment did not focus on human health. At the same time, the chemical-based approach to malaria control was failing because of genetic and social resistance. The Panel of Experts on Environmental Management, PEEM, was formed in 1981 to reintroduce the concept of environmental management in water resource development projects. It was, primarily, collaboration between the WHO and the Food and Agriculture Organisation (FAO) of the United Nations. Its members recognized the need for HIA and a series of publication followed [4–6]. There was still no policy to support such assessment.

The following three contrasting case studies, based on personal experience, illustrate some of the difficulties of HIA in developing countries. The piece-meal approach adopted diverges from the systematic procedures and methods that we advocate. This was dictated by local circumstances.

Dams in Sarawak

Sarawak is a small country on the north coast of Borneo that is owned and administered by Malaysia. The indigenous community are largely forest dwelling and subdivided into different tribes while the administrators are largely mainland Malaysians. The territory is naturally covered with dense tropical forest and there are a number of large river systems. There have been international protests for many years about the exploitation of the forest and proposals to drown large valleys in order to construct reservoirs [7]. The author visited Sarawak on behalf of the WHO [8]. Because the project was so sensitive he was refused permission to visit the hydropower construction site he intended but was allowed a two-day visit to the site of a previously completed hydropower project at Batang Ai.

The Batang Ai project was located in a remote area of Sarawak and involved construction of a large dam drowning several villages and creating a narrow reservoir some 40 km long. The inhabitants of the drowned villages had been resettled at a specially constructed site near the dam. These people traditionally live in longhouses near the river so that they can bathe and collect drinking water. Local pastimes include gambling on cockfights. Their traditional agricultural system is based on upland rice, which is the staple food and the spiritual heart of the community. The best place to grow crops is the alluvial soil in the narrow valley bottoms.

An EIA had been undertaken as part of the project feasibility study. This had included a special report by a consultant anthropologist, who had interviewed the local community before they were resettled. This report was regarded as sensitive and as often happens with such reports access to it was restricted.

The community had accepted the need for resettlement. They made clear their preference to continue growing their upland rice crop but also agreed to grow some cash crops. The impact assessment had recommended the construction of a community centre at the resettlement site that would include health care, school, administrative headquarters, and market.

Although malaria is an important disease in many parts of forested Southeast Asia, it is a relatively small problem in Sarawak. The local malaria vector has a restricted habitat and is easily controlled by spraying longhouses with insecticides. Records of reported cases kept by the malaria control team showed a substantial rise in the number of case reports during the period of resettlement and dam construction.

The resettled villagers had been provided with lump sum cash compensation. However they were subsistence agriculturalists with limited experience of the cash economy. Many of them had spent their cash compensation unwisely in ways such as gambling on fighting cocks. In the resettlement village the government had built longhouses similar to those that the people had left behind, but with modern materials, glass windows, rectangular doorframes, water supply, sanitation, and electricity. Villagers had used some of their compensation money to buy colour televisions and there were many aerials on the roof. However these no longer worked since the electricity had been cut off because the villagers had not paid their bills. The site for the community centre was bare ground, as the project had run out of money. Projects often overrun their budgets and elements considered unnecessary to the primary objective are cut. One old woman explained that she preferred her original longhouse because she had been able to bathe in the river everyday. The modern longhouse was too far from the river and the tap water was bad. The water was unpalatable because it was drawn from the river downstream of the dam site. Water released from dams is often anaerobic and smells of hydrogen sulphide.

The hillsides around the resettlement site were planted with unhealthy looking young cocoa trees. Government had ignored the wishes of the resettled community and insisted that their main crop should be cocoa and not rice. The new crop required large inputs of labour, fertilizer, pesticides, and agricultural machinery. There would be a delay of some years before the cocoa crop was ready for harvest and its value would vary with the world market. The people were upset by the loss of rice, their spiritual heart, and were attempting to grow it between the cocoa plants. Some of them had run away, deep into the forest, to set up new rice farms where the government would not disturb them or protect them from malaria.

This example illustrates a number of easily predictable failures on the part of the project planners. Highly trained and experienced engineers, who have profound understanding of the physical world, have little or no understanding of 'social engineering'. The problems encountered are attributable to a lack of commitment to impact assessment.

Cotton pesticides in Egypt

Agricultural development projects often depend on expensive inputs of fertilizer, seeds, machinery, infrastructure, and pesticides. Pesticides can cause acute or chronic, accidental or deliberate poisoning. This case study is based on a post-project evaluation of the health impacts of changing from conventional pesticides to a novel and much safer method of control using a pheromone to control the insect population by disrupting mating.

Cotton is a major export crop in Egypt and is cultivated on much of the arable land. Production is intense and relies on heavy use of conventional pesticides. Many agricultural pesticides are available in Egypt. The relative toxicity and persistency of each compound is reasonably well known [9]. The most persistent chemicals have been banned from use on cotton for many years. Residues are still detected in milk and meat at very low levels that are probably not due to current applications [10]. Newer compounds are not persistent, so the risk to the public by poisoning from this source seems very low, although the reuse of wastewater in agriculture has also led to concern about the accumulation of toxic compounds [11].

Most of the poisoning associated with the conventional pesticides is probably an occupational hazard of applicators and formulators. Safe use is difficult because of intense heat, poorly designed or used equipment, and lack of concern about safety [12]. As one field supervisor expressed it: 'better a healthy cotton crop and a sick labourer than the other way round'. Blood tests of applicators show high frequencies of cholinesterase depression, a sign of poisoning [13]. With up to 1 million hectares under cotton, there are some 100,000 adult male pesticide applicators. Some 10,000–100,000 children frequently work in the fields, picking caterpillars from the undersides of cotton leaves.

Pesticides in Egypt are formulated in a number of small factories. These factories employ between 1000 and 10,000 workers in total. Permanent neurological damage is commonplace [14]. In addition to the occupational hazard, acute pesticide poisoning occurs because of attempted suicide and accidents. Both of these categories are assisted by the ease with which pesticides may be purchased and stored in the domestic environment. There are some 1000–10,000 cases per year in Egypt.

The response to this serious public health problem in Egypt was greatly assisted by public pressure during the 1980s [15,16]. As a result of improved methods of pesticide application and alternative pest control techniques, the annual application of conventional pesticides has been reduced and this should benefit both the environment and human health. However, these improvements have been possible because pesticide application is centralized and not under the control of individual farmers. At the time of the assessment, there were plans for economic liberalization that might remove such centralized control. In this case, conventional pesticide usage would be expected to increase.

In conclusion, cotton growing in Egypt provides an example of the health impacts of conventional pesticides and of the need to assess the potential positive health impacts of new pest control technologies.

Wastewater in Syria

The world is facing a global water shortage and wastewater reuse is increasing. But wastewater reuse projects have many health impacts and should be assessed at the design stage, as the following example illustrates.

The city of Damascus, in Syria, is an ancient city that has traditionally used its river as a source of domestic water supply and as a method of disposing of waste. As the river leaves Damascus it divides into a number of distributaries that flow across the plains before disappearing into a shallow lake on the edge of the desert. The distributaries provided irrigation for the fertile agricultural plain, which supplied much of the city's food and fuel. The modern city consumed a large amount of domestic water and dumped a large amount of both domestic and industrial waste into the river. The sewage water was used for irrigation and the crops included fuel wood, fruit trees, salads, and vegetables. There were popular recreations sites among the fruit trees, where Damascenes took their children to play and eat during the spring months. Small towns downstream from Damascus had stopped using river water for domestic supplies and were using boreholes. There was no sewage treatment plant. Plans made in the 1970s had not been implemented because of the cost.

A large percentage of the inhabitants of Damascus suffer from infection with the large roundworm called *Ascaris*. The eggs are defecated by the human host and enter the sewage system. They have a long half-life of 1–2 years. They can adhere to food plants irrigated with sewage and so be ingested by a new human host. Each summer there were reports of epidemics of 'watery diarrhoea'—a common euphemism for cholera. In order to prevent this, regulations had been promulgated to stop farmers using untreated wastewater for salads and certain vegetable crops. Many farmers had complied and had sunk open boreholes on their land. But enforcement was limited. The use of wastewater meant that large quantities of nitrate rich water were recharging the ground water. The nitrate and nitrite content of ground water had risen and was affecting drinking water supplies.

Partly in response to geopolitical events during the early 1990s, Syria received a large loan for building a sewage treatment plant and wastewater reuse system. The plans drawn up during the 1970s were reactivated. The original site was now a suburb of Damascus and could not be used. Interceptor sewers would collect the domestic wastewater flowing out of Damascus and channel it to a treatment plant. This would be one of the largest in the region using activated sludge technology. The sludge would be collected and dried before being sold to farmers as fertilizer. The liquid component would be treated with chlorine and then delivered to a series of canals and pumping stations that would distribute it as safe wastewater for use in agriculture. Separate facilities would be developed for industrial waste.

The author was commissioned to undertake a rapid HIA of the project with about 10 days in the field and the help of a local research assistant, fluent in both Arabic and English [17]. As the treatment plant was already half constructed, the assessment had more to do with compliance than with healthy public policy. It highlighted a number of simple points of concern.

The WHO has published guidelines for the safe reuse of wastewater in agriculture and aquaculture in both English and Arabic [18,19]. It became apparent that these guidelines were not being used, and the only person (an occupational physician), who appeared to have one had not been consulted by the project. The principal recommendation was that a number of those guidelines should be purchased in Arabic and distributed to the project managers. The lack of availability of those guidelines and the associated lack of awareness about risks illustrates an institutional determinant of health that could seriously affect the health impact of a project. For example, neither the plans for drying and storing sewage sludge nor the plans for monitoring water quality took account of *Ascaris*.

There were other structural deficiencies. The most obvious one was the plan to use an activated sludge treatment plant rather than waste stabilization ponds. Activated sludge treatment plants are very expensive to construct and operate but require relatively small amounts of land. They are very profitable undertakings for construction companies. By contrast, stabilization ponds are relatively cheap to construct and operate although they require relatively large amounts of land. In Syria, there is plenty of land available in the desert and there are examples of pond systems elsewhere in the Middle East.

The plan to chlorinate the river of treated sewage also had difficulties. Syria did not manufacture its own chlorine. Large numbers of trucks would be required to transport it from the ports to Damascus. There are problems of storage. Supplies would vary and would be sensitive to both budget cuts and world prices. When the river of treated sewage left the treatment plant it would move from the jurisdiction of the plant operators to the jurisdiction of the Irrigation Department. The monitoring laboratory belonged to the plant operators and not to the Irrigation Department. There was no independent verification procedure. The Irrigation Department would be unable to satisfy themselves that the product was fit for purpose. The quality of the product can vary. Skilled managers are required to operate an activated sludge treatment plant correctly and low government salaries do not attract them. When the plant is operated correctly, the product looks clean and transparent rather than black and smelly. The suspended particles have been removed but it is still highly pathogenic. The retention time in the system is too short for several pathogenic organisms. Strict controls are still required on the crops that can be grown in the water. Elsewhere in the Middle East, farmers abstract such water immediately downstream of the treatment plant, for the illegal production of salad crops.

This assessment did not conform to the procedures that are now advocated for HIA. For example, there was no in-country steering committee to receive the report and evaluate it, so it is unlikely that any of its recommendations led to action. The author was a contracted consultant who came, wrote a report, and left and as is commonplace with such consultants had no further contact with the client and did not know the outcome.

Policy change

Over the last 10–15 years there has been a slow change in policy regarding HIA of development projects. This should ensure a more systematic and timely approach. Principle one of Agenda 21 placed people at the centre of development, justified the inclusion of health concerns in all development policies, and recommended EIA and HIA [20]. In Europe, the Maastricht Treaty, 1992, and Amsterdam Treaty, 1999, required that member governments of the EU shall ensure that their actions do not have an adverse impact on health, or create conditions that undermine health promotion [21]. Implementation of the Articles of the Maastricht Treaty was slow but progress is now being made. Many EC policies have adverse health impacts, such as agriculture and transport [22,23]. The European Policy for Health advocates multi-sectoral accountability through HIA for both internal and foreign policies [24,25]. The European Charter on transport, environment, and health, recognized the need for HIA [26].

The Organisation for African Unity has declared that malaria prevention and control should include EIA and HIA of development projects [27]. The Director

General of WHO emphasized the links between human health and environmental policies and practices [28]. The World Commission on Dams expressed concern about health impacts [2].

Health impact assessment is likely to pay an increasing role in international fora such as the World Trade Organisation (WTO), which has been subject to widespread popular criticism. Concerns have included low wages and child labour; both have negative health implications [29]. Trade agreements are made without explicit consideration of their health or other social impacts and this has been referred to as 'trade-creep' [30]. The global rules of trade and intellectual property rights can create perverse incentives that undermine the public health [31].

The British Department for International Development has funded programmes to provide guidance about health impacts [32,33]. International Development Banks such as the Asian Development Bank and the World Bank have issued guidance [34,35]. The African Development Bank and the European Investment Bank have plans. The International Finance Corporation recently expressed concern that its own safeguard policies neither covered the area nor provided systematic advice [36]. It noted that health issues were increasingly the concerns of sponsors and communities.

The WHO has initiated regional guidelines in the Eastern Mediterranean, European, and African regions but they have not been published. Some multinational industries are adopting a proactive approach. For example, the Royal Dutch/Shell Group has published minimum health standards [37]. These state that 'a HIA is to be made in conjunction with any Environmental and Social Impact Assessments that are required for all new projects, major modifications and prior to abandonment of existing projects where there is the potential to impact on the health of the local community, company and contract workers, or their families'. This is a powerful policy statement and it being backed up by draft guidelines [38].

The International Association for Impact Assessment (http://www.iaia.org) has included health alongside environmental and other impact assessments. A Memorandum of Understanding has been signed with WHO, a board position has been created, the mission statement has been modified, and papers on HIA now form a regular part of the annual conference.

International capacity-building

As governments, development agencies, and industries accept a policy of assessing health impacts, they will need to draw upon a trained cadre in government departments, universities, and consultancy companies of donor and recipient nations. At present they will not find one, as there are few training programmes. However, a number of pilot training projects have been implemented and lessons have been learned [39–41]. For example, a single course run on one occasion in a country is insufficient to build national capacity, although international funding opportunities often have this effect. Courses need to be designed, tested, and then transferred to national training institutions. Participants require empowerment by their line managers if they are to implement what they have been taught. Health impact assessment involves sharing information between professionals, sectors, and communities. But information is the currency of power. Specialists are not trained to share information with outsiders. So HIA courses must provide skills in intersectoral collaboration.

A number of companies and nations are embarking on their own capacity-building projects. For example, Shell is running courses for Nigerian consultants. The UK government has funded capacity building by the International Health Impact Assessment Consortium and other groups. Many courses have been held since 1999. There is a growing body of experienced UK and other European practitioners and increasing opportunities for them to transfer that experience to developing countries.

References

1 World Commission on Dams. *Dams and Development a New Framework for Decision-Making.* London: Earthscan, 2000.

2 **Colson E.** *The Social Consequences of Resettlement.* Manchester: Manchester University Press, 1971.

3 **Jobin W.** *Dams and Disease—Ecological Design and Health Impacts or Large Dams, Canals and Irrigation Systems.* London and New York: E& FN Spon, 1999.

4 **Birley MH.** (1991). Guidelines for Forecasting the Vector-borne Disease Implications of Water Resource Development. World Health Organisation, Panel of Experts in Environmental Management (PEEM) Guidelines No. 2, WHO/CWS/91.3, 2nd edn.

5 **Mills AJ and Bradley DJ.** Methods to Assess and Evaluate Cost-Effectiveness in Vector Control Programmes, Selected Working Papers prepared for the 3rd, 4th, 5th and 6th meeting of the WHO/FAO/UNEP PEEM, WHO, Geneva, 134–141, 1987.

6 **Tiffen M.** Guidelines for the incorporation of health safeguards into irrigation projects through inter-sectoral cooperation, with special reference to vector-borne diseases. WHO, 1989.

7 **Hong E.** *Natives of Sarawak: Survival in Borneo's Vanishing Forests.* Pulau Pinang, Malaysia: Institut Masyarakat, 1987.

8 **Birley MH.** Feasibility Test of Guidelines for Forecasting the Vector-borne Disease Implications of Water Resources Development in Two Countries in South-East Asia, Thailand and Malaysia. WHO, WHO/VBC/894, WHO/VBC/893, 1988.

9 WHO Public health impact of pesticides used in agriculture. Geneva: World Health Organisation in collaboration with the United Nations Environment Programme, 1990.

10 Dogheim SM, El-Shafeey M, Afifi AMH, and Abdel-aleem FE. Levels of pesticide residues in human milk samples and infant dietary intake. *Journal of the Association of Analytic Chemists* 1991; **74**(1): 89–91.

11 Safwat A-D and Abdel-Ghani M. *Concentration of Agricultural Chemicals in Drainage Water.* Drainage and Water Table Control, Nashville, American Society of Agricultural Engineers, 1992.

12 Loosli R. (1989). Toxicological hazard evaluation and field workers safety in the use of plant protection products. Pesticide intoxication in the third world countries. In Proceedings of the first third-world conference on environmental and health hazards of pesticides, Cairo University, Cairo, 1994.

13 El-Khatib NY, Farid MM, GMM, and El-Herrawi MA. *Effect of Exposure to Pesticides on Blood Cholinesterase in Agricultural Workers.* Cape Town, S. Africa: Pan African Environmental Mutagens Society, 1996.

14 Amr MM, El Batanouni M, Emara A, Mansour N, Zayat HH, Atta GA, and Sanad A. Health profile of workers exposed to pesticides in two large-scale formulating plants in Egypt, Forget G, Goodman T, and de Villiers A. (eds.), *Impact of Pesticide Use on Health in Developing Countries*, Ottawa: International Development Research Centre, pp. 118–130, 1993.

15 El-Gamal A. Persistence of some pesticides in semi-arid conditions in Egypt. In Proceedings of the International Conference on Environmental Hazards of Agrochemicals in Developing Countries, Alexandria, Alexandria University, 1983.

16 Amr MM. *Environmental and Health Hazards of Pesticides.* Cairo: Cairo University, 1988.

17 Birley MH. Environmental health risks associated with effluent and sludge re-use for inclusion in: Damascus Sewerage Project, environmental review of proposals for use of treated wastewater reuse and sludge for agricultural use in the Damascus Ghouta area. Howard Humphries and Partners Ltd with General Company for Engineering and Consulting, Damascus, Syrian Arab Republic, April 1997.

18 Mara D and Cairncross S. *Guidelines for the Safe Use of Wastewater and Excreta in Agriculture and Aquaculture.* Geneva: WHO and UNEP, 1989.

19 World Health Organization. Health Guidelines for the Use of Wastewater in Agriculture and Aquaculture. World Health Organization, Technical report series, 778, 1989.

20 United Nations. *The Global Partnership for Environment and Development a Guide to Agenda 21 Post Rio Edition.* New York: United Nations, 1993.

21 European Commission Europa (search for Amsterdam Treaty). 1999. Access Date: 23 August, 2000, http://:www.europa.eu.int

22 Dahlgren G, Nordgren P, and Whitehead M. (eds.) Health impact assessment of the EU common agricultural policy, Policy report from Swedish National Institute of Public Health, 1997.

23 Dora C. A different route to health: implications of transport policies. *British Medical Journal* 1999; 318: 1686–1689.

24 WHO. *Health 21—Health for all in the 21st Century, An Introduction.* Copenhagen: Europe, World Health Organization, 1998.

25 WHO European policy for health for all. 1999. Access Date: 23 August, 2000, http://:www.who.dk.

26 WHO (1999), *Draft charter on Transport, Environment and Health.* Copenhagen, Europe: World Health Organization, 1999.

27 Organization for African Unity Harare declaration on malaria prevention and control in the context of African economic recovery and development. Organization of African Unity Assembly of Heads of State and Government, Thirty-third ordinary session, AHG/Decl 1 XXXIII, 2–4 June, 1997.

28 **Brundtland G.** Speech to WHO. 1998. Access: http://:www.who.ch.int.

29 **Denny C.** Why the fight for union rights could harm the world's poorest. *The Guardian.* London: 11, 1999.

30 **Koivusalo M.** World Trade Organisation and Trade-creep in Health and Social Policies. National research and development centre for welfare and health, Occasional paper on global social policy, 4/1999, 1999.

31 **Mashelkar R.** (2002). Level playing field. *Orbit* 2002; **83**: 6–7.

32 **Birley MH.** *The Health Impact Assessment of Development Projects.* London: HMSO, 1995.

33 **Birley MH and Lock K.** The health impacts of peri-urban natural resource development. Liverpool: Liverpool School of Tropical Medicine, 1999. http://: www.liv.ac.uk/~mhb.

34 **Birley MH and Peralta GL.** Guidelines for the Health Impact Assessment of Development Projects. Asian Development Bank, Environmental Paper No 11, 1992.

35 **Birley MH, Gomes M, and Davy A.** Health aspects of environmental assessment. Environmental Division, The World Bank, Environmental Assessment Sourcebook Update, 18 July 1997.

36 Compliance Advisor Ombudsman of the International Finance Corporation. Review of IFC's safeguard policies. 2002. Access Date: 18 November 2002, http://:www.cao-ombudsman.org/ev.php.

37 Shell International (2001). Minimum health management standards. Shell International, 2, 2001.

38 Shell International Exploration and Production B.V. Health impact assessment guidelines. Shell International, EP 95 0372 (1 May 2002), 55.

39 **Birley MH, Bos R, Engel CE, and Furu P.** A multi-sectoral task-based course: Health opportunities in water resources development. *Education for Health: Change in Training and Practice*, 1996; **9**(1): 71–83.

40 **Engel C, Birley MH, Bos R, Furu P, and Gotsche G.** Intersectoral Decision-making Skills in support of Health Impact Assessment of Development Projects, report on the development of a Course in the Context of Health Opportunities in Water Resources Development 1988–1998. World Health Organisation, WHO/SDE/WSH/00.9, 2000.

41 **Bos R, Birley MH, Furu P, and Engel C.** Final reports and training material for WHO training course entitled, Health opportunities in water resource development. World Health Organization, draft 2002.

Chapter 33

HIA of agricultural and food policies

Karen Lock and Mojca Gabrijelcic-Blenkus

Background: public health and agricultural policy

The public health implications of agricultural practice and policy making became more prominent after the agricultural crisis in Europe caused by the discovery of bovine spongiform encephalopathy (BSE or mad cow disease). This was a new disease in cattle first seen in the mid-1980s in the United Kingdom. A retrospective independent public inquiry recognized that poor agricultural practices and bad intersectoral policy making in the United Kingdom, which had not taken public health into account, led to BSE not only becoming an animal epidemic but caused it to be transmitted to humans as a new fatal disease, variant Creutzfeld–Jakob disease (vCJD) [1]. As of April 2003, 127 people have died from this disease (probable or confirmed) in the United Kingdom. The financial costs have also been enormous: estimated to be £4.2 billion [2]. In the aftermath of the crisis, the independent inquiry called for health impact assessments (HIAs) to be used by the British government to help policy decisions [1].

Unsurprisingly since the advent of BSE, policy makers across Europe have reacted by creating more stringent food safety regulation, and overemphasizing chemical and microbiological content of food as the key health issue. Food safety continues to be perceived as the most important health problem in agriculture despite the relatively higher importance of food security, nutrition, and other risk factors in terms of actual burden of disease [3,4]. The basic aim of many agricultural policies is to provide adequate food for the population. In reality, the situation in each country is a much more complex combination of agriculture, food, trade, and health. Issues of food security and balanced nutrition compete for prominence in policies with environmental and food safety standards, agrichemical and biotechnology use, foreign investment, food processing and product branding, land ownership, rural development, and international trade agreements. The broader public health issues that are raised by these aspects of agriculture and food production are rarely considered by policy makers.

Health impact assessment is still a developing approach that has been used to consider the potential, or actual, health impacts of a proposed project or policy. There are now many examples of projects and programmes that have been subjected to HIA, but much less experience of applying HIA to national policy. In those countries that have used HIA at policy level the methods are more varied, and the stages are often less distinct. Agriculture and food programmes and policies worldwide are often subjected to environmental impact assessments (EIAs) [5], but to date there have been very few published studies of HIA applied to agriculture, particularly at the policy level. This chapter will look at how HIA has been applied to agricultural and food policies internationally and discuss whether it is a useful tool for raising broader public health issues on the agricultural and food policy agenda. It will briefly review the different approaches that have been used and draw some conclusions about the strengths and weaknesses of each for improving health considerations in this policy sector.

Environmental HIA in Canada

The Canadian government has published two HIAs of agricultural systems in Québec, as part of an integrated approach to HIA, incorporating health within the framework of environmental assessments [6]. The approach is presented in a three-volume manual, which includes discussion of the use of social impact assessment (SIA), epidemiology, health evaluation, economics, risk assessment, and the role of health professionals. Rather than looking at agricultural policies, the two published examples are HIA of agricultural practice, hog (pig) farming (see Table 33.1), and pesticide use in apple growing. These two health assessments have been conducted on discrete issues in single agricultural systems in response to public concern. Despite the theoretical integrated methods they present in the manual, the actual examples take a very quantitative approach, drawing on risk assessment methods and data on known health risks, mainly focusing on the issue of environmental pollution. There is very little consideration of broader psychosocial impacts on health.

Assessment of UK foot and mouth outbreak

Foot and mouth disease is a mild, mostly self-limiting disease of animals, which has almost never been known to be transmitted to humans. In 2001, the United Kingdom saw its first outbreak of the disease for over 30 years. In an attempt to control the spread of the outbreak and maintain livestock exports, the government ordered a mass slaughter policy. The main motivation for this decision was economic impact rather than animal or human health. As a result of the policy a very large number of carcasses had to be disposed.

Table 33.1 Main health risks considered in the HIA of large-scale hog farming in Québec

♦ Pollution caused by animal waste: inorganic and organic pollution (the latter focused on bacterial infections)

♦ Water pollution: nitrogen and phosphorous concentrations, formation of chlorination by-products

♦ Odours

♦ Occupational health risks: acute and chronic respiratory problems, zoonotic diseases

♦ Social and human impacts: in an attempt to analyze the impact on quality of life, the levels of mental distress relative to the density of hog raising operations were also considered, but no causal relationship was established

Very early in the process the Chief Medical Officer of the English Department of Health called for a qualitative risk assessment to examine the effect on public health of the disposal methods of the slaughtered animals [7]. There has been greater experience of using traditional health risk assessment methods in agriculture. It is a particularly useful approach when there is a single specific and well-defined health risk and it has been applied extensively to issues of food safety. Although the government study of carcass disposal methods was presented as an expert environment and health risk assessment it took a broader approach by also considering psychosocial health impacts of the policy, similar to the approach taken in an HIA. The health assessment proved to be an important tool to get other ministries to take account of wider public health issues that had previously not been considered. It influenced changes made to the animal disposal policy, and resulted in plans for the long-term environmental and health monitoring to be led by the Department of Health [8].

In the English county of Devon, a rapid HIA of the local effects of the foot and mouth disease crisis was conducted in May 2001 [9]. The HIA was carried out in partnership by the local county council and health authority, and unlike the Department of Health study used a recognized HIA methodological framework. Qualitative and quantitative data was collected over the course of three weeks. Stakeholders and key informants were interviewed, documents were obtained from local organizations, and the evidence base was reviewed. They also used available health data to ascertain whether the burning of the carcasses had a measurable effect on respiratory health, and primary care records to investigate changes in the prevalence of mental health problems. The most significant potential health impacts identified are summarized in Table 33.2. The HIA report made several practical recommendations including further mental health support services for farmers, monitoring of private water

Table 33.2 Potential impacts identified by the rapid HIA of foot and mouth disease in Devon

---

- Mental health: it was believed that there would be an increase in anxiety, depression, and suicides in the farming community. This will also affect the unemployed, which will include those in the tourist industry. However, the data systems were not in place to study this.

- Social structure and community: it was equivocal whether there would be any long-lasting impacts. In the short term, there was very little social contact in rural areas due to fear of spreading the disease.

- Economic: the effect on job losses was thought to be the biggest health impact. This would be greatest on the socially disadvantaged such as seasonal workers in both tourist and farming industries.

- Environmental health: the health data did not indicate any increase in acute respiratory illness or gastrointestinal disease due to air or water pollution from carcass disposal.

---

supplies, and health surveillance of primary care for respiratory diseases and mental health problems. It also recognized the financial burden of foot and mouth disease on the local council, and the potential indirect health impact on the wider community by its effects on reducing resources for other service provision.

Swedish HIA of the EU Common Agriculture Policy

The EU Common Agricultural Policy (CAP) is a key agricultural policy at a transnational level. The CAP provides various agricultural subsidies, the results of which have impacts not just in Europe but worldwide due to the distortion of world food prices, and hence trade. This has potentially adverse impacts on less developed nations in particular [10]. The Swedish Institute of Public Health conducted a review of the potential health impacts of the CAP in 1996 [11]. Although this was entitled an HIA it did not take a recognizable HIA approach and is actually a useful descriptive review of the potential health effects of four CAP regimes: dairy products, fruit and vegetables, tobacco, and alcohol. The report has had very little impact in the European Commission (EC) or on CAP reform. Since it was published, CAP negotiations have continued to marginalize the public health dimension. Clearly, if this had been an applied HIA it would have not been considered a success in effecting change.

The Swedish Institute of Public Health produced an updated analysis of the public health implications of the CAP in 2003. This is a much more detailed and critical analysis of the potential public health impacts of several commodity regimes and policy instruments [12]. This report was prepared by the Institute of Public Health, but since publication it has stimulated the start of

Table 33.3 Selected key recommendations from the National Institute of Public Health report on the health aspects of the EU CAP [12]

- Phase out all consumption aid to dairy products

- Limit the school milk measure to include only low-fat products

- Introduce a similar school measure for fruits and vegetables

- Redistribute agricultural support so that it favours increased fruit and vegetable consumption

- Improve the support to farmers who wish to cease wine production

- Plan to phase out subsidies for tobacco growing

---

meetings on the health effects of the CAP between the health sector and Ministry of Agriculture in Sweden. Although not HIA in the formal methodological sense, this report provides important evidence for use by policy makers across Europe on the public health impacts of the CAP. The analysis makes strong recommendations on the need for reform of several policy instruments of the CAP, which it suggests would lead to substantial improvements in public health across EU member states (see Table 33.3).

Slovenia and the CAP

As far as we were aware, Slovenia is the only country that has carried out a prospective HIA of new national agricultural policy. Although there were many reasons why an HIA of agricultural and food policy was believed to be important for Slovenia, the most significant reason was Slovenia's application to join the EU. In December 2002, the EU invited eight countries from Central and Eastern Europe (Czech Republic, Estonia, Hungary, Latvia, Lithuania, Poland, Slovakia, and Slovenia), plus Cyprus and Malta, to join the organization in 2004. Negotiations about enlargement of the EU had commenced in 1998 [13]. Since then there has been a period of rapid transition in Slovenia and across Europe. The HIA in Slovenia was proposed towards the end of a complex and bureaucratic negotiation process with the EC and member states during which candidate countries have been adopting thousands of pages of the EC legal framework, known as the *acquis communitaire* [14]. Each candidate country has to sign up to the *acquis* in its entirety and to accept that European law takes precedence over national law. This includes the influence of the EU CAP on national agricultural and food systems. All candidate countries, including Slovenia, had been experiencing problems in the negotiations for the national terms for adopting agriculture policies. The agriculture chapter of the *acquis* was still being negotiated when the HIA work was started in Slovenia.

In December 2001, the Slovenian Ministry of Health and the WHO European region proposed to undertake an HIA of agriculture, food, and nutrition policies. The HIA project in Slovenia was conducted as a pilot project to develop the methods of HIA and the evidence base in this sector.

The Republic of Slovenia is a small country of approximately two million inhabitants, and is bordered by Austria, Hungary, Italy, and Croatia. Formerly a constituent part of Yugoslavia, Slovenia declared its independence in 1991. The country is divided into 9 health and 12 statistical administrative regions. Agriculture contributes only 3.2 per cent of gross domestic product (GDP), and is dominated by dairy farming and animal stock, with the main crops being corn, barley, and wheat [15]. In addition to concerns as the effect that the CAP legislation would have on Slovenian agricultural policy, there were also national Slovenian issues that supported development of the HIA work including national development of an intersectoral food and nutrition action. There were marked differences in standardized mortality rates between the regions in the east and west of Slovenia [16]. The reasons for the differences had not been explained, but the northeast region, Promurje, which has the highest all-cause mortality, is also the region with the largest agricultural sector in the country. In Promurje, 20 per cent of the population are employed in farming or related industries, which are most likely to be affected by the CAP after accession.

Setting up the HIA of agricultural policy in Slovenia

HIA was proposed as an appropriate approach that could be used to investigate the health concerns in the multi-sectoral development of agriculture, food, and nutrition policy in Slovenia. This was particularly important in the agricultural sector, where public health was not on the agenda because it is not a directly negotiated factor within the CAP. The HIA basically followed a six-stage process: policy analysis; rapid appraisal workshops with stakeholders from a range of backgrounds; review of research evidence relevant to the policy; analysis of Slovenian data for key health-related indicators; a report on the findings to a cross-government group; and evaluation.

The major difficulty in the initial stages of the HIA was clarifying the policy options to be assessed. Although there were national proposals for new agricultural policy and a food and nutrition action plan, these were still at the stage of development rather than being firm government proposals. To complicate matters the HIA had to take into account the effect of adopting the CAP into Slovenian law. This could not be done with any degree of accuracy or certainty as there were ongoing negotiations with the EU about the nature and amount of CAP subsidies that Slovenia would be allocated on accession, and the date of accession had still not been confirmed. These issues were not resolved until December 2002, when the CAP subsidies were finally agreed between the EC and the Slovenian government. The complexities of European agricultural policy and how it will be applied in Slovenia made conducting a detailed HIA very difficult. The CAP is an enormous and relatively inflexible body of legislation. The HIA project involved agricultural economists at the University of Ljubljana who were important in modelling and interpreting potential policy scenarios that would be likely to occur in Slovenia when integrating the CAP requirements into Slovenian national policy [17]. Obviously, the adoption of the CAP has an enormous influence on national policy, and it was decided that the main focus of the HIA should be the broad effects of the CAP adoption. The HIA looked at the effects of some of the specific commodity regimes including the fruit and vegetable, wine, and dairy sectors, and the policy instruments for rural development. The policy analysis had to be balanced against the national proposals, which particularly promoted the rural development measures such as rural diversification and environmentally friendly policies. Although these national proposals were based on the CAP, it was widely believed that the EU negotiations would prevent them being adopted in full.

Stakeholder workshops

The most important part of an HIA is identifying and collecting information for health impacts that a policy might create. The HIA approach taken in Slovenia involved national and regional stakeholders. The first HIA workshops were held in March 2002 in the northeast region of Promurje. A total of 66 people participated, including representatives of local farmers, food processors, consumer organizations, schools, public health, non-governmental organizations, national and regional development agencies, and officials from several government ministries. These officials included Ministries of Agriculture, Economic Development, Education, Tourism, and Health, and a representative of the president of Slovenia [18]. The participants were asked to identify potential positive and negative health impacts of the proposed agricultural policies. This was achieved by conducting a series of rapid appraisal workshops, which were facilitated by using a semi-structured grid assessment framework. This prompted participants to consider the core policy issues and identify potential health impacts using the main determinants of health. As part of this, participants were asked to identify which population groups would be most affected by each policy area.

The qualitative information gained from the workshops enabled a picture of probable positive and negative health impacts to be constructed, including

Table 33.4 Results of stakeholder workshops: key determinants of health potentially affected by agricultural policy development in Slovenia

---

- Changes in income, employment, housing, and issues of social capital in rural areas
- Changes in the rural landscape and cultural impacts
- Increased food imports and effects on exports
- Nutritional value and food safety of produce and food products
- Environmental issues: farm intensification leading to soil and water pollution
- Potential benefits of organic agriculture and food
- Barriers to increasing organic production or small-scale on-farm industries (including knowledge of farmers and absorption capacity for EU money)
- Occupational health of farm workers and food processors
- Capacity of local services and institutions, including employment, education, health, and social services

---

Source: Stakeholder HIA workshops, Slovenia, March 2002.

areas of speculation and disagreement (see Table 33.4). The next step was to combine this information on potential health impacts with evidence from other sources in order to test the 'hypotheses' of health impacts proposed. For example, one theme from the workshops was the hypothesis that adoption of the CAP would result in larger farm sizes and intensified production methods, leading to loss of small family farms, increased rural unemployment, and a consequent increase in ill-health, including depression. This was in regions that already had high rates of alcohol-related deaths and suicide. The next stage set out to clarify whether evidence supported the links between adopting the CAP and loss of small family farms, links between farm intensification and increased rural unemployment, and links between either of these and increased rates of ill-health. This review produced recommendations aimed to identify policy instruments in the CAP, which could be applied to maintain small farms, such as conversion from grain to horticulture production more suited to smallholdings.

Review of research evidence

To plan the evidence review, an expert meeting was held to assess the strength of the evidence for the links between the policy issues identified in the workshops, and health determinants and health outcomes. Unsurprisingly, for several key areas the evidence was found to be patchy or not available in an up-to-date, easily synthesizable form. For the HIA to proceed, the next stage had to be mapping out a more detailed evidence base for how agriculture and food policies affect health. Evidence reviews were commissioned that linked relevant agriculturally related health determinants and health outcomes for the six policy topics identified in the stakeholder workshops. These policy topics were environmentally friendly and organic farming methods, mental health and rural communities, socio-economic factors and social capital, food safety, occupational exposure, and issues of food policy, including price, availability, diet, and nutrition.

The final aspect of the project collected health and social indicators in Slovenia. These indicators are determinants of health and were used in the HIA as measures of intermediate health outcomes. This allowed the interpretation of the literature review evidence for the Slovenian context. The National Institute of Public Health, Ljubljana, coordinated the national and regional data collection. As with many HIAs, the uncertainty of the extent of policy change after accession meant that for many indicators we were unable to quantify the health outcomes precisely and could only predict the direction of the effect.

The final HIA report was presented to the intergovernmental committee on Health at the launch of the National Food and Nutrition Action Plan in Slovenia in October 2003. This report presented the results and recommendations for the government of Slovenia on a range of agricultural issues including the fruit and vegetable, grain, and dairy sectors, and rural development funding.

Lessons learnt

As far as the authors are aware, this was the first project to attempt to estimate specific national health impacts of incorporating the CAP, and the first prospective HIA undertaken of national agricultural and food policy. Although a formal evaluation has not yet been undertaken, several important learning points have already arisen. The main problems encountered during the HIA were the complexity of the policies being assessed and the lack of evidence of health impacts. As the CAP is such a huge and difficult policy area it was essential to have effective cross-governmental working in place at a national and regional level to tackle the policy issues. Relatively good intersectoral relationships existed between the Ministry of Health and other ministries, including agriculture and economic development, before the HIA commenced. The HIA helped to develop new communication between the ministries on these issues. In common with many HIAs at project or policy level, this HIA was limited by pressures of time and human resources, as everyone involved had to work on the HIA in addition to carrying out their existing responsibilities. At the start of the work most people in Slovenia were unfamiliar with the methods or aims of HIA. The project initially failed to recognize the importance of this, and found that some data or evidence from sources was not tailored to a form best suited for use in the HIA.

Even though this was planned as a pilot project feeding into national policy development, the political time frames created pressure to provide support for the Slovenian government position during the EU negotiations on the CAP subsidies. Providing such support was often not possible. In 2002 the goal of accession had been a moveable target, and the proposed nature of EU subsidies changed regularly. Consequently, it proved very difficult to quantify or assess some outcomes with any certainty. However, the process of conducting the HIA has achieved some important intermediate outcomes that were not initially foreseen. The HIA involved experts from the Ministry of Agriculture who were negotiating the Slovenian policy position on subsidies with the EC. This not only put wider health and social issues on the agricultural policy agenda, but also resulted in agricultural experts arguing the case for 'healthy' agricultural policy in the Slovenian National Media. The end result was that the health and agricultural sectors have begun to support each other in some agriculture and food policies that they want implemented in Slovenia after accession.

There is a growing experience of HIA applied to agriculture and food policies worldwide. Various methods and approaches have been used, all of which aim to assess the impact of an agricultural practice or policy on public health. Despite this, there is still much uncertainty about what HIA can realistically do for policy making and how it can be used by national and regional governments.

In many respects the experience of HIA of agriculture and food policies is similar to that found in other policy contexts. The major benefits seem to result in strengthening policy-makers' understanding of the interactions between health and other policy areas, and creating new opportunities for improving intersectoral relationships [19]. For example, in Slovenia, the ability of HIA to involve a wide range of stakeholders was considered a very important part of the process. It broadened the issues and enabled them to be considered from different viewpoints. It also engaged other ministries and sectors in public health issues, which created shared agendas and goals in the future policy negotiations. However, stakeholder involvement may not always be necessary. The Swedish CAP analysis contributed to improved intersectoral working between health and agricultural sectors but was conducted as a desk-based expert-led study.

Feeding HIA into policy

In terms of achieving more specific outcomes, many problems still exist with the HIA process in such complex policy environments as agriculture. These include the often-discussed issues of the timing of an HIA, the evidence base for HIA, and how to embed HIA in organizational culture. It is still not clear when is the best time to conduct an HIA of any policy. In the HIA of agricultural policy in Slovenia, as has been the experience of national HIA in the Netherlands and Wales [19,20], if an HIA is attempted at too early a stage the policies may be still too vague or change too frequently to make a strong definitive assessment possible. Conversely, a HIA that feeds into the decision making too late will also have little or no ability to effect change. This was important with the health assessment of the foot and mouth disease disposal policy conducted by the English Department of Health. The rapid, early health assessment was crucial in influencing policy change and improving public health consideration during the foot and mouth disease outbreak.

All the approaches used in the HIAs of agriculture policy are broadly similar, using assessment based on broad determinants of health. By using health determinants in this way HIAs will always reveal large uncertainties in potential health impacts. In food and agriculture, the causal pathways are very complex, and the current evidence base is patchy and often not relevant for assessing specific policy options. However, this does not mean that there is no evidence for health impacts of a policy. The lack of an adequate evidence base is a recurrent problem in HIA at project or policy level [21]. There is an ongoing debate about how to assemble relevant evidence for HIA and policy making [22,23]. In the Slovenian HIA some new reviews of research evidence were commissioned relevant to the agricultural policy interventions being assessed. In most cases, there is neither the time nor money available to undertake such systematic reviews or synthesize evidence relevant to the specific policy context.

How HIA is applied by governments will affect its ultimate long-term influence on policy [24]. Those countries that have an effective HIA programme at policy level have institutionalized HIA in various ways [19,24,25]. No country has yet institutionalized HIA of agriculture or food policies. The HIAs in Slovenia, Sweden, and the United Kingdom were conducted as single projects. However, in both Slovenia and the United Kingdom there was at least a clear mechanism of how the HIA would feed into government strategy making. In Slovenia this was the Food and Nutrition Action Plan, and in the United Kingdom the Department of Health was part of the central government emergency response team for handling the foot and mouth disease outbreak. If HIA is not embedded in the organizational structure of decision-making bodies, benefits to intersectoral working may be lost. This was the case in British Columbia, Canada, where, owing to political changes, HIA fell off the policy agenda after previously having a central cabinet-level role [24].

A comparison of the HIA approaches shows that there are still many limitations with HIA application at an agricultural policy level. In the wider context of policy making, HIA should be seen as one useful tool that can be used to embed public health across policy sectors including agriculture. It is clearly not the only way to support effective intersectoral working or 'healthy' policy development. Its strengths include a structured approach, the flexibility of methods, and its involvement of stakeholders in the process [21]. The problem still remains that the public health sector has not yet reached a common understanding of HIA, and how it should be used in policy making. This is confusing to policy makers wishing to apply HIA. The experience gained in Slovenia shows that HIA has potential as a means of contributing to more integrated intersectoral policies, not only in agriculture but a range of policy areas. The agricultural experience from the United Kingdom shows that rapid HIA can also have a key role in creating public health consideration by other ministries in emergency situations. Further evaluation of the outcomes of such policy-level HIA should enable us to direct the development of HIA in the most practical way to support governments make healthier choices in the agriculture and food policy sector.

References

1 The BSE Inquiry. The BSE Inquiry: The Report. 16 volumes. London: The Stationary Office, 2000. http://www.bseinquiry.gov.uk/report/index.htm.

2 House of Commons. *BSE: The Cost of a Crisis.* London HMSO: House of Commons Select Committee on Public Accounts, 1999.

3 **Ezzati M, Lopez A, Rogers A, Van DerHoorn S, Murray C,** *et al.* Selected major risk factors and global and regional burden of disease. *The Lancet* 2002; **360**:1347.

4 WHO. *The World Health Report 2002: Reducing Risks, Promoting Healthy Life.* Geneva: WHO, 2002.

5 World Bank. *Agriculture and Rural Development. Environmental Assessment Sourcebook.* Washington DC: World Bank, 1999.

6 Health Canada. The Canadian Handbook on Health Impact Assessment. Vol. 1–3. Ottawa: Health Canada, 1999.

7 Department of Health. *A Rapid Qualitative Assessment of Possible Risks to Public Health from Current Foot and Mouth Disposal Options.* London: Department of Health, 2001.

8 Department of Health. *Foot and Mouth Disease: Disposal of Carcasses. Programme of Monitoring for the Protection of Public Health.* London: Department of Health, 2001.

9 **Herriott N.** Rapid health impact assessment of foot and mouth disease in Devon: North and East Devon Health Forum, 2001. http://www.hiagateway.org.uk/Resources/showrecord.asp?resourceid=76

10 ActionAid. *The Developmental Impact of Agricultural Subsidies.* London: ActionAid, 2002:32.

11 Dahlgren G, Nordgren P, and Whitehead M. *Health Impact Assessment of the EU Common Agricultural Policy*. Stockholm, Sweden: National Institute of Public Health, 1996.

12 Schafer Elinder L. *Public Health aspects of the EU Common Agricultural Policy: Developments and Recommendations for Change in Four Sectors: Fruit and Vegetables, Dairy, Wine, Tobacco*. Stockholm, Sweden: National Institute of Public Health, 2003.

13 European Commission. *European Union Enlargement: A Historic Opportunity*. Brussels: European Commission, 2000.

14 Maclehose L and Mckee M. Looking forward, looking back. Gateway to the European Union: health and EU enlargement. *Eurohealth* 2002; **8**(4):1–3.

15 Albrecht T, Cesen M, Hindle D, *et al.* Slovenia. *Health Care Systems in Transition* 2002; **4**(3):1–21.

16 Selb J and Kravanja M. Analiza umrljivosti vSloveniji vletih 1987 do 1996 (mortality rate analysis in Slovenia 1987 to 1996). *Zdrav Varst* 2000; **39**(Supplement):S5–S18.

17 Kuhar A and Erjavec E. Situation in Slovenian agricultural and food sectors and related policies with estimation of the likely future developments. Ljubljana: University of Ljubljana, Biotechnical faculty, 2002.

18 Lock K, Gabrijelcic M, Martuzzi M, *et al.* Health impact assessment of agriculture and food policies: lessons learnt from HIA development in the Republic of Slovenia. *Bulletin of the World Health Organization* 2003; **81**(6):391–398.

19 Breeze C and Hall R. *Health Impact Assessment in Government Policymaking: Developments in Wales*. Brussels: WHO Europe, ECHP Policy Learning Curve, 2001.

20 Council for Public Health and Health Care. Healthy Without Care. Zoetermeer, The Netherlands: Report to the Minister of Health, Welfare and Sport, 2000.

21 Lock K. Health impact assessment. *British Medical Journal* 2000; **320**:1395–1398.

22 Parry J and Stevens A. Prospective health impact assessment: pitfalls, problems and possible ways forward. *British Medical Journal* 2001; **323**:1177–1182.

23 Mindell J, Hansell A, Morrison D, *et al.* What do we need for robust quantitative health impact assessment? *Journal of Public Health Medicine* 2001; **23**(3):173–178.

24 Banken R. *Strategies for Institutionalising HIA*. Brussels: WHO Europe, ECHP Policy Learning Curve no. 1, 2001.

25 Varela Put G, den Broeder L, Penris M, and Roscam Abbing E. *Experience with HIA at National Policy Level in the Netherlands*. Brussels: WHO Europe, ECHP Policy Learning Curve no. 4, 2001.

Chapter 34

HIA and the National Alcohol Strategy for England

John Kemm

Background to the English Alcohol Strategy

From earliest times alcohol has been recognized as having good and bad effects on health. The harmful effects of alcohol include physical and mental ill-health, accidents, crime and antisocial behaviour, family disruption, loss of productivity, and economic loss. The beneficial effects include enjoyment, facilitation of social contacts, some health benefits, employment, and revenue for the government. It is therefore to be expected that any government should be concerned about positive and negative health impacts of policy and legislation concerning alcohol.

The most recent development of alcohol policy in England began in 1998 with the publication of a new health strategy in the white paper 'Our Healthier Nation'. This document contained the statement 'The government is preparing a new strategy on alcohol to set out a practical framework for a responsible approach' [1]. This promise of a national Alcohol Strategy for England was warmly welcomed by those working in the alcohol field and raised expectations that it would shortly appear. Alcohol Concern, the UK alcohol charity responded by publishing proposals for an alcohol strategy [2] in the hope that this would influence government thinking. However there was no apparent progress towards developing the strategy. The National Health service (NHS) Plan [3] clarified the intended timetable stating 'by 2004 we will be implementing a Strategy to reduce alcohol misuse'. Elsewhere progress was being made. The Home Office published an Alcohol Action Plan [4], proposals for a revision of licensing law [5], and also produced guidance on aspects of alcohol and offending [6]. Scotland undertook an extensive consultation process and in 2002 produced its own strategy 'An Alcohol Strategy for Scotland' [7].

Finally the Strategy Unit within the Cabinet Office was asked to develop a national alcohol harm reduction strategy ready for implementation in 2004. The first step was a consultation process undertaken in autumn 2002. They published a consultation document [8], which asked a wide-ranging set of questions about alcohol consumption, mechanisms of harm, and possible interventions and invited responses from all with an interest. The consultation was completed early in 2003 and at the time of writing this chapter the Strategy Unit is analyzing the response before developing policy options and producing a final report.

Throughout this protracted process no formal health impact assessment (HIA) has been undertaken though several parts of what happened had many of the characteristics of an HIA. This chapter considers why no formal HIA was undertaken, explores the assessment elements that did take place, and reviews what an HIA might have added to the process.

Why was there no HIA?

The 1998 white paper 'Our Healthier Nation' [1] stated the UK government's intention to apply HIA to its policies. Alcohol policy is an area that impacts on many aspects of life and touches numerous cross-cutting themes. It has implications for law and order, family, leisure, tourism, social inclusion, agriculture, employment, the economy, relations with the EU, and health. Alcohol consumption has both beneficial and harmful consequences as do nearly all measures intended to regulate consumption making alcohol policy an area where policy trade-offs are prominent. For these reasons alcohol policy might appear to be an obvious candidate for HIA. Furthermore the Department of Health that had responsibility for developing HIA also had lead responsibility for alcohol policy.

However alcohol policy has always been politically difficult. Measures to place downward pressure on alcohol consumption are not popular with voters or with the alcohol industry, which has considerable political influence. It is salutary to remember that the introduction of breath testing and tighter drink driving laws, a measure that was undoubtedly in people's best interests and a major contribution to public health, was extremely unpopular when it was first introduced in 1967. Furthermore the values of openness and participation associated with HIA may seem problematic when dealing with an issue with as many political dangers as alcohol. These considerations may explain why the opportunity to use HIA to assist in developing an English Alcohol Strategy was not taken.

There are two main elements in alcohol policy. 'The harm minimization' element seeks to reduce the harm associated with alcohol consumption without affecting the amount of alcohol consumed. This element focuses on the need to treat those who drink harmfully and to prevent drinking in contexts associated with harm. It is popular with all and raises few political problems but many authorities question whether it can ever be effective by itself. The other element, 'the risk reduction' element, emphasizes that alcohol-related harm in a population tends to increase when alcohol consumption increases. Its supporters argue that any strategy to reduce alcohol-related harm must include a downward pressure on alcohol consumption through measures such as regulation and taxation. This second element is politically difficult.

Alcohol Concern's proposals for a National Alcohol Strategy

Alcohol Concern is a national UK charitable body concerned to reduce the level of alcohol misuse and develop the range and quality of helping services available to problem drinkers and their families. It is independent but receives most of its income from government grants. In order to help development of an English strategy it produced a report 'Proposals for a National Alcohol Strategy for England' [2], which made proposals for an alcohol strategy covering the areas of taxation and prices, licensing, community safety, drink driving, advertising and promotion of alcohol, changing attitudes, and support and treatment. It reviewed the current situation and briefly discussed some of the ways in which its proposed measures could affect the future. The authors never used the term, but was it an HIA?

The report was undoubtedly intended to influence the government. Officials from the Department of Health were closely consulted and were observers on the advisory group. Although the final document was sent to the Department of Health, no progress was made on producing a strategy and it is unclear how much the report influenced their thinking. If the document was not influential this was probably due to the fact that the relevant minister changed just after the work was completed rather than to any weakness of the document.

The process by which the document was produced was reminiscent of an HIA. There was an advisory group made up of people with relevant expertise and experience. The problem was scoped and then evidence was collected through policy literature review and consultation. Over 200 stakeholders were identified and consulted through a written questionnaire. A large number of organizations and individuals responded including alcohol service providers, government departments, health authorities, local government, medical associations, police, probation, schools, colleges, and youth organizations. The trade associations of the alcohol industry were invited to take part but did not do so though it is likely that MAFF (Ministry of Agriculture,

Food and Fisheries) and the Department of Culture, Media and Sport would have put forward the industries perspective. The written consultation was followed up by interviews with selected key informants. Views were also sought from focus groups including groups of service users and groups of young people.

The report outlined five objectives

- reducing the level of alcohol-induced ill-health,
- reducing the number of alcohol-related injuries,
- reducing the rate of alcohol-related crime,
- reducing the number of alcohol-related road accidents,
- reducing economic loss in the workplace due to alcohol misuse.

While rather narrow these objectives might be health outcomes considered in an HIA. However, other than implying that its proposals would contribute to reaching the stated objectives, the report made little attempt to predict the consequences of implementing them or compare this with the alternative of doing nothing. For this reason it is probably more appropriate to consider the report as a lobbying exercise rather than an HIA. A later section of this chapter will consider how an HIA would have treated these issues differently.

Work of the Strategy Unit of the Cabinet Office

The process managed by the Strategy Unit in the Cabinet Office that will lead to a National Strategy is still underway and will undoubtedly include deep policy appraisal but will it be an HIA? The stage of the policy-forming process at which an HIA takes place is important. Before policy options have been roughly shaped is probably too early and after policy options have been chosen is certainly too late. The widespread consultation with which the strategy unit began their task used very open and wide-ranging questions. In an HIA there would be further extensive consultation once the options were clearer. The future process leading to the production of a final report remains to be seen.

One can be confident that health impacts and particularly costs imposed on the health services will be analyzed. It will surely follow the government's guide [9], which identifies inclusivity as one of nine features of (good) modern policy-making. However 'inclusivity' is not the same as 'openness' and within the UK civil service it is clearly understood that advice to ministers is confidential. One other feature that Strategy Unit process shares with HIA is that it is managed by experts in the process (HIA/policy analysis) rather than by specialists in the policy area.

What might an HIA have added

This next section offers the sort of analysis that an HIA might have brought to consideration of an alcohol strategy. It begins to map out the causal pathways, by which alcohol impacts negatively and positively on health. It starts to indicate the sorts of health trade-offs that have to be made in optimizing a strategy and the many areas where there is uncertainty.

Individual consequences of drinking

The impact of alcohol policy on the health of a population is ultimately mediated by the drinking behaviour of members of that population (Fig. 34.1). There are three ways in which an individual's drinking behaviour can harm themselves or others.

- Chronic consumption of large amounts
- Intoxication
- Drinking in inappropriate contexts (e.g., driving or when doing work requiring best physical and mental function)

Dose–response curves describe the relation between risk of particular outcomes and consumption of alcohol. Such curves have been published for harms such as cirrhosis, oesophageal cancer, breast cancer, stroke, head injury, alcohol dependence, suffering an assault, and 'adverse social consequences' [10,11].

In addition to those conditions for which alcohol consumption increases risk there are also some conditions for which alcohol appears to decrease risk. These include heart disease and thrombotic stroke. For these conditions the relationship is J shaped and light drinkers are at lower risk than non-drinkers [12]. The diseases for which alcohol appears to have a protective effect are common and any assessment of alcohol policy has to take this benefit into account. Calculations in the United States [13], the United Kingdom [14], and Finland [15] have suggested that the harmful health effects of alcohol are counterbalanced in part or in whole by these good effects. While the evidence for a protective effect is strong it is worth pointing out that ecological evidence suggests that any protective effect of alcohol is less important. In the United Kingdom during the latter half of the twentieth century both heart disease mortality (rose and then fell) and alcohol consumption (rose and then reached a plateau) showed dramatic

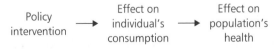

Fig. 34.1 The need to consider individual consumption.

changes. However the relationship between these two does not suggest that even large changes in alcohol consumption have any visible effect on changing heart disease mortality.

It must be remembered that there are important gaps in our understanding of dose–response curves for both harmful and beneficial effects of alcohol. There are many important harms such as involvement in crime, antisocial behaviour, family dysfunction, or lost productivity for which the dose–response relationships are largely unknown. Equally there are no dose–response curves for the most important beneficial outcomes such as enjoyment and facilitation of social networks. Often the curves are based on data from middle-aged and predominantly male cohorts (data from studies primarily designed to study heart disease and cancer) and very different relationships may apply to younger age groups [16]. The measure of exposure in dose-response curves is usually mean daily consumption, which may be an inadequate descriptor. Certainly for heart disease and hypertension there is evidence that the pattern of drinking is as important as the overall mean consumption. The health consequences of drinking in binges are different from the consequences of drinking the same amount regularly spread out over the week [17,18]. Most studies suggest that the type of beverage in which the alcohol is consumed is unimportant [19].

At a population level there is also a wealth of evidence that overall consumption is related to frequency of harm [20]. Prediction of the effect of any policy intervention on overall consumption must be a crucial part of any HIA.

Policy intervention and consumption

The relationship between individual consumption and harms or benefits is reasonably well understood but in order to predict consequences of policy interventions it is also necessary to understand the link between those interventions and consumption. The relationship between consumption and price, income and affordability has been extensively studied. Most alcoholic beverages are fairly price elastic, that is to say that their consumption is reduced when their price goes up [21]. Beer is the most price elastic and wine the least. However alcoholic beverages are also income elastic and their consumption tends to increase as incomes rise. Many alcohol strategies rely heavily on taxation to influence consumption through its effect on price. There is a solid evidence base for predicting the effects of fiscal interventions in HIAs.

Other methods of influencing consumption are control of production and marketing (as used in Sweden where alcohol sales are a state monopoly), regulation of sales through licensing, and regulation of advertising. The effect of these measures on consumption is uncertain. Experience in Scotland suggests that relaxation of licensing hours did not cause dramatic changes in consumption [22]. Studies on density of retail outlets and consumption produce equivocal results [23]. The effect of advertising on consumption is also contested [24,25] making it difficult to predict the impacts of regulation of advertising. Drinking patterns are strongly culture-specific making it difficult to transfer conclusions derived from interventions in one country to another.

Distribution of consumption

The distribution of consumption within populations is not uniform. The French statistician Lederman [26] suggested that distribution of consumption in populations followed a log normal distribution and for years there has been a rather sterile debate as to whether he was correct. There is no doubt that the distribution is not strictly log normal. However it is highly skewed with a median much below the mean and a long tail of heavier drinkers (Fig. 34.2). Furthermore when mean consumption increases there is a proportionate or greater increase in the number of heavier drinkers [27].

A mass of policy interventions attempts to alter the shape of this distribution seeking to reduce the long tail of heavier drinkers without affecting the mean. It is hard to find any example of a country in which this has been achieved. The rationale for public education and promotion of sensible drinking benchmarks is presumably that if people understand the relationship between alcohol consumption and risk of ill-health they would choose to restrict their consumption. While there is evidence that public education does change knowledge there is very little evidence that it changes consumption [28].

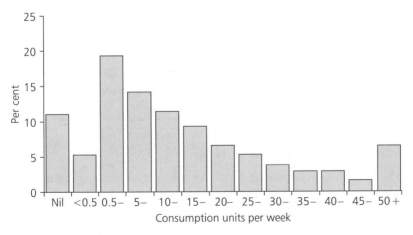

Fig. 34.2 Distribution of alcohol consumption among men.

Source: General Household Survey 2000.

Policy interventions based on public education alone seem unlikely to reduce alcohol-related harm though they are probably justified on the ground of producing an informed population.

An alternative approach is to try and alter the shape of the dose-response curves for harm so that the risk of harm at a particular level of individual consumption is reduced. There are examples of this being achieved. Drink driving offences and road deaths associated with alcohol have markedly reduced in the United Kingdom at a time when alcohol consumption was rising or static. A change in the law, firm enforcement, and sustained public education succeeded in changing attitudes and reducing the risk that people drove after drinking. Antisocial behaviour at football matches is another example of how risk has been changed. Banning drink inside football grounds and on trains going to matches has reduced the amount of alcohol-related violence in these settings without changing overall consumption. Policy interventions directed at licensed premises such as requiring improved design, better training of bar and door staff, and stricter enforcement of existing law may all reduce the risk of antisocial behaviour and nuisance [29]. Improved late night public transport from city centres may similarly reduce alcohol-related violence. A requirement for toughened glasses in licensed premises will reduce the risk of severe facial injuries (glassings in which broken bottles or glasses are used as weapons) [30]. These are all examples of attempting to reduce the risk of harmful consequences without influencing overall consumption.

'Culture' meaning the way in which people think about alcohol, drinking and intoxication, their expectations, beliefs, values, and norms is clearly an important factor in determining how alcohol will impact on a society. Many policy documents talk about the need to change the culture. Cultures do change but it is exceedingly difficult for policy makers to influence the direction or timing of those changes. Many of the interventions considered in this section will operate through changes in culture but because the causal relationships are so complex it is difficult to build them into the predictions of an HIA.

Policy interventions and the young

'The young' are a common focus of alcohol policy. Politically this is an easy area partly because protecting the young is self-evidently good and partly because those below voting age cannot register objections through the ballot box. In the United Kingdom young adults are the group who drink most heavily and suffer most frequently from certain forms of alcohol-related harm (road traffic injuries, assaults, suicide, acute alcohol toxicity). Children are particularly vulnerable to the acute toxic effects of alcohol. Drinking habits acquired when young tend to persist into later life. Those who drink heavily as young adults are more likely to drink heavily as older adults. These are all good reasons for considering young people in alcohol policy.

Attempts to prevent young people from purchasing alcohol are a feature of many policies. These usually centre on proof of age cards and enforcement of law on underage purchasing. It is not unreasonable to think they may reduce underage purchasing and the licensed trade generally favours them but there is little evidence on how much they reduce the problem. Regulation of alcohol advertising and marketing to prevent it being directed at recruiting young drinkers has been advocated. The justification for this lies in theory rather than any sound empirical base. Alcohol education in schools or directed at young people is another common intervention. This increases knowledge but there are very few examples where it can be shown to have reduced alcohol consumption or delayed the onset of drinking. It may be that these educational activities have a 'sleeper' effect and reduce the likelihood that the individual will become a heavy drinker in later life but there is no evidence to confirm or refute this suggestion.

Services for problem drinkers

Most accept that alcohol strategies should cover provision of services for problem drinkers and their families and consideration of the effectiveness of those services. The debate about the relative merits of different approaches to treatment is beyond the scope of this chapter but it should be noted that early intervention is widely accepted to give better results than later intervention. Some forms of intervention have been assessed as not only effective but also cost effective [31,32]. Treatment services are essential in a humane society and politically popular but have little impact on the overall size of the problem. Brief interventions directed at the heavy drinker rather than the drinker with established problems are effective [33] and may be one of the few effective ways of reducing the 'tail' of heavy drinkers without tackling overall consumption.

Economic benefits

The manufacture, distribution, and retailing of alcohol is a large financial sector in the United Kingdom and any major disruption of trade in alcohol would have negative impacts on the economy and on health. About 4 per cent of exchequer income is derived from alcohol [34]. If this income were lost revenue would either have to be raised from other sources or expenditure (probably including expenditure on health and welfare) would have to be cut.

Large numbers are employed in the various branches of the industry and the health damaging effects of loss of employment are well recognized [35,36]. For many more, while their employment does not depend on alcohol, loss of alcohol trade would significantly reduce their income and lower income is associated with poor health. The economic relations are complicated but can be modelled. If alcohol duties were raised the greater revenue per volume would probably more than compensate for the fall in volume so that exchequer revenue would rise as consumption falls.

Impact on health equity

Health impact assessment is concerned not only with the overall level of health in the population but also the distribution of health within that population. Alcohol-related harms affect different groups in society very unevenly. The use made of and cultural attitudes towards it differ widely between various ethnic and faith communities. Possible control measures such as increasing duty, changed enforcement practice, alcohol education, and provision of services would differentially affect these groups. In assessing the impact of different strategy components it would be necessary to examine the impact on different groups and attempt to ensure that equity between groups was increased.

Participation

So far this chapter has taken an epidemiological and technical approach (tight focus) to HIA. For issues such as frequency of cirrhosis, distribution of consumption levels, tax revenues, and numbers employed a tight focus approach will probably produce better predictions than a democratic participatory approach (broad focus). There is however another set of issues, such as self-esteem, happiness, social networking, and willingness to comply, where epidemiology and econometrics have less to offer. Any HIA that ignored these issues would be very inadequate and it is important that the 'voices' of the community should be added to the dry calculations of the technicians.

Good participation in HIA is challenging to arrange. It is particularly so at a national level. In England there are 50 million stakeholders in the alcohol strategy since the whole population will be affected by an alcohol strategy. Among them there are widely divergent views on alcohol, what constitutes acceptable behaviour, and what liberties should be constrained. The requirement for confidentiality within the civil service makes meaningful participation even more difficult to arrange. Representative democracy is an important path for participation but involvement in elections is low and policy is not always adequately influenced by the views of the population. There are no simple solutions but certainly thought must be given as to how an HIA of national policy can be fully owned by government and still achieve participation.

Conclusion

There are numerous causal paths by which the elements of an alcohol strategy could impact harmfully or beneficially on the health of the population (see Fig. 34.3). The map is complex and no one could predict all the ramifications and trade-offs within this field. However enough is known to predict many of the likely policy consequences. For some paths the size of the impacts can be fairly precisely predicted; for other paths the direction and rough order of magnitude can be predicted; for others even the direction of impact is uncertain. An HIA cannot tell what the alcohol strategy should be but it can inform and structure thinking and assist in making judgements.

However in the case of the development of an English alcohol strategy HIA does not seem to have contributed much. The brutal lesson is that unless HIA is seen to be useful by policy makers it will not contribute to healthy public policy (HPP).

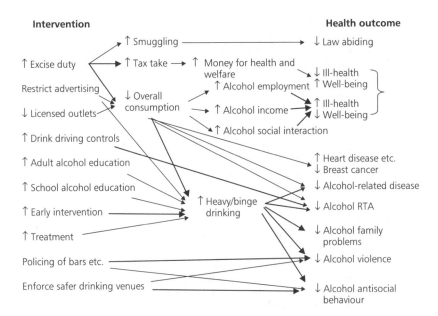

Fig. 34.3 Causal pathways in alcohol policy.

References

1 Department of Health. *Our Healthier Nation. A Contract for the Nation.* London: The Stationery Office, 1998.

2 Alcohol Concern. *Proposals for a National Alcohol Strategy for England.* London: Alcohol Concern, 1999.

3 NHS. *The NHS plan: a Plan for Investment, a Plan for Reform.* London: The Stationery Office, 2000.

4 Home Office. *Alcohol Action Plan.* London: Home Office, 2001.

5 Home Office. *Time for Reform; Proposals for the Modernisation of our Licensing Laws.* London: The Stationery Office, 2000.

6 Home Office. *An Alcohol Related Crime Toolkit.* London: Home office, 2001.

7 Scottish Executive. *Plan for Action on Alcohol Problems.* Edinburgh: Scottish Executive, 2001.

8 Department of Health. *National Alcohol Harm Reduction Strategy: Consultation Document.* London: Department of Health, 2002.

9 Cabinet Office Professional Policy making for the twenty First Century London Cabinet Office, 1999. www.cabinet-office.gov.uk/moderngov/1999/policy

10 **Edwards G, Anderson P, Babor TF, Casswell S, *et al.*** The individuals drinking and degree of risk. In *Alcohol Policy and the Public Good.* Oxford: Oxford Medical Publications, 1994.

11 **Corrao G, Bagnardi V, Zambon A, and Arico S.** Exploring the dose response relationship between alcohol consumption and the risk of several alcohol related conditions: a meta analysis. *Addiction* 1999; **94**:1551–1573.

12 Royal College of Physicians Working Party Alcohol and the heart in perspective: sensible limits reaffirmed London Royal College of Physicians, 1995.

13 **Thun MJ, Peto R, Lopez AD, Monaco JH, Henley SJ, Heath CW, and Doll R.** Alcohol consumption and mortality among middle aged and elderly US adults. *New England Journal of Medicine* 1997; **337**:1705–1714.

14 **Britton A and McPherson K.** Mortality in England and Wales attributable to current alcohol consumption. *Journal of Epidemiology and Community Health* 2001; **55**:383–388.

15 **Makela P, Valkonen T, and Poikolainen K.** Estimated number of deaths from coronary heart disease 'caused' and 'prevented' by alcohol: an example from Finland. *Journal of Studies of Alcohol* 1997; **58**:455–463.

16 **Romelsjo A and Liefman A.** Association between alcohol consumption and mortality, myocardial infarction and stroke in 25 year follow up of 49618 Swedish men. *British Medical Journal* 1999; **319**:821–822.

17 **McElduff P and Dobson AJ.** How much alcohol and how often? Population based case control study of alcohol consumption and risk of a major coronary event. *British Medical Journal* 1997; **314**:1159–1164.

18 **Kuahenen J, Kaplan GA, Goldberg DE, and Salonen JT.** Beer binging and mortality: result from the Kuopio ischaemic heart disease risk factor study, a prospective population study. *British Medical Journal* 1997; **315**:846–851.

19 **Rimm EB, Klatsky A, Grobbe D, and Stampfer MJ.** Review of moderate alcohol consumption and reduced risk of coronary heart disease: is the effect due to beer, wine or spirits? *British Medical Journal* 1996; **312**:731–736.

20 Edwards G, Anderson P, Babor TF, Casswell S *et al.* Population drinking and the aggregate risk of alcohol problems. In *Alcohol Policy and the Public Good.* Oxford: Oxford Medical Publications, 1994.

21 Godfey C. The influence of price, income and taxation on UK alcohol consumption and problems. In Raistrik D, Hodgson R, and Ritson B. (eds.) *Tackling Alcohol Together.* Abingdon Society for the Study of Addiction, 2000.

22 Duffy JC and Plant MA. Scotland's liquor licensing changes: an assessment. *British Medical Journal* 1986; **292**:36–39.

23 Stockwell T. Liquor outlets and prevention policy: the need for light in dark corners. *Addiction research* 1997; **92**:925–930.

24 Smart RG. Does alcohol advertising affect overall consumption.? A review of empirical studies. *Journal of Studies on Alcohol* 1988; **49**:314–323.

25 Calfee J and Scheraga C. The influence of advertising on alcohol consumption. A literature review and an econometric analysis of four European nations. *International Journal of Advertising* 1994; **13**:287–310.

26 Ledermann S. Alcool, alcoolisme. Alcoolisation. Donnees scientifiques de caractere physiologique economique et social. Institut National d'Etudes Demographiques Travaux et Documents. Cahier No 29. Paris: Presses Universitaires de France, 1956.

27 Colhoun H, Ben-Shlomo Y, Dong W, Bost L, and Marmot M. Ecological analysis of collectivity of alcohol consumption in England: importance of the average drinker. *British Medical Journal* 1997; **314**:1164–1168.

28 Gorman DM and Speer PW. Preventing alcohol abuse and alcohol related problems through community intervention: A review of evaluation studies. *Psychology and Health* 1996; **11**:95–131.

29 Graham K and Plant MA. Harm minimisation. In Plant M and Cameron D. (eds.) *The alcohol report.* London: Free Association Books, 2001.

30 Coomaraswamy KS and Shepherd JP. Predictors and severity of injury in assaults with barglasses and bottles. *Injury Prevention* 2003; **9**:81–84.

31 Fleming MF. Denefit cost analysis of brief physician advice with problem drinkers in primary care settings. *Medical Care* 2000; **38**:7–18.

32 Health Technology Board for Scotland Prevention of relapse of alcohol dependence Health Technology Assessment Advice 3 2002. www.htbs.co.uk/docs/pdf/HTA3fullversion.pdf

33 NHS Centre for Reviews and Dissemination. Brief interventions and alcohol use. *Effective Care Bulletin* 1993; 7.

34 British Beer and Pub Association Statistical Handbook 2002 (Table F1) London: Brewing Publications Ltd., 2003.

35 Mathers CD and Scholfield DJ. The health consequences of unemployment: the evidence. *Medical Journal of Australia* 1998; **168**:178–182.

36 Bartley M. Unemployment and health: understanding the relationship. *Journal of Epidemiology and Community Health* 1994; **48**:333–337.

Chapter 35

HIA in SEA and its application to policy in Europe

Carlos Dora

This chapter describes recent work to integrate health aspects into environmental impact assessments (EIAs), and notably into strategic environment assessments (SEAs) as required by the new international legislation in Europe. It outlines a rationale for this activity, describes its aims and what has been achieved, and points out the benefits and limitations. It discusses the potential role of health impact assessment (HIA) as part of SEA in helping to ensure that policies and projects in different sectors contribute to health and environment protection, and serve as tools for healthy public policy (HPP). It concludes by identifying the challenges that the health sector will need to address in order to realize that potential.

What is SEA?

A strategy defines the policy goals, the overall direction, and the types of action to achieve those goals. A project will fit within that strategic direction, and accomplish specific actions that contribute to the bigger picture. The broader decisions made at strategic level define and restrict the options and decisions to be made at the project level. Strategies are subject to SEA while projects are subject to EIA. Strategic environment assessments was developed to address and overcome some of the limitations of project-level EIA. It aimed to provide mechanisms to bring environmental considerations into the mainstream of higher and broader level of decision making, so as to achieve more ambitious goals. An EIA often cannot address the decisions made at a higher level (policies and strategies). There is some overlap between SEA and EIA, especially for very large projects with wide implications. A simple rule of thumb can be used to identify the type of SEA a policy requires. Policies that initiate or fix the type, form, or location of concrete projects can be subject to SEAs that follow similar steps to an EIA (e.g., engage with the specific stakeholders affected and document the expected impacts in quite some detail). Policies that focus on 'why' or 'if' questions in general need to be subject to a policy appraisal. Policy appraisal is broader than SEA, considers a wider range of scenarios and possibilities, and is less likely to lend itself to quantification [1]. Equivalent considerations apply when HIA is applied at strategic or project level and it might be helpful to distinguish between strategic or policy HIA and project HIA.

Rationale for including HIA in SEA

One of the greatest challenges for making HPPs is to establish processes, in which health issues can be considered as part of developing and implementing those policies. Existing knowledge of health determinants provides a good understanding of how policies impact on health. The need for 'cross-sectoral policy making' has often been expressed but proved difficult to translate into practice.

Health impact assessment is now increasingly used as a tool for achieving HPP. Experience with implementing HIA, for example, in British Columbia, Canada (see Chapter 16), demonstrate the importance of having clear rules that require HIA to be applied during the decision-making process [2]. If the decision to undertake HIA relies on the motivation of individuals, it is unlikely to be continued when political power or government changes. Institutionalization is therefore necessary to ensure that health impacts of policies are systematically considered as part of the process of decision making. Incentives are needed to encourage individuals and governments to include consideration of possible health impacts in the process of policy and decision making.

There is little experience on institutionalizing the use of HIA in policy making. Descriptions of how this can be done are only now beginning to be documented and evaluated. However there is considerable experience of institutionalizing the use of EIA and SEA in policy making where they are used to ensure environmental concerns are addressed. There are three reasons why this experience is relevant to the debate on HIA and policy making.

First, we can learn from comparing Impact Assessments (EIA and SEA) to other mechanisms used in the attempt to incorporate environment considerations into policies. Lessons for mainstreaming health through HIA can be drawn from this wider experience. Second, assessment of health impacts is formally required as part of EIA and SEA. These processes are already institutionalized and required by national and international law for policy decisions in many fields. There is already an extensive body of expertise in applying EIA to policy. Third, despite there being a requirement to include assessment of health impacts in EIA and SEA, it is hardly ever complied with. The reasons why this is so and why EIA and SEA so rarely include adequate assessment of health contain important lessons for the practice of HIA.

The foregoing considerations provided an important background for the negotiations for a new protocol on SEA as part of the convention on EIA. The countries of Europe plainly expressed a wish that health issues should be clearly addressed in this protocol. The negotiation created an opportunity to make progress on institutionalizing the assessment of health impacts of policies.

Coverage of health impacts in EIA

Formally health has been included in EIA, but in practice it has been conspicuously absent. Three surveys show how wide this gap is. A series of case studies in six countries of central and eastern Europe were recently carried out in collaboration with WHO. These studies investigated, using standard methodology, how health impacts had been considered in 11 EIAs. These EIAs covered a range of projects, including the construction of a highway, a new industry to produce chromic oxide, changes in an aluminium casting factory, changes in an incineration plant burning chlorinated hydrocarbons, a radioactive waste site; deepening of a sea port; two new waste landfill sites, a new oil extraction site, a new technology to produce pesticides, and a car painting facility.

In only one of the countries was there a requirement for health authorities to participate in the early stages of EIA. In that country the health authorities seem to have had a significant influence on the scope of the EIA and its coverage of relevant health issues. In other countries only health issues for which there was a legal limit were considered (e.g., water and air contamination or pollution). In consequence the definition of health used in the assessment was narrow, and several relevant health issues were excluded. Because of the narrow definition, groups affected by the proposal often had no opportunity to participate in the assessment or contribute to the debates. In one case this led to public outcry. The EIA legislation did not specify how health should be assessed, and in most cases no baselines on health were established. Reviews of the environment impact assessments (EISs) did not include health experts and the other inputs from health authorities were largely overlooked. Although the environment authorities and experts carrying out EIA were interested in assessing health impacts they had few incentives or resources to do so. Some of these case studies were presented at a meeting on HIA in SEA in 2001 [3] and a publication on the project is being prepared by the WHO.

A review of how health issues were treated in 28 EIAs of road projects in Sweden since 1990 made similar findings [4]. It concluded that health experts were rarely consulted and that the majority of assessments did not identify the populations affected by the plans or consider vulnerable groups. Reference to health was made only by consideration of compliance with environmental standards (e.g., for air pollution) that are set on the basis of health impacts. In consequence the road projects overlooked a number of important health determinants and impacts, such as those associated with physical activity and recreation. In addition the EIAs made no reference to relevant national health goals. As a result of these deficiencies it is not possible to assess the cumulative effects of these projects on health.

The WHO carried out a similar analysis for European countries, as background material for the debate about a possible legal instrument in Transport Environment and Health. The findings were similar to those of the Swedish study, and suggested that incentives were needed if health is to be included in EIA of transport policies and projects [5].

European moves to include health in SEA

European governments, in a number of inter governmental fora between 1998 and 2000, expressed a wish to see health impacts of policies examined as part of policy-making processes. This was clear in the Declaration adopted by 72 Ministries at the 1999 Environment and Health Conference in London [6]. Countries adopting the declaration undertook to 'carry out EIAs fully covering impacts on human health and safety . . . , invite countries to introduce and carry out strategic assessments of the environment and health impacts of proposed policies plans and strategies . . . , invite international finance institutions also to apply these procedures . . . '. The conference report states that ' . . . several countries supported the idea of a protocol on strategic environment and health impact assessment . . .'. This desire was also expressed at a meeting of the parties of the Aarhus convention on Access to Information, Public Participation and Access to Environmental Justice, in July 2000. A decision to prepare a legally binding protocol on SEA was made at the second meeting of the parties to the EIA Convention in 26–27 February 2001, attended by environment departments and non-governmental organizations [7].

These decisions fitted into a general trend towards identifying the implications of policies for health and the environment, creating safeguards, and engaging the public in decision making. These developments were taking place when a number of health scares such as the BSE crises in the United Kingdom, were making clear the high political and economic costs of failing to take account of health risks. The need to consider people's concerns, to assess risks, and to adopt precautions was evident [8].

Relevant legislation has been introduced including the UNECE EIA convention, which came into force in 1997, and the EC directives on EIA (97/11/EC–85/337/EEC) and on SEA (2001/41/EC). The EC Directorate General DG environment has introduced a new department to deal with environment and health issues.

Moves to include health considerations in public policy-making were also advancing. In 1997 the Amsterdam treaty of the EU called for 'a high level of human health protection to be insured in definition and implementation of all community policies and activities'. In this same period several national health departments developed HIA in national policy making. Both the WHO and SANCO,[1] have taken initiatives to promote HIA. Efforts to develop integrated impact assessment, that cover both environmental and health aspects are part of this continuing trend. Development of HIA came later and somewhat separately to that of EIA. The advantages and disadvantages of bringing the two assessments closer need to be discussed.

Preparation of the draft protocol on SEA

An *ad hoc* working group was established to negotiate the draft SEA Protocol, under the UNECE EIA convention. The group included representatives of the WHO and representatives from some Ministries of Health, who were invited to take part in the negotiations. A vice chair of this working group, from the Ministry of Health of the Czech Republic, was elected to facilitate preparation of health aspects of the draft. Three broad interest groups contributed to this preparatory process, those coming from EIA, those interested in health impacts, and those focusing on public participation and access to information. Before drafting commenced, a small meeting and then a larger expert workshop involving the three groups was held. The meeting and workshop built mutual understanding of the three groups' perspectives, and clarified how those issues might be reflected in the protocol draft.

The preparatory work on health impacts included preparation of a review document describing concepts, tools, methods, and experience in HIA and the links with SEA [9]. This review was based on an expert workshop led by the WHO, and reflected the views and practice of HIA in Europe [10]. The WHO prepared for health departments a short policy briefing and a brochure clarifying the potential benefits of having health well covered in the SEA protocol [11]. The WHO also gave presentations on the subject [11].

---

[1] The European Commission Directorate General for Health and Consumer Protection (DG SANCO).

Both health and environment groups needed to have the potential benefits of including health in SEA clarified and to be persuaded of its worth. The environmentalists tended to fear that a specific mention of health would detract attention from the ecological issues. Some in the health sector, who were less accustomed to working across sectors, feared that a link with environment assessment might limit HIA to quantitative assessment of environmental contamination. The fears of both sides had some foundation but were also exaggerated.

During the protocol negotiations the debate on health focused on fundamental issues of principles and values that should guide the health assessment as well as on accountability and control mechanisms. For example, what health impacts should be included, all health impacts or only those mediated by the environment? Who would assesses health impacts? What were the roles of health authorities and health agencies? Who sets the standards for health assessments? Would it be obligatory to include health impacts in every SEA?

The arguments for integrating HIA in SEA were the similarities in their goals and the parallel methodologies and procedures by which they were implemented. Both aim at evidence-based policy making, and both serve as devices for information sharing, consultation, public participation, and negotiation between administration and public. Both HIA and EIA analyze and document effects of proposed actions, identify alternative measures to mitigate adverse effects, and try to ensure that the decision-making process considers relevant findings. Both HIA and EIA/SEA share several key stages (screening, scoping, appraisal, etc.). The procedures for implementing each stage are also similar (policy appraisal, risk assessment, indicators for risk management, public participation). Some of the debates on different approaches to HIA among the health community have equivalents in debates within the EIA/SEA community.

The new SEA protocol

The draft protocol was finalized on 30 January 2003, adopted, and opened to signature at the Ministerial 'Environment for Europe' Conference, 21–23 May 2003 in Kiev, Ukraine. The phrase 'environment including health' was used throughout the protocol. This was broader than 'environment mediated health impacts' that some preferred, and at the same time clarified that the starting point for the impact assessment was environmental concerns, and that health issues were part of it. The final text of the SEA protocol refers to health in an unequivocal manner [12]. It requests that health be considered at the different stages of the SEA process, and that health authorities are consulted at those stages. The protocol provides a clear response to the requests from

European countries for health impacts to be clearly included. It is in that sense a landmark in the institutionalization of health assessments of policies, and in the inclusion of health into SEAs.

This formal requirement for the assessment of health impacts and for the involvement of health authorities in the different stages of SEA is new. When the protocol comes into force it will place demands on health systems but will also increase the opportunities for these systems to address health determinants and to promote health in a systematic and substantial way. The existing limitations of health assessments as part of EIA will need to be addressed. This will require a clearer reference to health, and consideration of the whole range of health determinants, covering positive as well as negative health impacts, and the impacts on specific population groups. Experience from EIAs and HIAs done separately suggest that it is not only desirable but also feasible for health issues to be adequately within the context of an EIA. Authorities are starting to recommend that this be done [13].

The inclusion of health issues in EIA and SEA is likely to increase demands on health systems and authorities in a number of ways. The will be asked to provide more evidence of health impacts of policies and better tools and methods for HIA. They will be asked to analyze their experience in implementing HIA. There will be a need to generate more commonality in the way that environment and health authorities understand health impacts and determinants. Greater awareness of what HIA is and what it can deliver is likely to be needed within the health sector.

There may be need for capacity building to increase the numbers of those who can undertake the assessments. Policy stakeholders, who are the users of SEA results, will have to analyze their experience when health assessments are included and determine the added value of doing this. Continued implementation of health assessment in SEA, documentation of the experience gained, and exchange of the lessons learnt are key to further protecting the environment and promoting public health. The need to respond to demands for health assessments in the context of SEAs, can be an incentive for health systems to strengthen their stewardship role. It will also stimulate them to increase their understanding of the determinants of health and contextual factors, to engage with the actors who influence those determinants, and to build public trust in the legitimacy of public health decision-making [14,15].

Overall one would hope the new SEA protocol, which institutionalizes assessment of health impacts as part of international law, could encourage further thinking on cross-sectoral policies. It could encourage bolder action and help to bring concern with intersectoral policy back into mainstream public health practice.

References

1 Sadler B and Verheem R. Strategic Environmental Assessment: Status, Challenges and Future Directions. Netherlands Ministry of Housing, Spatial Planning and the Environment (VROM), publication no 53. PO Box 351, 2700AJ Zoetermeer, Netherlands, 1996.

2 Banken R. Strategies for institutionalising HIA. ECHP Health Impact assessment Discussion Papers; no. 1. WHO Regional Office for Europe, 2001. Access Date: 16 May 2003, http://www.euro.who.int/document/e75552.pdf

3 WHO workshop on HIA in SEA. Orvieto Italy, 20 November 2001. Access Date: 16 May 2003, http://www.who.dk/healthimpact/NewsEvents/20020115_1

4 Alenius K. *Consideration of Health Aspects in Environmental Impact Assessment for Roads*. Stockholm: National Institute of Public Health, 2001. ISBN 91-7257-113-6.

5 UNECE and WHO. Overview of Instruments Relevant to Transport, Environment and Health. Recommendations for Further Steps—Synthesis Report Document, ECE/AC.21/2001/1–EUR/00/5026094/1, 2001. Access Date: 7 April 2003, http://www.euro.who.int/document/trt/advreport1.pdf

6 Declaration. Third Ministerial Conference on Environment and Health. London 1999. Access Date: 16 May 2003, http://www.who.dk/EEHC/conferences/20021010_2

7 Accessed 16 May 2003: http://www.unece.org/env/eia

8 Harremoës P, Gee D, MacGarvin M, Stirling A, Keys J, Wynne B, and Vaz S. Late Lessons from Early Warnings: the Precautionary Principle 1896–2000. Environmental Issue Report No. 22 European Environment Agency, 2002.

9 Breeze C and Lock K (eds.) Health Impact Assessment as part of Strategic Environment Assessment. WHO Regional Office for Europe, 2001. Access Date: 16 May, 2003, http://www.who.dk/healthimpact/MainActs/20030120_1

10 European Centre for Health Policy, WHO Regional Office for Europe. Health Impact Assessment: main concepts and suggested approach. Gothenburg Consensus Paper. WHO 1999. Accessed on 16 May 2003: http://www.who.dk/document/PAE/Gothenburgpaper.pdf

11 Key documents in HIA in SEA. Policy briefing, leaflet and one conference presentation, can be found in the following URL. Access Date: 16 May, 2003, http://www.who.dk/healthimpact/MainActs/20030120_1

12 The final draft of the SEA Protocol, and commentary on the references to health in the Protocol focus can be found in the following URL. Access Date: 16 May 2003, http://www.unece.org/env/eia

13 Levett-Therivel. Draft guidance on the SEA Directive. Office of the Deputy Prime Minister, UK, October 2002. http://www.planning.odpm.gov.uk/consult/sea/04.htm

14 Saltman RB and Ferroussier-Davis O. The concept of stewardship in health policy. Bulletin of the World Health Organization. 2000; **78**: 732–739. www.who.int/bulletin/pdf/2000/issue6/bu0614.pdf

15 Travis P, Egger D, Davies P, and Mechbal A. Towards better stewardship: concepts and critical issues. WHO EIP 2002. www3.who.int/whosis/discussion_papers/pdf/paper48.pdf

Chapter 36

Future directions for HIA

Jayne Parry and John Kemm

The past decade has witnessed much activity taking place under the health impact assessment (HIA) banner. Initially there was disagreement as to what HIA was, uncertainty regarding how it should be done, and doubt about its value. The chapters in this book and other work demonstrate that progress has been made towards establishing a theoretical base and appropriate methods for predicting health impacts. Practical experience of conducting assessments in different settings has been gained. This book has set out to survey the work undertaken, and its chapters demonstrate the diversity of activity and ideas. The different contributors are far from unanimous on all points, but it is possible to identify areas of emerging agreement as well as areas where there are still questions to be answered.

Mackenbach and colleagues (Chapter 3) write in the field of HIA 'methods development may simply not have reached a stage in which consensus would be useful'. This is true but as consensus can be uncomfortably close to stagnation it is also reassuring. As with any field subject to active research numerous bipolarities have emerged. McCarthy and Utley (Chapter 6) identify three of these 'quantitative or qualitative', 'health or disease', and 'participative or expert'. Others such as 'rapid or in-depth' and 'separate or integrated with other impact assessments', could also be included. In the early days of HIA, debate as to the merits of these differing camps was the norm; now however we seem to be moving to a position whereby any approach to undertaking HIA can be justified so long as it is 'fit-for-purpose'. Whether this is essential pragmatism in the world of dynamic policy-making, or the default Cartesian response ('I do not see it therefore it does not exist') to difficult methodological challenges remains to be seen.

What is HIA?

It seems that there can now be agreement on the question what is HIA. The majority of work described as HIA in this book has two common characteristics, which might be regarded as the necessary and sufficient features for an HIA.

- It attempts to predict the consequences of adopting different options.

- It is intended to influence and assist decision makers.

Accepting a definition of HIA based on these two criteria has certain corollaries. First if HIA is concerned with prediction then all HIA is 'prospective'. The practice of referring to 'retrospective' or 'concurrent' studies as HIA is confusing and the time has come to cease using these terms. The terms evaluation and surveillance adequately cover most of the activities sometimes described as 'retrospective' and 'concurrent' HIA. Second it follows that many activities, which describe themselves as HIA, are not, and that many activities, which do not describe themselves as HIA, are. In some parts of the world where there is apparently little activity in HIA it is present but given some other label such as comparative risk assessment, social impact assessment (SIA), or the health component of EIA. In seeking to remove the description 'HIA' from retrospective evaluative and community development projects one is in no way devaluing their worth. Third the statement that HIA is intended to influence decisions has clear implications as to how it should be done. As Mathers and Parry (Chapter 32) argue, the process needs to be undertaken in a timescale that permits influence on the decision-making process.

There are dissenters: for example, Mittelmark and colleagues (Chapter 13) argue to extend the definition of HIA but the value of the work they describe is in no way reduced if it is called 'involving citizens in health planning' or community development rather than HIA-CD. The task of thinking about, discussing, and improving HIA is made easier if it has a precise and somewhat narrow meaning. Petticrew and colleagues (Chapter 7) also include retrospective HIA in their review of evidence.

The meaning of HIA may be coming to be agreed on within the HIA community but outside it still causes problems. As Breeze notes elsewhere [1], 'the term "health impact assessment" itself has some inherent difficulties . . . There is a still a tendency for "health" to be too narrowly interpreted . . . the meaning of "impact" is open to debate on what can be measured, while the term "assessment" seems to imply to some that it is a highly technical process that is in the domain of experts only'.

One area in which terminological confusion is still present is the description of small-scale HIA, which has been variously described as mini HIA, rapid appraisal, and/or desktop HIA. In the Netherlands, the rapid appraisal is termed 'screening' (Chapter 16). Ison (Chapter 11) clarifies the differences between the different types of small-scale HIAs as well as the difference between them and the first screening step in the 'classic' HIA process. Her reasoning provides a possible basis for a terminology of small-scale HIAs: it may take a short time (mini HIA), it may not involve participation (desktop

HIA), and it may not involve new data collection or fresh literature searching (rapid appraisal). These characteristics are not mutually exclusive.

Participation and stakeholder involvement

Nearly all contributors agree that stakeholders should be involved in HIA but doubt remains as to how this involvement contributes to the HIA. McCarthy and Utley (Chapter 6) describe communities as 'not necessarily well informed about potential health impacts' and likely to assess proposals 'from their own subjective viewpoint'. But others argue that that subjective viewpoint is an essential component of HIA and that technocratic approaches risk failing to address the concerns of key stakeholders. Elliott and colleagues (Chapter 8) point out that 'lay knowledge' is essential if one is to focus on the 'determinants of the determinants' and understand how a proposal may be modulated by interaction with social structures and human contexts. Furthermore, participation may bring additional benefits. Mittelmark and colleagues (Chapter 13) suggest how certain ways of doing HIA can increase community understanding. Other contributors (Breeze, Chapter 18; Roscam Abbing, Chapter 16; Bowen, Chapter 21; Lock, Chapter 33) emphasize how drawing stakeholders into HIA fosters new partnerships. Although community participation is intuitively appealing and (perhaps) theoretically appropriate, in practical terms it is extremely difficult to arrange adequately. In this regard, Ison's (Chapter 11) suggestions on how to conduct a participatory HIA are particularly useful.

The authors in this book demonstrate the divisions within the HIA community on the importance attached to participation. Many claim to have undertaken participatory HIAs but frequently one has to ask was there true participation or merely tokenism, which reinforced community division and perpetuated the difficulties experienced by marginalized and hard-to-reach groups? Achieving genuine participation appears to be particularly difficult at a policy level where the number of stakeholders is large and there may be many conflicting interests. In the future one must certainly hope to see further developments in the theoretical understanding of how participation contributes to impact prediction and other goals of HIA, and in the practice of when and how to use participation in HIA.

The need to consider distribution (equity) of impacts

Within the HIA community, consensus is apparent in the recurring call for HIA to consider the distribution of impacts within a population. While noting this, Mackenbach and colleagues (Chapter 3) comment that in the majority of HIAs, the assessment of inequalities has been in conducted in a relatively unstructured manner, and their suggestions for a future research agenda in this area make compelling, if challenging, reading. In this regard, Chapter 3 has resonance with the findings of a recent study commissioned to explore progress in reducing inequalities in the United Kingdom since the Acheson Report [2]. The study notes, 'health inequalities impact assessments should be applied extensively within (government) departments' but that 'evidence of such assessments is currently sparse'. The authors go on to recommend: 'there could be greater and/or more sensitive application of health inequalities impact assessment (especially across central government), through for example developing methodology, improving skills and capacity, refining data collection, conducting assessments prior to implementation and changing the scope of performance management systems' [5].

It is clear that at present there is a mismatch between the aspirations of HIA and the reality of assessments in terms of considering health inequalities in an adequate manner. Certainly, future HIA should strive to explicitly set out the estimation of the differential distribution of effects arising from the policy under investigation. Given that the aim and methods adopted for an HIA may differ according to context, it is not possible at present to agree on a uniform approach and methods for inequalities assessment. That being said, it may be possible to achieve consensus on the minimum required within any HIA—for example consideration of the effects of intervention stratified by sex, age, ethnicity, and socio-economic status relative to the 'whole' population.

Bases for prediction in HIA

Petticrew and colleagues (Chapter 7) point us towards important sources of information that might assist risk estimation but they suggest that 'the real problem remains that there is little sound evidence of any sort—either qualitative or quantitative—of the health impacts of social interventions to aid sound, scientifically based HIAs'. They are right to warn that 'in the pressure to engage in predictive HIA we should not forget there is a basic need for evaluation and monitoring of actual impacts'.

What comprises 'evidence' opens another well-rehearsed debate and the existing precedence given to 'research' and 'expert' evidence has relevance for HIA. As Elliott, Williams and Rolfe (Chapter 8) note, in seeking to identify those risks that may be measured we may miss the 'fine grain' information inherent in lay knowledge that is only uncovered through more qualitative approaches to fieldwork.

This need both to synthesize evidence from a variety of sources and to provide access to such work are key challenges facing HIA. An ideal situation would see a database of well-organized information on how the different determinants of health impact on populations. The content of these databases would where possible be expressed in a way that is easy to use in quantitative predictions (if this population of 1000 people is exposed to an increase of $PM_{2.5}$ particulate levels of 10 $\mu g/M^3$ there are likely to be x additional admissions to hospital with asthma each year). The database would also cover 'social' interventions but since the outcome of these is frequently very context dependent, they would also have to contain guidance on the contexts in which they were relevant. Finally, the database would be easily accessed by the web or other methods so that every HIA could have the benefit of high-quality literature reviews and the assessment team would be able to concentrate on working out how the general information on determinants applied to the specific case of their HIA.

Relation of HIA to EIA and other impact assessments

Increasingly there is recognition of the overlap between HIA and other forms of impact assessment. The arguments for integrating HIA with EIA are reviewed by Bond (Chapter 12). The practice of approaching health in the context of EIA is well established in Australia (Wright, Chapter 20) and Canada (Kwiatowski, Chapter 27) and will be required as part of SEA in Europe (Dora, Chapter 35). This approach is not without its dangers and problems but it may be the route through which HIA becomes 'institutionalized' (i.e., has a legislative base).

Integrating HIA with other assessments may go some way to increasing utilization of the process by organizations already overwhelmed by requirements for risk management (Milner, Chapter 22). Attempts are already being made in international (EU [2]) and some national governments (England [3]) to introduce integrated policy appraisal tools. The danger is that this exercise could degenerate into a tokenistic check box exercise. The promise is that it could be a mechanism that makes all policy makers aware of health and other cross-cutting issues and a trigger for increase partnership and working between departments. The challenge for the HIA community is to give away ownership of health impacts, become more aware of other cross-cutting issues and allow integrated impact assessment to develop in a way that benefits the health of the population.

Institutionalizing HIA

Many contributors talk about institutionalizing HIA, by which they mean making it required by law or otherwise obligatory. Some see this as a way of avoiding the British Columbia experience in which HIA passed from favour to neglect (Banken, Chapter 15). Others see it as a quick way to make reluctant officials give a higher priority to health. However, legislation before there is general willingness to comply rarely produces the desired change in behaviour.

Institutionalization implies that methods are at a stage where they can be standardized and specified by law and many would question that that stage has been reached. Many aspects of HIA are already institutionalized through legislation on EIA and SEA (Wright, Chapter 20; Dora, Chapter 35). It may be wiser to study and learn from those areas where HIA has already been institutionalized than to press for more.

Capacity for HIA

Douglas (Chapter 17) and several other authors discuss the problem of lack of capacity for HIA. Progress with HIA depends on more HIAs being done, which in turn depends on more people being prepared to try one. It is clear that if HIA is to be widely applied then the process must be adopted and undertaken by people outside the small community currently interested in HIA. It is unfortunate that HIA has acquired a mystique (and there is a danger that this book will add to it) of being difficult, the domain of experts, and very *recherché*. A brief glance at the majority of HIA reports would serve to demonstrate the falsity of this view—but it is still widespread. In order to move on there must be a dramatic increase in the number of people who feel they know enough about HIA to undertake one. In order to do this many more people must be familiarized with what HIA is and a few more must learn enough about it in order to act as a resource for their organization and help their colleagues to undertake HIA.

How to assess the added value of HIA

Discussion of 'evaluation' brings us to HIA's weakest point. A discussion of evaluation of HIA has been published [4] but there is some confusion between evaluation of the HIA and evaluation of the decision it was meant to inform. While some attempts have been made at evaluating how well the process of HIA has been conducted, there is little information about the utility of HIA against its primary objective—changing the development and implementation of public policy so as to improve health and reduce inequalities. To understand this requires knowledge of whether the findings of an HIA influence the decision-making process. The Greater London Authority is commissioning an assessment of their HIA programme (Bowen, Chapter 21) and similar work is going on in Sweden (Berensson, Chapter 19). These investigations may shed some light on the utility of HIA. However, Lehto (Chapter 5) reminds us that policy making is a complex process and HIA needs to become much more sophisticated in its understanding of how policies are developed.

In future many more evaluations of HIA are needed. In order to make evaluation easier it should be standard practice for reports to clearly state what decisions they sought to inform and who would make those decisions. Discussion with the decision makers should then reveal in what ways the HIA was useful to them and in what ways it influenced their thinking.

Accounts of HIA also make it clear that HIA may have other benefits, which are real but difficult to evaluate. Many accounts describe building of partnerships as a result of HIA (Breeze, Chapter 18; Bowen, Chapter 19; Aziz and colleagues, Chapter 25; Barnes, Chapter 26). Others talk of changing the way that policy makers approach their task and raising health issues on their agenda. Evaluation must not lose sight of these incidental outcomes.

Though the true costs of doing HIA are commonly 'lost' in other budgets, HIA is not a cost-free activity. Once its benefits have been established decision makers will go on to ask is it cost effective? Several authors in this book give crude costings for some of their HIA activities. An economist reading them would be struck first by their wide range and second by the lack of sophistication displayed in costing. It is progress that costs are being discussed but HIA has to get much better at estimating cost as well as benefits.

Conclusion

The work presented in this book demonstrates the impressive variety and range of work that has taken place in the field of HIA. Health impact assessment has indeed come a long way in the past 10 years, but it needs to improve further both in terms of methodological technique and practical application if it is to truly fulfil its promise and become a useful adjunct to decision making.

References

1 Breeze CH and Hall R. Health impact assessment in government policymaking: developments in Wales. Policy learning curve series Number 6. Brussels: WHO European Centre for Health Policy, 2002.

2 Commission of the European Communities. Communication from the commission on impact assessment COM(2002) 276 final Brussels, 2002.

3 DEFRA. Sustainable development strategy: Foundations for our future Chapter 2 new thinking better solutions. 2000. www.defra.gov.uk/corporate/sdstrategy/chapter2.htm

4 Taylor L and Quigley R. Guide to evaluation of HIA. London: Health Development Agency, 2003.

5 Exworthy M, Stuart M, Blare D, and Marmot M. Tackling health inequalities since the Acheson Inquiry. The Policy Press, 2003.

Index